Lecture Notes in Mechanical Engineering

Lecture Notes in Mechanical Engineering (LNME) publishes the latest developments in Mechanical Engineering—quickly, informally and with high quality. Original research reported in proceedings and post-proceedings represents the core of LNME. Volumes published in LNME embrace all aspects, subfields and new challenges of mechanical engineering. Topics in the series include:

- Engineering Design
- Machinery and Machine Elements
- Mechanical Structures and Stress Analysis
- Automotive Engineering
- Engine Technology
- Aerospace Technology and Astronautics
- Nanotechnology and Microengineering
- Control, Robotics, Mechatronics
- MEMS
- Theoretical and Applied Mechanics
- Dynamical Systems, Control
- Fluid Mechanics
- Engineering Thermodynamics, Heat and Mass Transfer
- Manufacturing
- Precision Engineering, Instrumentation, Measurement
- Materials Engineering
- Tribology and Surface Technology

To submit a proposal or request further information, please contact the Springer Editor of your location:

China: Ms. Ella Zhang at ella.zhang@springer.com
India: Priya Vyas at priya.vyas@springer.com
Rest of Asia, Australia, New Zealand: Swati Meherishi at swati.meherishi@springer.com
All other countries: Dr. Leontina Di Cecco at Leontina.dicecco@springer.com

To submit a proposal for a monograph, please check our Springer Tracts in Mechanical Engineering at https://link.springer.com/bookseries/11693 or contact Leontina.dicecco@springer.com

Indexed by SCOPUS. All books published in the series are submitted for consideration in Web of Science.

More information about this series at https://link.springer.com/bookseries/11236

José Machado · Filomena Soares ·
Justyna Trojanowska · Vitalii Ivanov ·
Katarzyna Antosz · Yi Ren ·
Vijaya Kumar Manupati ·
Alejandro Pereira
Editors

Innovations in Industrial Engineering II

Springer

Editors
José Machado
Department of Mechanical Engineering
University of Minho
Guimarães, Portugal

Justyna Trojanowska
Poznan University of Technology
Poznan, Wielkopolskie, Poland

Katarzyna Antosz
Faculty of Mechanical Engineering and Aeronautics
Rzeszow University of Technology
Rzeszow, Poland

Vijaya Kumar Manupati
Department of Mechanical Engineering
National Institute of Technology Waranga
Hanamkonda, India

Filomena Soares
Department of Industrial Electronics
University of Minho
Guimarães, Portugal

Vitalii Ivanov
Department of Manufacturing Engineering, Machines and Tools
Sumy State University
Sumy, Ukraine

Yi Ren
School of Reliability and Systems Engineering
Beihang University
Beijing, China

Alejandro Pereira
Faculty of Industrial Engineering
University of Vigo
Galicia, Pontevedra, Spain

ISSN 2195-4356 ISSN 2195-4364 (electronic)
Lecture Notes in Mechanical Engineering
ISBN 978-3-031-09359-3 ISBN 978-3-031-09360-9 (eBook)
https://doi.org/10.1007/978-3-031-09360-9

This Springer imprint is published by the registered company Springer Nature Switzerland AG
The registered company address is: Gewerbestrasse 11, 6330 Cham, Switzerland

Preface

This volume of Lecture Notes in Mechanical Engineering gathers selected papers presented at the Second International Scientific Conference (ICIE'2022), held in Guimarães, Portugal, during 28–30 June 2022. This conference was organized by the School of Engineering of University of Minho, throughout MEtRICs and Algoritmi Research Centres.

The aim of this conference was to present the latest engineering achievements and innovations, and to provide a chance for exchanging views and opinions concerning the creation of added value for the industry and for the society. The main conference topics include (but are not limited to):

- Innovation
- Industrial Engineering
- Mechanical Engineering
- Mechatronics Engineering
- Systems and Applications
- Societal Challenges
- Industrial Property.

The organizers received 139 contributions from 16 countries around the world. After a thorough peer-review process, the committee accepted 81 papers written by 335 authors from 15 countries for the conference proceedings (acceptance rate of 58%), which were organized in three volumes of the Springer Lecture Notes in Mechanical Engineering.

This volume, with the title "Innovations in Industrial Engineering II", specifically focuses on innovative technologies associated with and strategies for the development of Industry 4.0. The chapters discuss new ways to improve industrial production and supply chain management by applying mathematical and computational methods. They also cover important issues relating to sustainability, education, and collaborations between industry and universities, and national developments. Last but not least, it analyses important issues proposing a good balance of theoretical and practical aspects. This book consists of 25 chapters, prepared by 96 authors from 7 countries.

Extended versions of selected best papers from the conference will be published in the following journals: Sensors, Applied Sciences, Machines, Management and Production Engineering Review, International Journal of Mechatronics and Applied Mechanics, SN Applied Sciences, Dirección y Organización, Smart Science, Business Systems Research, and International Journal of E-Services and Mobile Applications.

A special thanks to the members of the International Scientific Committee for their hard work during the review process.

We acknowledge all that contributed to the staging of ICIE'2022: authors, committees, and sponsors. Their involvement and hard work were crucial to the success of ICIE'2022.

June 2022

José Machado
Filomena Soares
Justyna Trojanowska
Vitalii Ivanov
Katarzyna Antosz
Yi Ren
Vijaya Kumar Manupati
Alejandro Pereira

Contents

The Meaning and Directions of Development of Personalized Production in the Era of Industry 4.0 and Industry 5.0

Jakub Pizoń[1] and Arkadiusz Gola[2](✉)

[1] Department of Enterprise Organization, Faculty of Management, Lublin University of Technology, ul. Nadbystrzycka 38, 20-618 Lublin, Poland
j.pizon@pollub.pl

[2] Department of Production Computerisation and Robotisation, Faculty of Mechanical Engineering, Lublin University of Technology, ul. Nadbystrzycka 36, 20-618 Lublin, Poland
a.gola@pollub.pl

Abstract. Because of changing market requirements and uncommonly fast development of computerization, automation and robotization the paradigms concerning methods and techniques of manufacturing are still evaluating to make production processes more and more flexible and effective. The paper presents the issue of the importance of personalized production in the perspective of contemporary customer requirements (including personalization of available products) and technological solutions that fit into the trends of modern industrial revolutions. In particular, the article focuses on the conditions and methods of organizing the work in a modular approach with the use of solutions that are key features of Industry 4.0 and 5.0, so that they effectively carry out production orders formulated as part of individual customer orders, thus implementing mass production. As an example illustrating the issue at hand, the model of production of personalized doors and windows in a production plant belonging to a company from the SME sector was presented.

Keywords: Personalized production · Mass customization · Mass personalization · Industry 4.0 · Industry 5.0

1 Introduction

We live in times when the life of a product is significantly shortened, and new things, concepts or technologies will appear and disappear at a pace unknown in other epochs [1]. As shown in the literature, the time between the first three industrial revolutions spanned centuries or decades, while since the third industrial revolution the period of transformation has significantly accelerated [2]. The intensity of the change and the emergence of new opportunities open up the prospect of solutions to both the already known problems and the new ones. Moreover, it motivates to consider what possibilities it opens up for the current issues, concepts and methods of work organization [3, 4]. Nevertheless, in order to be able to formulate new solutions or analyze the existing ones

J. Machado et al. (Eds.): icieng 2022, LNME, pp. 1–13, 2023.
https://doi.org/10.1007/978-3-031-09360-9_1

- conceptual work is needed, which will allow to consider the possibility of applying solutions and obtaining perceived benefits within real production systems [5, 6].

Modern industrial revolutions announced under the slogans of Industry 4.0 and Industry 5.0 (together with the technology solutions they refer to) offer a completely new perspective in terms of the requirements and possibilities of organizing production systems and processes. The aim of this paper is to analyze the possibility of transforming the approach to the issues of organizing production in manufacturing companies in such a way that it corresponds to modern market needs and technological possibilities, ensuring high efficiency, but at the same time maintaining the personalized nature of products delivered to customers.

2 Literature Review

2.1 Modern Industrial Revolutions

The concept of "Industry 4.0" was proposed for the first time in 2011 at the Hanover (Germany) fair as a synonym for the concept of intelligent production [7]. The fourth industrial revolution is identified with the concept of Industry 4.0, and this concept results from the numbering adopted to identify successive technological breakthroughs that have had a key impact on economic development [8]. According to the literature, in the first industrial revolution, people got rid of the limitations of many physical properties and used water and steam as energy for production and transport [9]. The second industrial revolution brought about the development of electricity, which replaced water and steam as energy for production, and machines were used for production for the first time. The third industrial revolution focused around the era of production automation, using electronic equipment and information technologies to increase production capacity [10].

Basically, Industry 4.0 is an intelligent combination of machines and industrial processes based on new information and communication technologies. It builds on emerging information and artificial intelligence technologies, including cyber-physical systems (CPS), the Internet of Things (IoT), cloud computing, Big Data, machine learning, service-oriented architecture (SOA), and industrial information. By delivering digital solutions through the integration of these complex, interconnected technologies, Industry 4.0 transforms production processes and systems from a centralized decision-making instance to a decentralized structure [11]. Therefore, one of the key characteristics of Industry 4.0 is the implementation of vertical integration, i.e. the integration of IT systems from various hierarchical levels in the company - from the production hall to the middle and upper management level - ensuring real-time data flow and supporting decision-making. Vertical integration enables intelligent machines to create a self-organizing system that can be dynamically reconfigured to accommodate different products. Thanks to such integration, companies obtain a transparent, predictable and adaptable intelligent manufacturing system, which is vertical integration and therefore considered the first step towards Industry 5.0 at the factory level [12].

After discussions between participants from research and technology organizations and funding agencies from all over Europe in two virtual workshops organized by the Directorate "Prosperity" of Directorate-General for Research and Innovation, on 2 and 9 July 2020, in 2021, the European Commission formally called for a fifth industrial revolution (Industry 5.0) by issuing on 4 January 2021 an official document entitled "Industry 5.0: towards a sustainable, human-centered and resilient European industry" [13].

Industry 5.0 is a human-centered design solution where the ideal human companion and cobots work together with human information sources to enable personalized autonomous production through the enterprise's social networks. This in turn allows people and machines to work hand in hand [14]. Cobots are not programmable machines, but they can sense and understand human presence. In this context, cobots will be used to perform repetitive tasks and work that require a lot of work, while a human will take care of customizing products to customer needs and critical thinking (thinking outside the box) [15].

Industry 5.0 is considered to be the next industrial evolution, with the goal of harnessing the creativity of human experts in collaboration with efficient, intelligent and accurate machines to obtain resource-efficient and user-preferred manufacturing solutions compared to Industry 4.0. [16]. Industry 5.0 is an evolving concept. There are currently two visions of Industry 5.0 emerging. The first is "human-robot collaboration," where the premise is that robot and humans will collaborate whenever and wherever possible. People will focus on creative tasks and robots will do the rest. The second vision of Industry 5.0, on the other hand, is related to the bioeconomy. Intelligent use of biological resources for industrial purposes will help to achieve a balance between ecology, industry and the economy [17].

2.2 Mass Personalization Versus Mass Customization

Modern industrial revolutions, starting with Industry 4.0 and the later announced Industry 5.0, have changed the approach to the organization of production in enterprises. The tools and technologies that modern industrial revolutions bring with them open up new perspectives for improving production processes and creating new value in an effective way. Industry 5.0 reminds you that social needs should be taken into account, but also points out that people and their work are a source of breakthrough innovations that cannot be achieved with the use of analytical methods.

It should be noted that, apart from discoveries in the field of new technological solutions, the driving force behind the ongoing changes in the organization of production is the growing market demand for personalized products. Although the need for personalized products has always existed, achieving this goal until a certain moment was impossible due to the limitations of production automation technology. In fact, it was only since 2010 that mass personalization became a realistic (achievable) goal (Fig. 1). Currently, producers are striving to be able to simultaneously manufacture highly personalized products in dynamically changing batch sizes, while maintaining the lowest possible production costs (similar to those achieved in the case of serial production) [18].

As indicated in the literature, the fourth industrial revolution made it possible to carry out orders as part of mass product matching [19]. The client can place an order and adjust it to his needs by means of a limited selection from the list. Thanks to this, despite the standard manufacturing process, it is possible to make products more attractive by adjusting them (to a certain extent) to specific needs. On the other hand, with the advent of the Fifth Industrial Revolution, where man becomes the center of attention again, there is the prospect of mass personalization, so as to (in a sense) return to the artisanal nature of production. Customers want to buy personalized products, i.e. those that emphasize the individualism and sometimes even the personality of a given customer. At the same time, there is an expectation that the price of personalized products (even if higher than standard ones) will be acceptable in the long term to a wide range of customers.

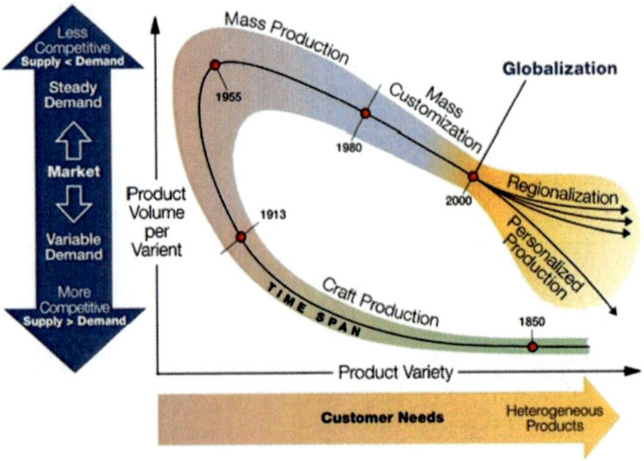

Fig. 1. Changes in the production paradigm [18].

Therefore, Industry 5.0 focuses on advanced analytical systems (Big data - see e.g. [20]) and collaborative robots (co-bots - see e.g. [21]), which allow for cost reduction using only one - at the same time, the potential of human-machine cooperation. Cobots are able to notice, understand and feel not only a person, but also the goals and expectations of a human operator. Like the student, the cobots will observe and learn as the person completes the task. Once learned, cobots will perform the desired tasks just like their human operators [22].

The difference between customization and personalization is that something unique is created through personalization. This artisanal nature of the manufactured goods and its uniqueness constitute an added value for the customer in relation to standard, matched products. In addition, the client actively participates in the process of designing a given product, has an influence on its form, selection of materials, or other [23].

2.3 Modular Production

The current changes that take place within the production systems allow for more and more frequent generation of product innovations. Therefore, with the advent of modern technological solutions, there are opportunities to develop the existing concepts of production organization. Technologies of modern industrial revolutions open up new perspectives for both new and existing solutions. In this perspective, the problem of organizing production by means of production modules, also referred to as modular production, draws particular attention. Modularity essentially refers to the ability to reconfigure the system on a plug-n-play basis, allowing it to respond quickly and efficiently to changes in customer requirements. The key advantages of this system are [24]:

- ongoing resolution of any problems that may arise in the supply chain, without stopping production completely,
- no need for employees to adapt to the speed of the assembly line and to move with it,
- the opportunity to learn from previous experiences and possible improvisation thanks to the methods and techniques of artificial intelligence,
- the ability to adapt and introduce other changes without incurring additional costs.

Modularization enables manufacturers to respond more flexibly to customer requirements. Thanks to modularization, it is possible to break down production lines into decentralized production modules with individual functionalities. To support shop floor workers in decentralized, modular production environments, they need independent access to information. Each employee has their own tasks and a specific role in the company. For these tasks, they need information about production modules and the production process [25].

The advent of Industry 4.0 technology and advanced distributed production control systems has provided additional opportunities for a plug-and-produce approach in manufacturing systems, further reducing reconfiguration time and effort. Following the concept of plug-and-produce modular production systems, production modules can be rearranged in line during the production system operations. Integrated in the system are special connection nodes where mechatronic modules can be added and removed with limited effort. These features significantly shorten the system reconfiguration time as well as the skills required during reconfiguration, enabling multiple adaptations of the system architecture during a production shift. Despite their advantages, modular "plug-and-produce" production systems entail additional challenges at the system design and management level that limit their prevalence in real production contexts [26].

Modular production is possible with the use of autonomous transport systems, i.e. a solution based on AMR vehicles. AMRs are the latest solution in the sector of unmanned mobile robots for material handling. Most often, AMR is used in large warehouses or inside production plants to carry out transport operations in the production hall and to transfer materials, semi-finished products, finished products or goods [27].

3 Application Case Study

Taking into account the needs of modern consumers and the existing possibilities of technology, this chapter presents the concepts of modular production in a company from the furniture industry. The analyzed example shows how to use individual solutions of industry 4.0 and 5.0 to meet consumer preferences and implement production in a personalized manner. The production system of a furniture manufacturing company was used as the basis for the implementation of the application case study.

3.1 Current State and Company's Needs

The existing production system consists of one production line. Such organized production allows for the production of a defined set of cabinets and shelves (the company has a specific product catalog). Production is carried out on request. Standard orders are placed by large retail chains specializing in the sale of furniture and interior finishing accessories. To a small extent, production is made to the warehouse. What's more, production is very rarely or sporadically carried out as personalized products - adequately to the reported needs of customers.

The analyzed company (in order to diversify the product portfolio) wants to be able to flexibly start the production of new products. That is why the company wants to obtain the possibility of producing personalized furniture, while maintaining the current production volume, identical to the implementation of previous orders for large retail chains. It is assumed that the production of personalized furniture would consist in the fact that the customer, in contact with a sales representative or via a web application, designs kitchen furniture adequately to the needs of his kitchen. It also has the option of adding personalized patterns to decorate cabinets created with the help of the web wizard. Moreover, the cutting of the cabinets and their arrangement inside are in line with the customer's needs and requirements. The company also wants to give customers the opportunity to choose whether the created furniture is to be delivered packed or already assembled, ready for assembly.

The need of the enterprise is also emphasized that due to the problems with obtaining qualified employees, the enterprise wants to maintain production capacity even in the case of limited staff. The intralogistics tasks should be automated. At the same time, the company emphasizes the need to employ highly qualified and experienced employees.

3.2 Current and Planned Manufacturing Process

The furniture production process used so far begins with cutting chipboards into cabinet construction elements, such as fronts, sides, rims, shelves, strips for frames or pedestals (Fig. 2).

This process requires proper layout of the elements on the plate sheet. In the next stage, the process of wrapping is carried out. It needs to take the edge from the warehouse, installing it in an edge banding machine and then carrying out the edgebanding process. The taped elements go to drills that make mounting holes as well as for handles and hinges. Such prepared batch of elements goes to the warehouse of semi-finished products. In the next step, packing is carried out. In this operation, appropriate furniture accessories (such as hinges, handles, actuators, etc.) are added to the furniture package. The accessories are distributed in separate plastic bags, securing these elements against loss during transport. After packing, the package goes to the finished goods warehouse. If a given type of furniture contains glazing, the glass pane is added separately to the finished package. Such a set is collected by transport from the shipping warehouse. It should be noted that the panes are sanded before they are packaged. However, in the next stage, they are packed and labeled. Transport between the described stages is carried out by means of platform transport trolleys. The company does not use a plate labeling system. A given batch of elements is glued or sanded (in the case of glass panes) in accordance with the production plan of a given stand. The production process realized in this way achieves the assumed goals, but it is not flexible enough and it is practically impossible to implement personalized production. Supervision over such production in the present system would require a significant increase in staff or the introduction of a system for marking elements. Marking would require the implementation of a marking system that allows easy removal of marks after the production process. Therefore, from the perspective of the possibility of personalized production, there are a number of problems, such as:

- no possibility to control the execution of individual orders,
- no possibility of carrying out personalized orders,
- production stability directly dependent on the workforce,
- a large amount of waste related to problems in the implementation of intra-logistics tasks, shortages of individual elements in the process of completing packages,
- the need to maintain a warehouse of semi-finished products, which increases the value of work in progress.

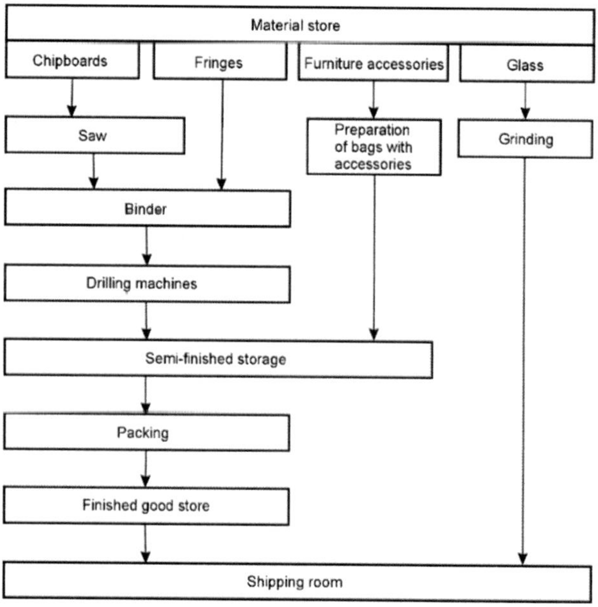

Fig. 2. Scheme of the manufacturing process for the manufacture of kitchen cabinets

It should be added that the method of organizing the flow based on the possibility of marking, although simple, is unfortunately not very flexible. The production is still carried out within one line, which emphasizes its nature dedicated to the implementation of large orders. Such a method enables quick implementation of new employees (even with small competences) and relatively simple management of the production process. So far, standard management methods have been used, focusing on planning orders in accordance with the MRP loop. Work progress is recorded after the production is completed based on the stock levels. This, however, confirms the existence of a significant information gap during the process, which closes any possibility of its optimization. Possible optimizations are reported by employees and are mostly a consequence of rationalization conclusions.

3.3 The Concept of the Production Process Implementation Model for the Purposes of Personalized Production

The organization plan of the production process has been defined in relation to the needs of the enterprise in order to achieve the expected benefits, while reducing the risk and solving the current problems. Therefore (taking into account the possibilities of modern information technologies and the flexibility of the modular production approach), it was decided that these techniques and means will be used in the development of the concept of the organization of the production process combining the features of mass production and personalized production.

While developing the process, a preliminary analysis of the company's needs was carried out. So that the new system meets key needs in the first place. Therefore, both key needs were identified and defined, as well as those that the enterprise requires, but are not crucial for the organization of the new system, as presented in Table 1.

Table 1. Initial analysis of the company's needs.

Need	Key	Not key
Diversification of the product portfolio, the possibility of flexible launching the production of new products	Yes	–
The possibility of producing personalized furniture	Yes	–
Possibility to produce standard orders	Yes	–
Possibility to choose whether the created furniture is to be delivered packed or already assembled	–	Yes
Possibility of production even in the case of staff reduction	Yes	–
The need to employ highly qualified and experienced employees	–	Yes

When planning the target process, it is necessary to focus on the key needs. In order to meet the requirements of flexibility and personalization in the first place, in the perspective of modular production the means of production should be compiled adequately to the product (so that a given module performs a given task from the production process). However, taking into account the need to implement the current production volume, and therefore a large number of products produced by pipelines - a tool for ongoing monitoring and control of the process is needed. As there are problems with employees, an IT tool is needed to carry out this task. Planning the process flow will therefore belong to the implemented IT system. Moreover, in order to become independent from staff limitations, intralogistics operations should be performed regardless of the availability of employees.

Secondly, non-key needs should be met, and the employment of highly qualified employees should be independent of the planned system. On the other hand it is possible to plan the possibility of choosing the form of the delivered order - folded furniture or packed furniture.

As a result of the above considerations, an organization concept was proposed using the modular production approach using modern IT solutions - so that they could effectively execute production orders formulated as part of individual customer orders (thus implementing mass personalized production as well as the production of standard orders). The solutions constituting the basis of the developed concept use the possibilities of technologies of modern industrial revolutions such as AMR (autonomous mobile robots), collaborative robots (cobots), internet of things, multidimensional data sets and industrial artificial intelligence.

The concept assumes that the production will be carried out within the framework of dynamically organized workplaces created adequately to the ordered production. Paying attention to the fact that standard production as part of standing orders will be carried out in the largest possible part by automated robots. On the other hand, personalized production will be handled by operators.

The basis for such organization of dynamic positions will be the AMR platform. Production tools (machines, robots), workbenches (dedicated to operators) or production materials (cabinet construction elements, finishing materials) will be installed on the platforms (Fig. 3). The organization of the stands depends directly on the structural elements for a given product (cabinet), which is the result of the first phase, i.e. cutting. Construction elements will be loaded on the leading AMR trolleys, the path of which in the production process begins with cutting and ends with the finished products warehouse.

Fig. 3. An example of autonomous mobile robot that can be used for an organized production process [27].

The leading trolley will contain all the structural elements required to complete a given order. It will contain all the structural elements so that when the finished products are delivered to the warehouse - the finished product is delivered in the form ordered by the customer (it is possible to create multi-trolley platforms). After the completed transport operation, the leading trolley returns to the beginning of the process, where it is serviced in a designated space and the trolley's batteries are charged.

At the post-cutting stage, with the advent of the leading trolley with an order for a personalized product, at a given production stage, a workshop stand will be dynamically organized, which will enable the production of a given product. However, if standard production is carried out, the position organized within the AMR platforms will be based on the work of robots performing standardized activities.

The task of the platforms will also be to provide the necessary materials for production. In the case of the considered system, these will be edgings, glass or furniture accessories. They will be loaded by loading robots in the right amount. In order to improve the stage of packaging implementation for those orders for which the furniture assembly was selected - a collaborative robot will be introduced (Fig. 4). The organization of this stand consists in planning cooperation, where the robot indicates and gives the fitter the next elements required for assembly. The operator (fitter) sets up the elements based on experience in order to ensure the best possible quality of assembly.

Fig. 4. An example of collaborative robot that can be used for an organized production process [own photo].

Summing up, it should be noted that the solutions described in the proposed concept use the possibilities of modern industrial revolution technologies such as AMR (autonomous mobile robots), cobots, internet of things, multidimensional data sets and industrial artificial intelligence.

Such organizational order will allow the implementation of personalized orders. These orders will be performed on positions managed by experienced operators.

4 Summary and Conclusions

Contemporary market requirements impose an increasing need to look for solutions enabling the production of personalized products in a highly efficient manner with the simultaneous reduction of the direct participation of people in the production process. New technologies come with help in this respect, opening new perspectives not only for quality improvement, streamlining the supply chain, but also the basis for the transformation of the organization of production processes.

This article focuses on the presentation of the concept of the organization of the production process in a furniture industry company enabling the simultaneous high-performance production of standard products and flexible production of personalized products. The elements enabling the implementation of this concept are the latest technological solutions belonging to the achievements of the so-called Industry 4.0 and Industry 5.0. in the form of autonomous mobile robots, cobots, internet of things, multi-dimensional data sets and industrial artificial intelligence. In accordance with the adopted assumptions, the developed concept will be a response to the current production needs, such as:

- the possibility of diversifying the product portfolio and the possibility of flexible launching the production of new products,
- the possibility of producing both personalized furniture and standard orders,
- possibility to choose whether the created furniture is to be delivered packed or already assembled,
- the possibility of production even in the case of staff reduction,
- the need to employ highly qualified and experienced employees.

The assumptions of this concept and the possibility of its implementation were initially discussed with the management of the analyzed enterprise. As part of further research, actions will be taken to detail the adopted assumptions and to develop a detailed plan for the organization of the production system taking into account these assumptions.

Acknowledgments. The project/research was financed in the framework of the project Lublin University of Technology-Regional Excellence Initiative, funded by the Polish Ministry of Science and Higher Education (contract no. 030/RID/2018/19).

References

1. Ciesla, B., Mleczko, J.: Practical application of fuzzy logic in production control systems of engineer to order SMEs. Appl. Comput. Sci. **17**(1), 17–25 (2021)
2. Hamid, M.S.R.A., Masrom, N.R., Mazlan, N.A.B.: The key factors of the industrial revolution 4.0 in the Malaysian smart manufacturing context. Int. J. Asian Bus. Inf. Manage. **13**(2), 1–19 (2022)
3. Antosz, K., Pasko, L., Gola, A.: The use of intelligent systems to support the decision-making process in Lean Maintenance Management. IFAC PapersOnline **52**(10), 148–153 (2019)
4. Lapunka, I., Pisz, I., Wittbrodt, P.: Stochastic scheduling of production orders under uncertainty. Adv. Intell. Syst. Comput. **649**, 348–358 (2018)

5. Gola, A.: Economic aspects of manufacturing system design. Actual Prob. Econ. **156**(6), 205–212 (2014)
6. Jasiulewicz-Kaczmarek, M., Zywica, P., Gola, A.: Fuzzy set theory driven maintenance sustainability performance model: a multiple criteria approach. J. Intell. Manuf. **32**(5), 1497–1515 (2021)
7. Lin, B., Wu, W., Song, M.: Industry 4.0: driving factors and impacts on firm's performance: an empirical study on China's manufacturing industry. Ann. Oper. Res. 1–21 (2019). https://doi.org/10.1007/s10479-019-03433-6
8. Antosz, K., Jasiulewicz-Kaczmarek, M., Paśko, Ł., Zhang, C., Wang, S.: Application of machine learning and rough set theory in lean maintenance decision support system development. Eksploatacja i Niezawodnosc – Mainten. Reliab. **23**(4), 695–708 (2021)
9. Kłosowski, G., Gola, A., Thibbotuwawa, A.: Computational intelligence in control of AGV multimodal systems. IFAC-PapersOnline **51**(11), 1421–1427 (2018)
10. Lu, Y.: Industry 4.0: a survey on technologies, applications and open research issues. J. Indust. Inf. Integr. **6**, 1–10 (2017)
11. Zhang, C., Chen, Y., Chen, H., Chong, D.: Industry 4.0 and its Implementation: a Review. Inf. Syst. Front. 1–11 (2021). https://doi.org/10.1007/s10796-021-10153-5
12. Tabim, V.M., Ayala, N.F., Frank, A.G.: Implementing vertical integration in the industry 4.0 journey: which factors influence the process of information systems adoption? Inf. Syst. Front. 1–18 (2021). https://doi.org/10.1007/s10796-021-10220-x
13. Xu, X., Lu, Y., Vogel-Heuser, B., Wang, L.: Industry 4.0 and Industry 5.0—inception, conception and perception. J. Manufac. Syst. **61**, 530–535 (2021)
14. Breque, M., De Nul, L., Petridis, A.: Industry 5.0 Towards a sustainable, human-centric and resilient European industry. Publications Office of the European Union, Luxembourg (2021)
15. Maddikunta, P.K.R., et al.: Industry 5.0: a survey on enabling technologies and potential applications. J. Indust. Inf. Integr. **26**, 100257 (2021). https://doi.org/10.1016/J.JII.2021.100257
16. Madsen, D.Ø.: An exploratory bibliometric analysis of the birth and emergence of Industry 5.0. Appl. Syst. Innov. **4**(4), 87 (2021)
17. Demir, K.A., Döven, G., Sezen, B.: Industry 5.0 and Human-Robot Co-working. Procedia Comput. Sci. **158**, 688–695 (2019)
18. Lu, Y., Xu, X., Wang, L.: Smart manufacturing process and system automation – a critical review of the standards and envisioned scenarios. J. Manuf. Syst. **56**, 312–325 (2020)
19. Aheleroff, S., Philip, R., Zhong, R.Y., Xu, X.: The degree of mass personalisation under Industry 4.0. Procedia CIRP **81**, 1394–1399 (2019)
20. Jasiulewicz-Kaczmarek, M., Antosz, K., Żywica, P., Mazurkiewicz, D., Sun, B., Ren, Y.: Framework of machine criticality assessment with criteria interactions. Eksploatacja i Niezawodnosc-Maint. Reliab. **23**(2), 207–220 (2021)
21. Castillo, J.F., Ortiz, J.H., Velásquez, M.F.D., Saavedra, D.F.: COBOTS in Industry 4.0: safe and efficient interaction. In: Ortiz, J.H., Vinajamuri, R.M. (eds.) Collaborative and Humanoid Robots. https://www.intechopen.com/books/10417. Accessed 02 Mar 2022
22. Nahavandi, S.: Industry 5.0—a human-centric solution. Sustainability **11**(16), 4371 (2019)
23. Mourtzis, D., Doukas, M., Psarommatis, F., Giannoulis, C., Michalos, G.: A web-based platform for mass customisation and personalisation. CIRP J. Manuf. Sci. Technol. **7**(2), 112–128 (2014)
24. In-depth: Industry 4.0 2021. https://www.statista.com/study/66974/in-depth-industry-40/. Accessed 02 Mar 2022
25. Birtel, M., Ruskowski, M.: Developing and implementing human-centered information services in a modular production environment. Procedia Manufac. **51**(11), 592–597 (2020)

26. Colledani, M., Angius, A.: Integrated production and reconfiguration planning in modular plug-and-produce production systems. CIRP Ann. **68**(1), 435–438 (2019)
27. Stączek, P., Pizoń, J., Danilczuk, W., Gola, A.: A digital twin approach for the improvement of an autonomous mobile robots (AMR's) operating environment - a case study. Sensors **21**(23), 7830 (2021)

A Digital Twin Platform-Based Approach to Product Lifecycle Management: Towards a Transformer 4.0

Henrique Silva[1,2](✉), Tomás Moreno[1,2], António Almeida[1,2], António Lucas Soares[1,2], and Américo Azevedo[1,2]

[1] INESC TEC, Rua Dr. Roberto Frias, 4200-465 Porto, Portugal
henrique.d.silva@inesctec.pt
[2] Faculty of Engineering, University of Porto, Rua Dr. Roberto Frias 378, 4200-465 Porto, Portugal

Abstract. Recently, we have been observing a significant evolution in products, machines, and manufacturing processes, towards a more digital and interoperable reality. In this sense, the power transformers sector has also been evolving to develop smart transformers for the future, capable of providing the digital capabilities to leverage new services and features that follow its entire life cycle, from the design and manufacturing to the use and dismantling/recycling. In this sense, this paper aims to present and demonstrate how an innovative digital twin platform can be used in a secure and trustable way for the enhancement of the power transformers' performance and potential lifespan, enabling, at the same time, the promotion of new business models. A real use case is also presented to demonstrate the applicability of Asset Administration Shells (AAS) for power transformer life cycle management, as well as the use of the International Data Spaces (IDS) for the secure and trustable horizontal interoperability along with the different actors of the value chain, from the manufacturers to the power network and maintenance services companies.

Keywords: Smart transformers · Asset administration shell · Digital Twin · Product lifecycle

1 Introduction

Industry 4.0 is designated as the fourth industrial revolution and is often associated with smart manufacturing [1] and is supported by innovative digital technologies such as the Industrial Internet of Things, Cyber-Physical Systems, Big Data, and other future-oriented technologies. [2]. Such types of technologies lead to the integration, automation, and optimization of manufacturing processes [3, 4]. This revolutionary context of integrated industry [1] enabled the concept of the Digital Twin (DT) as a means to connect the physical power transformers (PT) in the manufacturing shop floor with the digital space [5, 6] and its capabilities, therefore enabling digital data-driven services.

J. Machado et al. (Eds.): icieng 2022, LNME, pp. 14–25, 2023.
https://doi.org/10.1007/978-3-031-09360-9_2

As methods of energy generation and distribution change, transformer design and manufacturing must adapt to meet evolving global standards, reduce coil losses and improve transformer performance. Transformer design today is affected by several changing conditions in our electricity distribution networks. Energy generation, for example, is increasingly being achieved by renewable sources like wind and solar power, replacing more traditional forms of energy generated through power plants. Applications such as electric mobility are gaining popularity, meaning increased pressure on the transmission and distribution network.

The influence of the digital in this sector is also growing as seen in smart grid development, in the impact of complex monitoring and automated processes enabled by the increasing prevalence of cyber-physical systems, and last, but not least, the importance of global energy efficiency regulations.

The development and production of power transformers as is done today, is no longer compatible with this 'Energy 4.0' trend, due to a series of reasons:

- Technical information flow is not sufficiently open, shared, and efficient, e.g., in an internet of things era where all information is potentially shared, product and process design information or testing data still predominantly follows in a unidirectional and sequential flow from the project to the suppliers, customers, shopfloor and other stakeholders;
- Relevant information from product operation and usage during the lifecycle remains low, limiting not only the failure mode and effects and root cause analysis but also the continuous product improvement through an expansion of the design knowledge base;
- Designers and other stakeholders which make the creative workforce behind a product/service are often stuck on repetitive tasks, that, while crucial to its development, add little to no value to the final product, and as such industry is not making the most of its talent basis and failing to implement true open innovation practices.

Driven by the development of new digital concepts, power transformers have noteworthy opportunities, creating new affordances, through innovative data-enabled functions, leading to more competitive products for a fast-evolving sector that strongly impacts our quality of life and economic competitiveness. The different technical concepts identified and articulated in this paper are to serve as the underlying infrastructure for the development of a next generation PT that, grounded into the digital era of the DT, place the PT itself as the core informational asset for the development of new and innovative product-service systems.

2 Literature Review

2.1 Digital Twin

The DT can be perceived as an opportunity to optimize the performance of a manufacturing process. In the Industry 4.0 context, manufacturing is evolving from knowledge-based manufacturing to smart manufacturing, in which "smart" refers to the use of data-driven technologies [7]. On these service-oriented technologies, data and information can be perceived as the common ground and the key element [8]. In order to take advantage of the digital capabilities of cloud computing and artificial intelligence, it is crucial to connect the digital to the physical entity [7]. Thus, requiring the use of sensors to harvest data (on the physical entity) [7] in real-time and transmitted through IIoT (the communication between the digital and physical) [9, 10] throughout the product's lifecycle [7].

Furthermore, the DT can be separated into three different instances, the Digital Model (DM), Digital Shadow (DS) and Digital Twin (DT). *Digital Model* is the digital representation (model) of a non-existing physical entity, meaning that the digital entity will be the projection of a future physical entity [11]. This means that at the DM phase there is no communication between the physical and digital entity and all the processes and integration will be emulated. *Digital Shadow* differs from the DT in the sense that it does not support bidirectional communication. In the DS the data will flow automatically from the physical entity to the digital but manual from the digital entity to the physical [11]. *Digital Twin* is formed by three instances, the digital object, the physical object, and the communication between both objects [2, 12]. The digital entity is the representation of the physical entity in a digital environment and can monitor, emulate and control the behavior of the physical. In the DT, the data will flow continuously, bidirectional, and in real-time between the entities.

2.2 RAMI4.0 and Asset Administration Shell

Impacted by Industry 4.0, manufacturing industries were revolutionized and globalized, this stressed the development of a Reference Architecture Model for the Industry 4.0 (RAMI 4.0) [3, 5, 13, 14] to ensure standardization between enterprises that use I4.0 [15]. RAMI 4.0 is a service-oriented architecture that combines the fundamental elements and IT of Industry 4.0 in a structured three-dimension layer model [5, 15]. RAMI 4.0 is composed of three-axis, the product lifecycle, hierarchy levels in the factory, and the architecture. The product lifecycle is divided into type and instance. Type refers to the construction plan (i.e., development, construction, maintenance software…), covering the product lifecycle from the idea of the product to its manufacturing process [15, 16].

Within the RAMI4.0 architecture, the term Asset Administration Shell (AAS) starts to have its relevance in the technological world and can be related to DT implementation. According to [17], an "asset" can be categorized in everything that requires some sort of "connection" for an Industry 4.0 solution (e.g., machines and their components, contracts, orders, and supply materials). Moreover, the AAS will be the identification card of an asset in Industry 4.0. Furthermore, it will be responsible to provide controlled access to said information and provide a representation of the entire lifecycle of the asset [17]. The AAS aim to "enable partners in value creation networks to exchange meaningful information by conforming to a specified set of standardized elements" [18].

Besides, the AAS provides a standardization within digitized industrial production where the formatting of data/information will be according to a specified set of standardized elements [17]. Thereby, it will be composed of common standards of communication structures, rules for cyber-security, data protection, and language [17]. Furthermore, the AAS can be distinguished into three types: passive, reactive and proactive [19, 20].

- The *passive AAS* is a file written in JSON, XML, RDF, or AML that contains the catalog data of a particular asset that can be used to exchange within different partners.
- The *reactive AAS* supports the use of API to create a client/server connection, thus the content is made available using an interface.
- The *proactive AAS* in addition to supporting the API, client/server connection can ensure the desired behavior of the autonomous systems by containing decision-making skills.

2.3 International Data Spaces

International Data Spaces (IDS) initiative has the goal to create a safe data space in which industry enterprises can safely share their data assets [21] with different enterprises within the IDS ecosystem. IDS is defined by the IDS Association as a "Distributed network of Data Endpoints (i.e., instantiations of the IDS Connector), allowing a secure exchange of data and guaranteeing Data Sovereignty" [22]. Moreover, IDS allows establishing a secure connection between IDS connectors in which data is shared safely and can only be accessed according to the terms defined by the participants (typically the data owner and data user), guaranteeing data sovereignty for the data owner [22]. Therefore, with IDS, the implementation of smart devices can be done in a safe manner allowing innovative business processes to be created [21].

3 Power Transformer Digital Twin and Lifecycle Management

Figure 1 presents the initial design for the proposed DT to be used in this paper. This design is composed of the three main components of the DT namely, the physical object, the digital object, and the communication between both objects. Regarding the *physical object*, this component resides at the location of the PT and requires additional elements; the AAS deployments and the manager software. Whereas the data is generated from the PT and the AAS deployment will be responsible for the creation and storage of proactive AAS. Furthermore, providing an API to which the manager software will be able to request said AAS files. On this API it should also be possible to distinguish different types of AAS. Moreover, it should be possible for different types of request routes to the API, namely, requesting maintenance AAS or operation AAS. The software manager will then be responsible for pushing the AAS to the IDS connector from the provider of data that will afterward publish the contents to the IDS connector from the consumer of data. The *communication layer*, according to the DT requirements, must support bidirectional communication. Once the AAS contents are published to the IDS connector from the consumer, the secure data exchange between the data provider and consumer is completed, ensuring data sovereignty and ownership according to the terms defined by both participants. Thereby the AAS contents are ready to be processed by the digital services provider, used for emulation and preventive maintenance plans, improving the PT overall decision-making process.

Referring to the DT components, this will refer to the *digital object*, in which data will be processed and emulated onto a copy of the transformer in a digital environment capable of ensuring predictive maintenance and improving the overall performance of the transformer. Data will be displayed in interactive dashboards. Thus, supporting the decision-making process of the machine performance, capable of remotely and automatically controlling the transformer processes.

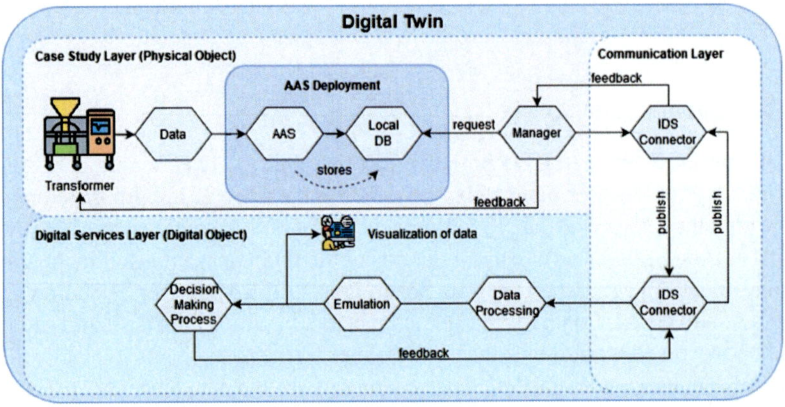

Fig. 1. Digital Twin design for a power transformer

Therein surges the premise of following the PT along its entire lifecycle supported by digital data-driven services built on top of the digital twin platform including the beginning of life, from conceptualization, definition and commissioning, middle of life, covering PT operation, usage and maintenance, and end of life, with disassembly, recycling and remanufacturing. It is then possible to distinguish the concept on two lifecycles connected between each other through stages: PT lifecycle & DT lifecycle.

1. In the *design* stage, through the *development of a digital model*, it is possible to visualize and materialize PTs in a more efficient and informed manner by leveraging historical data from previous designs, and from PTs in different stages of the lifecycle.
2. *Production stage*, the *digital model instantiation* can validate the PT concepts and designs in a safe digital environment and provide high fidelity models to optimize the product assembly process.
3. *Commissioning* stage, the *physical and digital object connection* can be eased with the DM to test the integration of the PT in the manufacturing process, reducing the risk of an unsuccessful implementation and the need for physical testing procedures.
4. In the *middle-of-life* stage, the *data-driven digital services* are provided by a set of data-driven tools that, in a product-service business model, generate added value for customers.
5. *End-of-life* refers to the tail end of the PT lifecycle and includes the *product status assurance* of the different components that can be reclaimed for refurbishing or remanufacturing and, finally, disposal.

Data, information and knowledge recovered throughout all these phases by the DT should be integrated and analyzing fueling other instances of the DT, in different stages of the lifecycle with historical data, crucial for continuous improvement (Fig. 2).

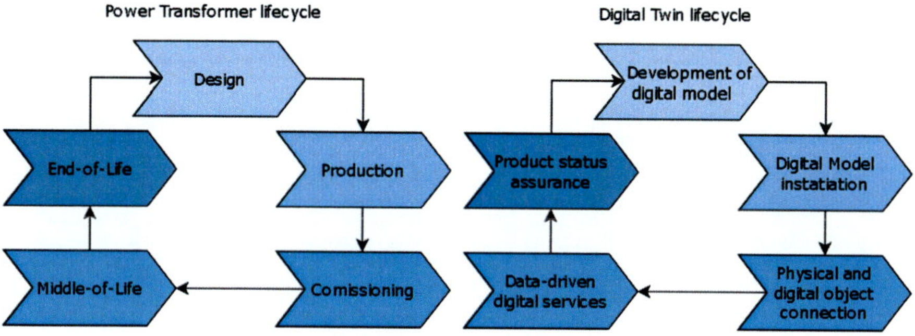

Fig. 2. Relationship between the PT & DT lifecycle

4 Framework Presentation

4.1 Digital Twin Platform

Digital platform (DP) is a term that is used in many fields of research. Conceptualizations range from technical point of view to socio-technical focused, defining platforms as a set of technical elements along with its associated organizational processes and standards [23, 24]. Starting from a socio-technical perspective [25] summarizes the characteristics that differentiate platforms into four properties: (i) compressed evolution as the capacity of platforms to shorten the period required to observe different market dynamics; (ii) evolution as a predictor of suitability; (iii) the ability to harness and leverage external disruptions; and (iv) the ability of architecture and governance to shape evolution.

Along with these properties, [26] see DP an innovative architecture for sharing assets such as data, models and services with ecosystems of end-users, enterprises, and other platforms. Our vision for a DT platform sees DPs as the infrastructure to support the DT as the core element for (i) managing product data collected from the multiple instances of DTs, and (ii) integrating design information (models, specifications, design data, and others) [27]. By leveraging a multi-sided platform model this structure, from an intra-organizational perspective provides DT-based tools to support the complete PLM; from an inter-organizational perspective, it can generate a marketplace of products and services that can be used to leverage the development of new business models.

Figure 3 presents an architecture that realizes our vision of the DT as the platform's central component and illustrates its essential components while comparing it to the reference ISO/DIS 23247-2 DT framework for manufacturing.

4.2 Interoperability Strategy Based on IDS

Data is the fundamental element of smart manufacturing and thus crucial when intending to implement a DT platform. In order to achieve this, it is mandatory to create a secure

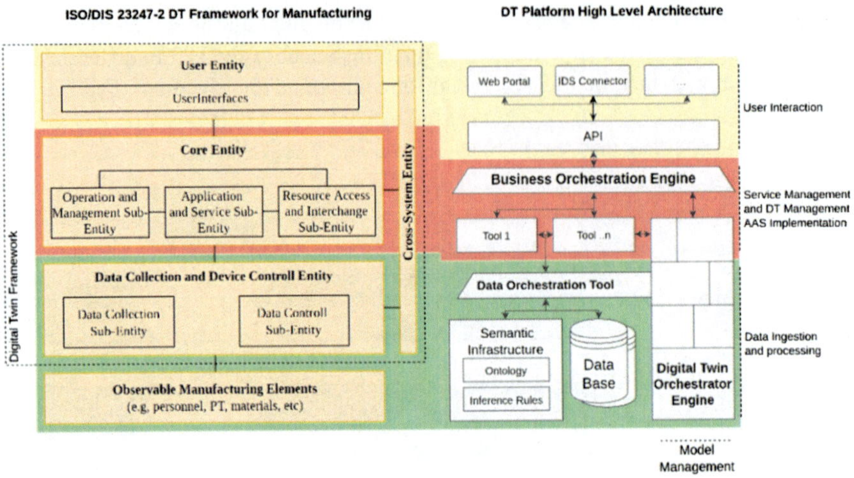

Fig. 3. Comparison of DT platform architecture and the ISO/DIS 23247-2 DT framework for manufacturing

and trusted channel of communication between the manufacturing shop floor of PT owners and the manufacturing industry of the PTs that will be responsible to provide the digital data-driven services built on top of the digital twin platform. Thereby with this statement, two major challenges arise. (i) Ensuring data sovereignty whereas the data assets generated by the PT remain in control of the PT owner even when used by the PT manufacturer to provide the digital data-driven services. (ii) Ensuring data security means that the data assets exchanges between the PT owner and the PT manufacturer are secure and protected. Based on this, this paper conducts an interoperability strategy based on the IDS to face these two major challenges. IDS allows establishing a secure connection between IDS connectors. Based on this, it is possible to adapt the communication layer of the DT with the inclusion of IDS connectors, and thus the innovative digital data-driven digital services are supported by a secure and reliable communication infrastructure.

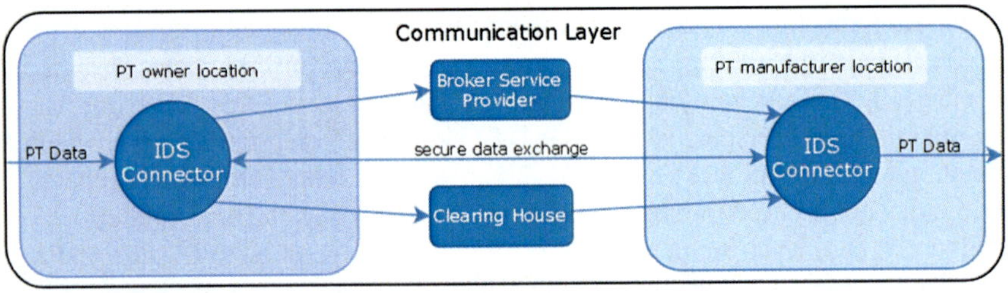

Fig. 4. Reformulated communication layer leveraging the IDS initiative

As shown in Fig. 4 data assets generated from the PT are pushed to the provider IDS connector (PT owner). There, the provider IDS connector publishes the data assets to the consumer IDS connector (PT manufacturer) in which the data assets are ready to be used by the digital data-driven services. The identity provider is capable of validating the identity of the desired target IDS connector, thus, avoiding unauthorized access to data.

Furthermore, in the Clearing House, the PT owner and the PT manufacturer log details regarding transitions of data assets. The usage of this module enables the evolving entities to resolve conflicts and provide financial reports for billing purposes. On top of that, the data exchanges within the IDS ecosystem are supported by a contractual agreement between the evolving entities, thereby, the PT owner can limit how their data assets are used by the PT manufacturer when providing the digital data-driven services. Data sovereignty is therefore ensured for the data owner.

Moreover, through the use of two IDS connectors, it is possible to establish trusted communication channels built in a secure ecosystem that enables enterprises at different locations to safely share their data assets while ensuring data ownership, sovereignty, and security. Thus, enabling new types of services and business models to be created. Through the use of the IDS initiative, PT manufacturers in addition to selling the physical product can now sell a set of digital data-driven services (on a subscription fee basis) that will support the PT performance.

4.3 Transformer AAS and Ontologies

The AAS will serve as the identification card of the power transformer and is responsible for providing a digital representation of the PT throughout its entire lifecycle. Furthermore, by utilizing the AAS it is possible to achieve a common understanding within digitized industrial production enterprises. The AAS also supports the use of external dictionaries such as cCl@ss and IEC CDD easing the process of assigning semantic descriptions to the AAS properties. Figure 5 displays the simplified AAS metamodel, in sum, an asset administration shell can be composed of multiple submodels, each submodel refers to a specific domain, and each submodel can be composed of multiple sub model elements and each sub model element can have one or more qualifier.

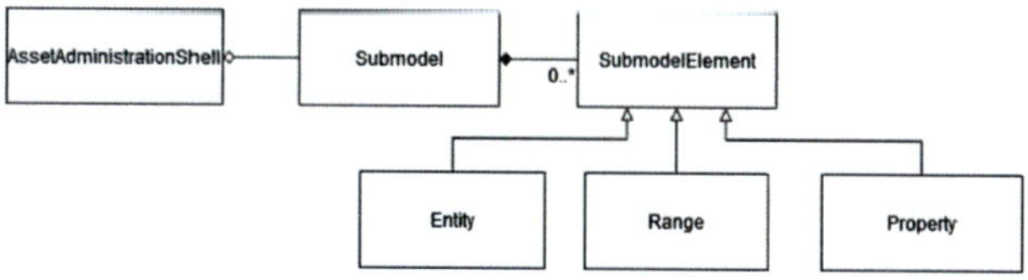

Fig. 5. Simplified asset administration metamodel

At the time of this writing, there are no frameworks developed capable of deploying proactive AAS, so as proof-of-concept was developed a passive AAS of a power transformer created with the IEC CDD dictionary in the software AASX Package Explorer as shown in Fig. 6.

This AAS is composed of three sub models, FixedPowerTransformers contains the technical data, identification contains data regarding the manufacturer of the PT, and

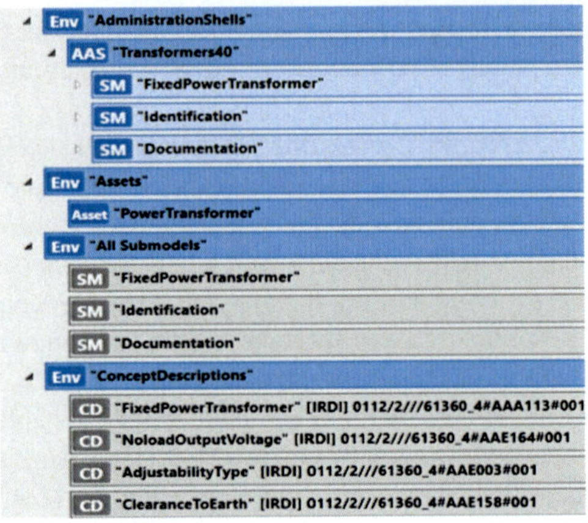

Fig. 6. Passive AAS for a fixed power transformer

documentation containing documents associated with the PT. Furthermore, ConceptDescriptions includes the definition of variables according to the IEC CDD, thus, ensuring semantic interoperability.

5 Use Case Presentation

Current power transformer development processes remain traditional in nature, relying on document-based information exchange and a set of PLM and simulation tools that are not interconnected [27]. We believe that our DT and the DT enabled DP visions described above when adopted and applied to the PT lifecycle can effectively shape its technologies and processes towards the Industry 4.0 standards. In partnership with a Portuguese company in the energy field, the platform is under active development to deliver to key company stakeholders a set of core tools and services to support the different stages of the PT lifecycle.

Supporting the conceptual design phase, the DT Platform can provide the engineering departments with reference PT designs that can serve as blueprints for developing new PT that fit a new customer's specific needs and requirements. Going further and leveraging the semantic descriptions of all the components of the PT and the datasets associated with all the components themselves, this new design can be tested and validated using simulation tools that leverage data gathered from real-world conditions of applications in previous projects. In the detailed design phases, the ability to detect design non-conformities based on operational insights, requirements, and rules simulation of critical operating conditions can be managed through the platform, providing the organization with critical information that will further optimize future PT development processes.

From an inter-organizational perspective, new value propositions will be leveraged from the DT platform, enabling the development of a new business model supported by it. Supported by this DT platform, the enterprise will be able to streamline PT development, resulting in reductions in cost and lead time, improve the design by certifying that it meets requirements and expectations, improve knowledge capture and sharing, and finally, add additional value to the product which will result in a better offering altogether.

6 Conclusions

Power transformers sector has also been evolving to develop smart transformers for the future, capable of providing the digital capabilities to leverage new services and features that follow its entire life cycle, from the design and manufacturing to the use and dismantling/recycling. In this sense, there is a strong opportunity to develop and adopt the Digital Twin concept to implement this vision. Nevertheless, there are a series of issues that arise to guarantee its full implementation. The first issue is related to data and knowledge management, in a comprehensive and effective way. The second issue is related to data governance and data sovereignty, between the different actors of the value chain and along the distinct stages of the product life cycle. To cope with these challenges, this paper proposes a new architecture for the management of digital twin and its instances. For this architecture, both Asset Administration Shell (AAS), the International Data Spaces (IDS) and an ontological data representation of the power transformer, based on standards, is proposed to deal with the issues previously presented. Finally, a real use case from a Portuguese power transformer producer is presented, pointing to potential benefits of this innovative approach for its business model and services enhancement.

Future work follows two streams. The development of technical intra-organizational components specific for the PT use case is ongoing, while research for the development of general reference models and architectures able to fit other market sectors is currently in progress.

Acknowledgments. The first author was funded by the Ph.D. Grant UI/BD/152565/2022 from Portuguese funding agency, FCT - Fundação para a Ciência e a Tecnologia.

The project TRF4p0 - Transformer 4.0 represented this work is cofinanced by the European Regional Development Fund - ERDF, through COMPETE - Operational Program Competitiveness and Internationalization (POCI) and by the Foundation for Science and Technology under the MIT Portugal Program under POCI-01-0247-FEDER-045926.

References

1. Luthra, S., Mangla, S.K.: Evaluating challenges to Industry 4.0 initiatives for supply chain sustainability in emerging economies. Process Saf. Environ. Prot. **117**, 168–179 (2018). https://doi.org/10.1016/j.psep.2018.04.018
2. Rolle, R.P., Martucci, V.D.O., Godoy, E.P.: Digitalization of manufacturing processes: proposal and experimental results. In: 2019 IEEE International Workshop on Metrology for Industry 4.0 and IoT, MetroInd 4.0 and IoT 2019 – Proceedings, pp. 426–431 (2019). https://doi.org/10.1109/METROI4.2019.8792838
3. Tissir, S., Fezazi, S.El., Cherrafi, A.: Industry 4.0 impact on lean manufacturing: literature review. In: 2020 IEEE 13th International Colloquium of Logistics and Supply Chain Management LOGISTIQUA 2020, pp. 2–4 (2020). https://doi.org/10.1109/LOGISTIQUA49782.2020.9353889
4. Piccarozzi, M., Aquilani, B., Gatti, C.: Industry 4.0 in management studies: a systematic literature review. Sustain **10**, 1–24 (2018). https://doi.org/10.3390/su10103821

5. Aheleroff, S., Xu, X., Zhong, R.Y., Lu, Y.: Digital Twin as a Service (DTaaS) in Industry 4.0: an architecture reference model. Adv. Eng. Inf. **47**, 101225 (2021). https://doi.org/10.1016/j.aei.2020.101225
6. Hermann, M., Pentek, T., Otto, B.: Design principles for Industrie 4.0 scenarios. In: Proceedings of the Annual Hawaii International Conference on System Sciences 2016, pp. 3928–3937 (2016). https://doi.org/10.1109/HICSS.2016.488
7. Tao, F., Zhang, M., Nee, A.Y.C.: Digital Twin and Big data. Digit. Twin Driven Smart Manuf. 183–202 (2019). https://doi.org/10.1016/b978-0-12-817630-6.00009-6
8. Bazaz, S.M., Lohtander, M., Varis, J.: Availability of manufacturing data resources in digital twin. Procedia Manuf. **51**, 1125–1131 (2020). https://doi.org/10.1016/j.promfg.2020.10.158
9. Rosen, R., Von Wichert, G., Lo, G., Bettenhausen, K.D.: About the importance of autonomy and digital twins for the future of manufacturing. IFAC-PapersOnLine **28**, 567–572 (2015). https://doi.org/10.1016/j.ifacol.2015.06.141
10. Qi, Q., Tao, F.: Digital Twin and big data towards smart manufacturing and Industry 4.0: 360 degree comparison. IEEE Access **6**, 3585–3593 (2018). https://doi.org/10.1109/ACCESS.2018.2793265
11. Kritzinger, W., Karner, M., Traar, G., et al.: Digital Twin in manufacturing: a categorical literature review and classification. IFAC-PapersOnLine **51**, 1016–1022 (2018). https://doi.org/10.1016/j.ifacol.2018.08.474
12. Grieves, M.: Digital Twin: Manufacturing Excellence through Virtual Factory Replication - A Whitepaper by Dr. Michael Grieves. White Paper 1–7 (2014)
13. Hofmann, E., Rüsch, M.: Industry 4.0 and the current status as well as future prospects on logistics. Comput. Ind. **89**, 23–34 (2017). https://doi.org/10.1016/j.compind.2017.04.002
14. Cai, Y., Starly, B., Cohen, P., Lee, Y.S.: Sensor data and information fusion to construct digital-twins virtual machine tools for cyber-physical manufacturing. Procedia Manuf. **10**, 1031–1042 (2017). https://doi.org/10.1016/j.promfg.2017.07.094
15. Schweichhart, K.: RAMI 4.0 reference architectural model for Industrie 4.0. InTech **66**, 15 (2019)
16. Federal Ministry for Economic Affairs and Energy: Plattform Industrie 4.0 - RAMI4.0 – a reference framework for digitalisation. Plattf Ind 40 (2019)
17. Bader, S., Barnstedt, E., Bedenbender, H., et al.: Details of the asset administration shell. Plattf Ind **40**(1), 473 (2019)
18. Belyaev, A., Block, C., Boss, B., et al.: Modelling the Semantics of Data of an Asset Administration Shell with Elements of ECLASS, vol. 54 (2021)
19. Motsch, W., Sidorenko, A., David, A., et al.: Electrical energy consumption interface in modular skill-based production systems with the asset administration shell. Procedia Manuf. **55**, 535–542 (2021). https://doi.org/10.1016/j.promfg.2021.10.073
20. Ocker, F., Urban, C., Vogel-Heuser, B., Diedrich, C.: Leveraging the asset administration shell for agent-based production systems. IFAC-PapersOnLine **54**, 837–844 (2021). https://doi.org/10.1016/j.ifacol.2021.08.186
21. Data, I., Association, S.: International Data Spaces Ids – a Standard for Data Sovereignty and an, pp. 1–4 (2019)
22. Otto, B., Steinbuß, S., Teuscher, A., Lohmann, S.: Reference Architecture Architecture (2019)
23. Sun, R., Gregor, S., Keating, B.: Information technology platforms: conceptualisation and a review of emerging research in IS research. In: Australasian Conference on Information Systems, pp 1–17 (2015)
24. de Reuver, M., Sørensen, C., Basole, R.: The digital platform: a research agenda. J. Inf. Technol. 1–12 (2017). https://doi.org/10.1057/s41265-016-0033-3
25. Tiwana, A.: Platform Ecosystems: Aligning Architecture, Governance, and Strategy. Newnes (2013)

26. Hunter, R., Coleman, M.: Competing for top talent: build the talent platform. Gart Res ID G 308714 (2016)
27. Silva, H.D., Azevedo, M., Soares, A.L.: A vision for a platform-based digital-twin ecosystem. IFAC-PapersOnLine **54**, 761–766 (2021). https://doi.org/10.1016/j.ifacol.2021.08.088

Value Creation in Technology Service Ecosystems - An Empirical Case Study

Agostinho Silva[1,2,3](✉) and Arminda Pata[4]

[1] CEFAGE-UE, Center for Advanced Studies in Management and Economics, Évora, Portugal
a.silva@zipor.com

[2] MARE, Marine and Environmental Sciences Centre, Polytechnic Institute of Leiria, Leiria, Portugal

[3] CIIC, Computer Science and Communication Research Centre, Polytechnic Institute of Leiria, Leiria, Portugal

[4] D. Dinis Higher Institute (ISDOM), Marinha Grande, Portugal

Abstract. Efficiency and market performance improving from technology-enabled service ecosystems is a hot topic in the literature. What outcomes on Value Creation Processes when companies evolve from the current state to the next level of the technology-based service ecosystem? Moving away from the conventional conception to a deeper and clear-cut understanding of digital-enabling, this research aimed to evaluate the efficiency and market performance of technology-enabled service ecosystems by using an interpretive research methodology. As an alternative to the traditional goods-dominant logic (GDL), the service-dominant logic (SDL) offers a theoretical foundation that elucidates the performative nature of markets, technologies, and business models, overcoming unbounded views of rationality in business model development through unification rather than dividing. An observational study on Portuguese ornamental stone sector was conducted. It was empirically concluded that innovation processes on value creation, improve positively when ornamental stones companies evolve operations from their current state to the next level of technology-enabled operations. Therefore, technology-enabled operations are recommended to the stone companies.

Keywords: Industry 4.0 · Service ecosystems · Service-dominant logic · Service · Value creation first section

1 Introduction

The literature highlights the digital transition as a driver to improve business models, reduce dependence on local support [1], enable effective regulation, and establish more equitable, fair, and socially responsible global markets [2].

From residual activities in the 1930s, services nowadays represent the main activity in developed economies and the fastest growing activity in these economies. In the migration to service economies, labor productivity has increased by transforming knowledge into value within increasingly specialized technological companies [3]. In other words, labor once used in the fields has migrated in these economies to public and research

J. Machado et al. (Eds.): icieng 2022, LNME, pp. 26–36, 2023.
https://doi.org/10.1007/978-3-031-09360-9_3

institutions, technological companies, and many other employers of specialized labor, supporting agricultural and industrial activities.

Although they were relatively adequate for a world less connected, according to some authors, the mental models for thinking about business, markets, marketing and value creation, in this "services" current time, in general, seems not appropriated [4, 5]. For these authors, this traditional focus on units of output, that once produced, the end-users consume is designated in the literature, as goods dominant logic (GDL) [6]. Schematically, in a GDL view, the producer creates value, puts the products on the loading dock ready for a waiting market that's demanding [7]. Despite its simplicity on supply chains and production [8], this model seems to elevate the firm to the creator of value, and the consumer to the destroyer of value [9].

Opposed to GDL, the service-dominant logic (SDL) perspective is based on the premise that broad sets of actors dynamically integrate and apply resources through the service-for-service exchange to cocreate value [10]. Moreover, according to SDL literature, (1) service is the fundamental basis of exchange [11]; (2) value is always cocreated by multiple actors, including the beneficiary; (3) all economic and social actors are resource integrators [5]; (4) value is always unique, and phenomenological determined by beneficiary [6], and (5) ecosystems coordination arises from actor-generated institutions and institutional arrangements. These institutions are signs, symbols, rules, social norms, regulations, languages, all those shortcuts that we have agreed upon to use to be able to create value. That is, the heuristics jointly agreed to be used by ecosystem's members [12]. From this institutional view, a service ecosystem is defined as a relatively self-contained, self-adjusting systems of resource-integrating actors connected by shared institutional arrangements and mutual value creation through service exchange [13].

Looking at the world through this service ecosystem view, the producer and the consumer are just two actors in a network of interacting actors, accessing, and giving access to a variety of resources to cocreate value and coordinated by institutions and institutional arrangements [14]. From the literature, this leads to a value creation narrative [15], where the actors are entities involved in resource integration and service exchange enabled and constrained by endogenously generated institutions and institutional arrangements, establishing nested and overlapping service ecosystems, which is the context for future resource integration service exchange [16]. This SDL view gives an understanding of services in the ever-widening space of increasingly connected objects: people and entities have become an important new challenge for academics, practitioners, and policymakers [17].

This SDL structural perspective makes the micro-level phenomena more understandable in which the most important feature consists of institutions, defined as rules, norms, meanings, symbols, practices, and similar aides to collaboration and, more generally, institutional arrangements, as interdependent assemblages of institutions [13]. Moreover, from this service perspective in value creation, a technology-enabled structured network becomes a service ecosystem [18] and therefore, its improvement and study can be supported on SDL philosophical axioms [10]. This research starts with a (1) resource-based institutional service perspective foundations on service ecosystems, followed by the (2) theoretical innovation sources identifications, (3) the theoretical innovation processes,

(4) the innovation processes on value creation, (5) observational study operationalization, and (6) data collection and analysis; and value creation assessment.

2 Technology-Enabled Institutionalization of the Ecosystems

Under the SDL institutional view, the literature highlights the role of the innovation sources in service ecosystems as a driver to: (1) improve business models; (2) reducing dependence on local support [1]; (3) enabling an effective regulation; and (4) establishing more equitable, fair and socially responsible global markets [2]. Moreover, these Innovation sources are related to the service system entities reconfiguration of resources in the ecosystem [19]. In other words, the network perspective inherent to SDL value cocreation and service systems [20] suggests new opportunities to configure all the resources needed to solve a given problem at a given time and place [21]. More specifically, innovation sources seem to play a determinant role in the company's transition from its current state to the next technology-enabled level and therefore, from the institutional perspective, we may argue that the technology-enabled level is related to the institutional arrangement (IA) in a particular moment of the company's life.

The SDL literature also highlights the role of innovation processes in the service ecosystems as drivers for actors to integrate the resources in social contexts that multiple actors also actively and collectively construct [11]. This boosts the actors performance by liquification of their resources [19], as it is referred to dematerialization through the separation of information from the physical side of the resource [22].

Moreover, the literature points out digital-enabled customer-oriented actions as innovation processes on value creation for marketing intensity, and from this, the following question may be stated: What outcomes on Value Creation Processes when companies evolve from the current state to the next level of the technology-based service ecosystem? To find an answer to this question, an observational empirical study will be conducted to evaluate whether, value creation processes improve or worsens when companies migrate their operations from the current state to the next technology-enabled institutional level on the ecosystem.

3 Methodological Approach

According to some authors, observational studies causation can be inferred if the researcher imposes restrictive assumptions, such not third constant variables [23]. Moreover, an observational methodological approach can ever be "determinate" [24], and requires that two or more concepts are linked to a causal chain, consisting of assertions about the relations between concepts.

Since the positivism worldview privileges an interpretive research methodology, the researcher starts from the general to the particular [25], implying the need to conceptualize a model (Fig. 1), followed by empirical testing.

This empirical test requires quantitative and qualitative constructs, variables, and concepts. These data must be collected from measuring instruments, observable checklists or even numerical records, among others [26].

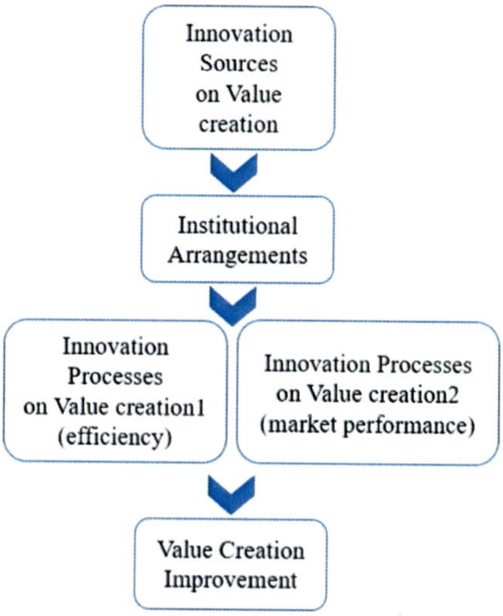

Fig. 1. Framework for observational study.

3.1 Observational Study and Application Field

Consisting mainly of Small and Medium-sized Enterprises (SMEs), the ornamental stone Portuguese companies (OS) operate either upstream or downstream. Representing more than 16,600 direct jobs and being one of the main generators of private employment in inland regions, the OS sector has economic relevance in the Portuguese economy. Although the contribution of exclusive upstream activities is significant to national wealth comes from the downstream activities [27].

As the result of multiple collaborative Research and Development (R&D projects, this sector has undergone successive waves of digitalization, which, according to some authors, are the driver that places Portugal as the eighth country in the international trade of OS and the second country in the world international trade per capita [28].

The first digital wave resulted from the Jetstone R&D Project: a partnership among ten entities, from which, it was developed technologies to respond to the market the raw materials in flexibly, obtain significant results in the reduction of raw material waste, and respond quickly to small orders. The second digital wave resulted from the Inovstone R&D Project. A more comprehensive partnership involving nineteen companies and scientific entities: recent studies point out that Inovstone technologies brought to OS companies a significant approach towards Industry 4.0 operations [29]. The third digital wave resulted from the Flexstone R&D Project: allowing OS companies to bring their operations close to the industry 4.0 level. The fourth digital wave resulted from the Inovstone4.0 R&D Project: through the digital technologies developed under Inovstone4.0 scope, OS companies seem to be taking a significant step towards collaborative manufacturing/smart manufacturing, oriented towards the co-creation of value in a digital marketplace, with Building Information Modelling (BIM) users.

There were identified 802 OS companies operating in downstream activities in mainland Portugal and employing at least five people.

These companies represent the population to be used in this research, from which a sample of 669 companies were surveyed along with this research, corresponding to 83.4% of the total population. To collect the independent variables, a local survey by the authors of this research was made on each one of these companies' facilities to verify their technology-enabling level in loco and objectively, from January 2019 to December 2019.

3.2 Theoretical Innovation Sources

According to the SDL literature, a connectable environment tends to be a source of innovation for interactions in the ecosystems. Therefore, the first Innovation Source (*InnovSource1*) can be defined as the operant resources directly involved in the production, enabling the transition to the first technology-enabled service ecosystem level. This $InnovSource_1$ is an innovation driver mostly related to the Jetstone innovations, reducing waste, increase production flexibility, and respond quickly to small orders.

Moreover, the literature refers to network Interactions as innovation sources for service exchange. Therefore, the second Innovation Source (*InnovSource2*) is defined as the combination of *InnovSource1* with the digital operand resources, integrating the operations management technologies. This $InnovSource_2$ is an innovation driver mostly related to the Inovstone innovations, technology-enabling companies to respond in an optimized and integrated way to medium scale projects, obtaining a significant reduction waste and an increasing capacity to respond to international market medium-size orders.

The literature refers to service-exchange as a source of innovation technology knowledge share [12, 30]. Therefore, the third innovation source (*InnovSource3*) is defined as the combination of InnovSource1 and InnovSource2 with the digital operational resources, for the integration of the company's activities with the digital production equipment. This $InnovSource_3$ is an innovation driver, mostly related to the Flexstone, technology-enabling companies to respond in a flexible, optimized, and integrated way to medium scale international projects.

The literature also refers the technology-knowledge-share as a source of innovation for proposals through technology-enabled symbols (Market) [4, 31, 32]. Therefore, the fourth innovation source (InnovSource4) is defined as the combination of InnovSource1, InnovSource2 and InnovSource3 with the digital operand interoperable resources for full integration into digital marketplace customers' BIM systems. The $InnovSource_4$ is an innovation driver, mostly related to the Inovstone4.0 technology, enabling companies to operate collaboratively into digital marketplaces, co-creatively customizing products with each BIM customer in an optimal and integrated way and responding to large-scale inquiries.

3.3 Theoretical Innovation Processes

As stated above, from SDL perspective all the actors entities are resource integrators and part of social contexts that multiple actors also actively and collectively construct [11]. Therefore SDL foundations support liquification [19], that is, the dematerialization through the separation of information from the physical side of the resources [22], is referred as a integration driver on resource density [33]. Keeping this view, technological innovation (such as the digital-enabled connectivity) is referred as the base for resource integration in a service ecosystem environment, ensuring the resource density increasing, boosting impartiality and trust in the ecosystem.

The SDL literature refers digital-enabling as Innovation Processes, i.e., innovation drivers for operant resources integration [34]. Therefore, it may be defined the first innovation process (*InnovProcess1*) is defined as a dependent variable, assessed by the labour productivity given by the relation between the company outcome and the operant resources.

$$InnovProcess_1 = Average(\frac{Outcome}{Operant_{Resources}}) \tag{1}$$

Moreover, the resources liquification through technology is refereed as innovation processes for Technological Combinatorial Evolution on sustainable relationships among stakeholders [34–36]. Therefore, it may be defined the second innovation process ($InnovProcess_2$) as a dependent variable, assessed through the relation between the company exports and the operant resources. That is, the company performance in international markets.

$$InnovProcess_2 = Average(\frac{Exports}{Operant_{Resources}}) \tag{2}$$

3.4 Innovation Processes on Value Creation

For SDL, as all economic actors are resource integrators, value is cocreated by multiple actors, through dematerialization mechanisms [37]. Resources liquification play a central role in the market actions, towards the value maximizing including the beneficiary and ensuring the ecosystem sustainability [19]. That is, the digital-enabled customer-oriented actions are innovation processes on value creation for marketing intensity [38]. Therefore, the first Innovation Process on Value Creation ($InnovProc\_on\_VCreation_{(1,IAn\text{-}1,n)}$) is defined as a dependent variable, assessed in terms of productivity gains, when companies evolve on their current institutional arrangement to the next one.

$$InnovProcess\_on\_ValueCreation_{(1,IAn-1,n)} \frac{InnovProcess_{1\,(IAn)}}{InnovProcess_{1\,(IAn-1)}} \times 100\% \tag{3}$$

$$(\mathrm{n} = 1,\ 2,\ 3 \text{ or } 4)$$

Moreover, the technology-enable institutions and institutional arrangements are innovation processes on value creation for market collaborative processes [39].

Therefore, the second Innovation Process on Value Creation ($InnovProc\_on\_VCreation_{(2,IAn\text{-}1,n)}$) is defined as a dependent variable, assessed in terms of the market performance gains, when companies evolve on their current institutional arrangement to the next level.

$$InnovProcess\_on\_ValueCreation_{(2,IAn-1,n)} = \frac{InnovProcess_{2\,(IAn)}}{InnovProcess_{2\,(IAn-1)}} \times 100\% \tag{4}$$

$$(\mathrm{n} = 1,\ 2,\ 3 \text{ or } 4)$$

4 Observational Study Operationalization

An observational approach intends to confirm if two concepts are linked. Therefore, the purpose of this research is to evaluate the innovation processes, when companies evolve operations from their current state to the next level of technology enabled operations.

To collect data to test observational study, each one of the 669 companies were surveyed in site, from January 2019 to December 2019. To each company, the following survey guidelines was used, (1) number of direct employees in 2018; (2) turnover in 2018; (3) 2018 exports; (4) current computerized production machines (5) current Enterprise Resource Planning System (ERP) (6) current ERP level of integration, and (7) current interoperable systems with BIM platforms.

From this survey, the quantitative data was collected: *Operant_Resources* – representing the number of human resources, and *Outcomes* – representing the turnover. For qualitative data collection, the current technology-enabled level of each company was evaluated: The *IA#0* is the level attributed to those companies not using computerized machines for production - qualitative level *IA#0* means the company did not start on the path of technology-enabling its operations; *IA#1* is the level attributed to those companies using productive computerized equipment, since from at least one year - qualitative level *IA#1* qualifies companies that, boosted by the $InnovSources_1$, already initiated on the path of technology-enabled service ecosystem; *IA#2* is the level attributed to those companies integrating digital production technologies together with management systems.

Boosted by the $InnovSources_2$ technologies, these *IA#2* classified companies have evolved their technology-enabled network to service ecosystem; *IA#3* is the level attributed to companies integrating information from computerized machines, together with operations management systems. Boosted by the $InnovSources_3$ an *IA#3* classified company evolved the technology-enabled internal network to a global business network; and, boosted by $InnovSources_4$ the *IA#4* level is attributed to companies integrating collaboratively the data generated by computerized production machines, together with digital marketplaces.

5 Data Collection and Analysis

From the 669 companies surveyed, 410 companies were found without computerised production machines and therefore, no technology-enabled service ecosystem was initiated yet. From those companies, 125 were qualitatively classified as *IA#1*: boosted by $InnovSources_1$ technologies, this second group of companies already started the way towards technology-enabled operations. A third group of 94 companies were classified as *IA#2*: boosted by $InnovSources_2$ technologies, these companies are already integrating production technologies together with management technologies. A fourth group of 25 companies was qualitatively classified as *IA#3*: boosted by $InnovSources_3$ technologies, these companies can integrate the information generated by computerised production and logistic equipment with the company's management and industrial engineering, that is, have evolved to a technology-enabled service ecosystem. Finally, a group of 15 companies wase qualitatively classified as *IA#4*: boosted by $InnovSources_4$ technologies, these companies collaboratively integrate the data generated by computerised machines on the shop floor together with industrial management and engineering and together with the digital marketplace.

Table 1. KPI's assessment from data collected.

KPI's	IA#0	IA#1	IA#2	IA#3	IA#4
InnovProcess1 (€)	73,964	83,091	87,331	90,838	96,149
InnovProcess2 (€)	10,862	33,971	61,212	81,676	93,542

From this representative sample of 669 companies, 61.3% of were qualitatively classified as *IA#0*, 18.7% as *IA#1*, 14.1% as *IA#2*, 3.7% as *IA#3*, and 2.2% as *IA#4* (Table 1).

6 Value Creation Assessment

By using four complementary analyses, this observational objective was to evaluate the innovation processes, when companies evolve operations from current state to the next level of technology enabled operations, by using the Eqs. (3) and (4) (Fig. 2).

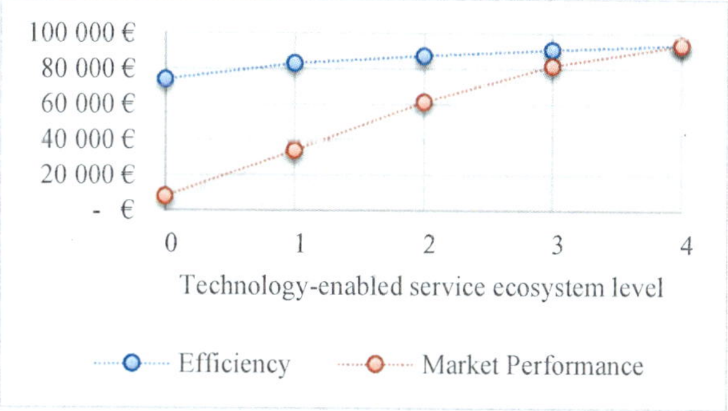

Fig. 2. Companies' performance and technology-enabled service ecosystem improvement.

Complementary analyses 1: shifting the operations **from the IA#0** to IA#1, the average outcomes on value creation processes were assessed: $InnProcValCreation_{(1\text{-}IA0,1)} =$ 112.3% and $InnProcValCreation_{(2\text{-}IA0,1)} =$ 312,8%.

Complementary analyses 2: shifting the operations **from the IA#1** to IA#2, the average outcomes on value creation processes were assessed: $InnProcValCreation_{(1\text{-}IA1,2)} =$ 105.1%; $InnProcValCreation_{(2\text{-}IA1,2)} =$ 180,2%.

Complementary analyses 3: shifting the operations **from the IA#2** to IA#3, the average outcomes on value creation processes were assessed: $InnProcValCreation_{(1\text{-}IA2,3)} =$ 104.0%; $InnProcValCreation_{(2\text{-}IA2,3)} =$ 133,4%.

Complementary analyses 4: shifting the operations **from the IA#3** to IA#4, the average outcomes on value creation processes were assessed: $InnProcValCreation_{(1\text{-}IA3,4)} =$ 105,8%; $InnProcValCreation_{(2\text{-}IA3,4)} =$ 114,5%.

Moreover, when a stone company evolves from the level ia#1 to ia#4, *innovprocess1* increases 22,185€, that is one employee contributes with more than 22,000€ per year to the company outcome.

7 Conclusions

From a resource-based institutional service perspective, the following research question stated: what outcomes on Value Creation Processes when companies evolve from the current state to the next level of the technology-based service ecosystem?

To find an answer to this question, an observational study was made along 2019, in a representative sample of 669 Portuguese ornamental stone companies. Along this observational study, four complementary analyses were conducted regarding the behavior of stone companies in the Innovation Processes on Value Creation when operations move from their current state to the next level of technology-enabled service ecosystem:

(1) analyzing the company's outcome when they evolve operations from the lowest institutional arrangement (IA#0) to the first level of a technology-enabled service ecosystem (IA#1), the average efficiency improves 12.3% and market performance improvement is 3,1 times; (2) analyzing the company's outcome when they evolve the operations from the IA#1 to IA#2, the average efficiency improves 5.1% and Market Performance improves 1.8 times; (3) analyzing the company's outcome when they evolve the operations from the IA#2 to IA#3, the average efficiency improves 4.0% and the market performance improves 1.3 times; (4) and analyzing the companies outcome when they evolve the operations from the IA#3 to IA#4, the average efficiency improves 5,8% and the market performance improves 1.5 times.

Moreover, it was found that each human resource working in stone companies operating in collaborative service ecosystem mode, in average, contributes with more 22,185€ per year to the company outcome, when compared to a human resource, working in a company that did not start the technology-enabled ecosystem way. From these findings, we empirically conclude that the Innovation Processes on Value Creation improve positively when ornamental stones companies evolve operations from their current state to the next level of technology-enabled operations. Therefore, technology-enabled operations are recommended to the stone companies.

References

1. Giraldo-Diaz, J.M., Fuerst, S.: Shipstra: the lean start-up of a digital freight-forwarding marketplace. Glob. Bus. Organ. Excell. **38**(5), 6–17 (2019). https://doi.org/10.1002/joe.21946
2. Bandara, R., Fernando, M., Akter, S.: Addressing privacy predicaments in the digital marketplace: a power-relations perspective. Int. J. Consum. Stud. **44**(5), 423–434 (2020). https://doi.org/10.1111/ijcs.12576
3. Demirkan, H., Spohrer, J.: T-shaped innovators: identifying the right talent to support service innovation. Res. Manag. **1**(6), 12–15 (2015). https://doi.org/10.5437/08956308X5805007

4. Akaka, M., Vargo, S., Schau, H.: The context of experience. J. Serv. Manag. **26**(2), 206–223 (2015). https://doi.org/10.1108/JOSM-10-2014-0270
5. Vargo, S.L., Akaka, M.A., Wieland, H.: Rethinking the process of diffusion in innovation: a service-ecosystems and institutional perspective. J. Bus. Res. **116**, 526–534 (2020). https://doi.org/10.1016/j.jbusres.2020.01.038
6. Vargo, S.L., Lusch, R.F.: Inversions of service-dominant logic. Mark. Theory **14**(3), 239–248 (2014). https://doi.org/10.1177/1470593114534339
7. Bettencourt, L.A., Lusch, R.F., Vargo, S.L.: A service lens on value creation: marketing's role in achieving strategic advantage. Calif. Manage. Rev. **57**(1), 44–66 (2014). https://doi.org/10.1525/cmr.2014.57.1.44
8. Siltaloppi, J., Koskela-Huotari, K., Vargo, S.: Institutional complexity as a driver for innovation in service ecosystems. Serv. Sci. **8**(3), 333–343 (2016). https://doi.org/10.1287/serv.2016.0151
9. Vargo, S.L.: On marketing theory and service-dominant logic: connecting some dots. Mark. Theory **11**(1), 3–8 (2011). https://doi.org/10.1177/1470593110393715
10. Vargo, S.L., Lusch, R.F.: Institutions and axioms: an extension and update of service-dominant logic. J. Acad. Mark. Sci. **44**(1), 5–23 (2015). https://doi.org/10.1007/s11747-015-0456-3
11. Edvardsson, B., Tronvoll, B., Gruber, T.: Expanding understanding of service exchange and value co-creation: a social construction approach. J. Acad. Mark. Sci. **39**(2), 327–339 (2011). https://doi.org/10.1007/s11747-010-0200-y
12. Akaka, M., et al.: The role of symbols in value cocreation. Mark. Theory **14**(3), 311–326 (2014). https://doi.org/10.1177/1470593114534344
13. Vargo, S.L., Lusch, R.F.: Service-dominant logic 2025. Int. J. Res. Mark. **34**(1), 46–67 (2017). https://doi.org/10.1016/j.ijresmar.2016.11.001
14. Michel, S., Vargo, S.L., Lusch, R.F.: Reconfiguration of the conceptual landscape: a tribute to the service logic of Richard Normann. J. Acad. Mark. Sci. **36**(1), 152–155 (2008). https://doi.org/10.1007/s11747-007-0067-8
15. Vargo, S.L., Lusch, R.F.: It's all B2B…and beyond: toward a systems perspective of the market. Ind. Mark. Manag. **40**(2), 181–187 (2011). https://doi.org/10.1016/j.indmarman.2010.06.026
16. Hsu, Y.: Integrating service science and information system technology: a case study. Int. J. Organ. Innov. **9**(1), 158–173 (2016). https://doi.org/10.4236/ti.2011.22015
17. Maglio, P., Kieliszewski, C., Spohrer, J., Kelly, L., Patrício, L., Sawatani, Y.: Handbook of Service Science, vol. II (2019)
18. Vargo, S., Wieland, H., Akaka, M.: Innovation through institutionalization: a service ecosystems perspective. Ind. Mark. Manag. **44**(2013), 63–72 (2015). https://doi.org/10.1016/j.indmarman.2014.10.008
19. Michel, S., Vargo, S.L., Lusch, R.F.: Reconfiguration of the conceptual landscape: a tribute to the service logic of Richard Normann. J. Acad. Mark. Sci. **36**(1), 152–155 (2008). https://doi.org/10.1007/s11747-007-0067-8
20. Maglio, P.P., Vargo, S.L., Caswell, N., Spohrer, J.: The service system is the basic abstraction of service science. Inf. Syst. E-bus. Manag. **7**(4), 395–406 (2009). https://doi.org/10.1007/s10257-008-0105-1
21. Simmonds, H., Gazley, A., Daellenbach, K.: Theorising change and stability in service ecosystems: a morphogenetic approach. Serv. Ind. J. **38**(3–4), 250–263 (2018). https://doi.org/10.1080/02642069.2017.1389908
22. Viriyasitavat, W., Hoonsopon, D.: Blockchain characteristics and consensus in modern business processes. J. Ind. Inf. Integr. **13**, 32–39 (2019). https://doi.org/10.1016/j.jii.2018.07.004

23. Radicic, D., Pugh, G., Douglas, D.: Promoting cooperation in innovation ecosystems: evidence from European traditional manufacturing SMEs. Small Bus. Econ. **54**(1), 257–283 (2018). https://doi.org/10.1007/s11187-018-0088-3
24. Della Porta, D., Keating, M.: Approaches and Methodologies in the Social Sciences. Cambridge University Press, Cambridge (2008)
25. Ghauri, P., Kjell, G., Ivar, K.: Research Methods in Business Studies. A Practical Guide. 4th edn. (2010)
26. Creswell, J.: Research Design: Qualitative, Quantitative, and Mixed Methods Approaches (2014)
27. Silva, A., Gil, M.: Industrial processes optimization in digital marketplace context: a case study in ornamental stone sector. Results Eng. **7**, 100152 (2020). https://doi.org/10.1016/j.rineng.2020.100152
28. Silva, A., Rabadão, C., Capela, C.: Towards Industry 4.0 | A case study of BIM deployment in ornamental stones sector. Int. J. Innov. Technol. Explor. Eng. **67**, 24535 (2020)
29. Vilas-Boas, J., Mirnoori, V., Razy, A., Silva, A.: Outlining a new collaborative business model as a result of the green building information modelling impact in the AEC supply chain. In: Camarinha-Matos, L.M., Afsarmanesh, H., Antonelli, D. (eds.) PRO-VE 2019. IAICT, vol. 568, pp. 405–417. Springer, Cham (2019). https://doi.org/10.1007/978-3-030-28464-0_35
30. Marx Gómez, J., Mora, M., Raisinghani, M.S., Nebel, W., O'Connor, R.V. (eds.): Engineering and Management of Data Centers. SSRISE, Springer, Cham (2017). https://doi.org/10.1007/978-3-319-65082-1
31. Kleinaltenkamp, M., Corsaro, D., Sebastiani, R.: The role of proto-institutions within the change of service ecosystems. J. Serv. Theory Pract. **28**(5), 609–635 (2018). https://doi.org/10.1108/JSTP-12-2017-0241
32. Müller, J.M., Kiel, D., Voigt, K.-I.: What drives the implementation of Industry 4.0? The role of opportunities and challenges in the context of sustainability. Sustainability **10**(1), 247 (2018). https://doi.org/10.3390/su10010247
33. Ng, I., Maglio, P.P., Spohrer, J., Wakenshaw, S.: The study of service: from systems to ecosystems to ecology. In: The SAGE Handbook of Service-Dominant Logic, pp. 230–240, SAGE Publications Ltd, Ed. 1 Oliver's Yard, 55 City Road London EC1Y 1SP (2018)
34. Lusch, R., Nambisan, S.: Service innovation: a service-dominant logic perspective. MIS Q. **39**(1), 155–175 (2015). https://doi.org/10.25300/MISQ/2015/39.1.07
35. Pohlmann, A., Kaartemo, V.: Research trajectories of Service-Dominant Logic: emergent themes of a unifying paradigm in business and management. Ind. Mark. Manag. **63**, 53–68 (2017). https://doi.org/10.1016/j.indmarman.2017.01.001
36. Wieland, H., Hartmann, N.N., Vargo, S.L.: Business models as service strategy. J. Acad. Mark. Sci. **45**(6), 925–943 (2017). https://doi.org/10.1007/s11747-017-0531-z
37. Normann, R., Ramirez, R.: From value chain to value constellation: designing interactive strategy. Harvard Bus. Rev. **71**(4), 65–77 (1993). http://search.ebscohost.com/login.aspx?direct=true&db=bth&AN=9309166477&site=eds-live
38. La Rocca, A., Caruana, A., Snehota, I.: Measuring customer attractiveness. Ind. Mark. Manag. **41**(8), 1241–1248 (2012). https://doi.org/10.1016/j.indmarman.2012.10.008
39. Breidbach, C., Maglio, P.: Technology-enabled value co-creation: an empirical analysis of actors, resources, and practices. Ind. Mark. Manag. **56**, 73–85 (2016). https://doi.org/10.1016/j.indmarman.2016.03.011
40. Silva, A., Dionisio, A., Almeida, I.: Enabling cyber-physical systems for Industry 4.0 operations: a service science perspective. Int. J. Innov. Technol. Explor. Eng. **9**(8), 838–846 (2020). https://doi.org/10.35940/ijitee.H6804.069820
41. Da Silva, A., Dionísio, A., Coelho, L.: Improving Industry 4.0 through service science. Int. J. Serv. Sci. **7**(2), 58 (2020). https://doi.org/10.1504/IJSSCI.2020.113189

Industry 4.0 Technologies for the Sustainable Management of Maintenance Resources

Stanislaw Legutko(✉)

Poznan University of Technology, 3 Piotrowo Street, 60-965 Poznan, Poland
stanislaw.legutko@put.poznan.pl

Abstract. Application of modern maintenance approaches such as Maintenance 4.0 is highlighted as one of the prevailing smart & sustainable manufacturing topics. The goal of this paper is to describe the latest trends within the area of maintenance management from the perspective of the challenges of the fourth industrial revolution and the economic, environmental and social challenges of sustainable development. The four stages of the machine maintenance approach related to the four industrial revolutions are characterized together with benefits and weaknesses of each approach to machine maintenance. Knowledge-Based Maintenance evolution has been shown. Comparison of maintenance data in different maintenance ages has been characterized. Less than three decades ago the concept of Green Maintenance (GMn) was proposed as a support for the implementation of the Green Manufacturing paradigm. Digitalization can empower maintenance services by making use of collected data and advance technologies to monitor the equipment health, diagnose faults, predict and troubleshoot failures well before it could happen and even optimize the performance. Intelligent and sustainable maintenance has been considered in three perspectives.

Keywords: Industry 4.0 · Knowledge Based Maintenance · Maintenance 4.0 · Predictive maintenance · Sustainability

1 Introduction

A few decades ago, maintenance was mostly seen as something that had to be done after such a failure, but it was also something that was difficult to manage. Nowadays, maintenance is widely recognized as an essential business function and a critical element of asset management. Organizations increasingly realize that they can improve their efficiency and reliability by planning maintenance interventions more effectively. This results in more preventive maintenance actions that are also better aligned with other business functions, such as production scheduling and spare parts control. Over the past 80 years, maintenance has evolved from an activity that is essential to repairing resource failures into embedded asset technology, into an essential part of generating profit and ultimately into a strategic component of every organization's cooperative partnership. Koochaki et al. [1] state that in the process industry unplanned downtime costs up to US $250,000 per hour. Recently, according to [2] maintenance processes have become very important for industry as their impact represents a total of 15–60% of the total operating costs of all manufacturing.

J. Machado et al. (Eds.): icieng 2022, LNME, pp. 37–48, 2023.
https://doi.org/10.1007/978-3-031-09360-9_4

2 Industrial Maintenance from Time Perspective

The industrial revolutions have significantly influenced the way of defining, organizing and implementing the maintenance function in the enterprise (Fig. 1).

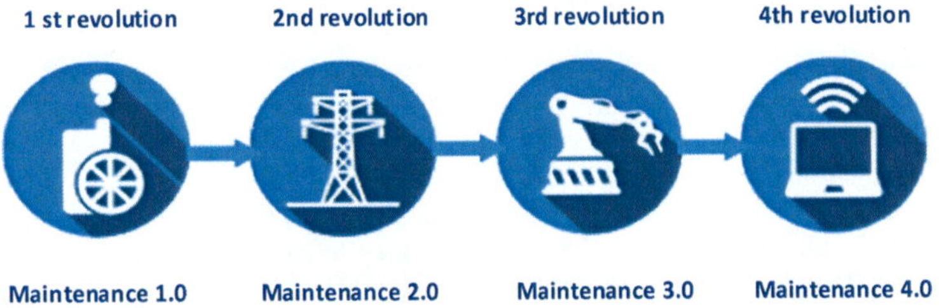

Fig. 1. Industrial revolutions and subsequent concepts for machine maintenance.

In the reactive approach to production equipment maintenance (**Maintenance 1.0**), the goal is to restore the damaged machine to normal operating conditions. The high costs of unplanned production downtime and the growing requirements for the efficiency of the use of the machine park have resulted in a change in the approach to maintenance aimed at preventing the occurrence of a failure, not just its quick removal. The concept of a preventive scheduled repair system (**Maintenance 2.0**) has emerged. Its essence is to carry out maintenance activities on machines and devices at fixed intervals, often using the checklist of the original equipment manufacturer's recommendations. Automation of production processes and even greater complexity of machines paved the way for the next generation - **Maintenance 3.0** (also known as Condition Based Maintenance - CBM). The expectations for the maintenance function in the enterprise were greater availability and reliability of the equipment, better product quality, long service life of the machines and high cost efficiency. CBM is a maintenance program that recommends maintenance activities (decisions) based on information obtained in the process of monitoring the technical condition of machines. The next generation of CBM (**Maintenance 4.0**), referred to in the literature as Predictive Maintenance (PdM), is related to the development of ICT. The use of digital technologies in maintenance gave rise to the concept of eMaintenance [3], which is further developed using Industry 4.0 technologies [4–6]. In the context of Industry 4.0, the maintenance function is often referred to as Smart Maintenance, Intelligent Maintenance, Maintenance 4.0. According to [7], "Smart Maintenance stands for an intelligent and learning maintenance management focusing on permanent improvement". According to [8] "Maintenance 4.0 does predictive analytics and suggests feasible solution, with major application in Industry 4.0 and especially on those maintenance aspects that deals with collection of data, its analysis and visualization and asset decision-making". According to [9] Smart Maintenance is "an organizational design for managing maintenance of manufacturing plants in environments with pervasive digital technologies".

One of the operational strategies most often referred to in the context of Maintenance 4.0 is Predictive Strategy (PdM), which uses condition monitoring data to detect anomalies (i.e., identify deviations from normal operating conditions) in production processes, production equipment and products, and diagnosing (i.e. characterizing the existing state of deviations) and forecasting (i.e. predicting the future evolution of deviations from normal operating conditions to failure). The main task of PdM is to forecast the Remaining Useful Life (RUL) of a technical facility in order to effectively use the intervals between maintenance and optimize maintenance activities.

The use of PdM has a number of benefits, and also involves overcoming several challenges. On the one hand, the benefits of PdM include improved productivity, reduction of system failures, minimization of unplanned downtime, increased efficiency in the use of financial and human resources, and optimization of maintenance planning [10, 11]. On the other hand, overcoming challenges involves the need to: integrate data from different sources and systems within a facility, which is important for gathering accurate information to develop predictive models. Another important aspect is the use of artificial intelligence and overcoming challenges such as: acquiring training data, dealing with dynamic operating environments, choosing an ML algorithm that better fits a given scenario, and the need to obtain contextual information related to the production process implementation environment [12]. Moreover, it should also be noted that PdM is not always the best solution [10]. The opportunity offered by Industry 4.0 technologies in maintenance consists rather in the possibility of defining the optimal operating strategy for each component of the machine, taking into account its specificity within the system, e.g. applicable safety regulations and environmental protection, quality standards, importance for business, physical features and functional, etc.

At the highest level of maturity, Maintenance 4.0 not only predicts the probability of failure, but also retrieves standard maintenance tasks from the library to determine the best actions to avoid failures. The concept of prescriptive maintenance goes beyond just predicting failure. Based on the analysis of historical and real-time data, the system predicts the required maintenance actions and determines the course of action. The main goal of this strategy is to reduce operational risk, eliminate a preventive approach based on planned renovations in favor of a proactive approach and intelligent maintenance planning [13, 14]. According to Grijalvo Martín et al. [15] prescriptive maintenance is primarily about sharing value with customers through more services, and with stakeholders in the internal ecosystem of a company through service delivery capabilities.

A new approach to maintenance requires the mastery of both new technologies and organizational issues related to their planning, implementation and maintenance. From a technology perspective, each of the nine pillars of Industry 4.0 (Autonomous/Collaborative Robots (A/CR), Augmented Reality (AR), Incremental Manufacturing (AM), Big Data & Analytics (BDA), Cloud Computing (CC), the Internet of Things (IoT), horizontal/vertical integration (HVI), Simulations (S) and Cybersecurity (CS) can play an important role in maintaining the movement of machines and devices in the enterprise [14, 16]. The use of these technologies has a significant impact on the maintenance processes and other areas of the company involved in maintenance, including logistics (management of spare parts and warehouse can be based on PHM results), health, safety and work environment (information about an intelligent system can be used for current risk monitoring), design (the use of intelligent components can lead to an increase in the reliability of solutions), etc. [17]. Summarizing, it is possible to formulate the benefits and weaknesses of the various approaches to machine maintenance (Table 1).

Table 1. Benefits and weaknesses of each approach to machine maintenance.

	Intention	Benefits	Weaknesses
Maintenance 1.0	Only undertake maintenance when a complete equipment failure occurs	No upfront planning or scheduling	Unpredictable equipment availability, shorter equipment lifetime, increased energy costs, and potentially lower production yield due to partial malfunctions
Maintenance 2.0	Carrying out regular services of the maintenance to avoid partial or complete failure of the equipment Preventive maintenance can be performed at specified intervals or when a certain condition is met	Increasing the availability of production equipment by performing maintenance at specified intervals and within a specified range	Premature component replacement and maintenance can increase costs
Maintenance 3.0	Actions carried out as needed, based on information from the monitoring system	Reduction of maintenance activities, only the necessary activities are implemented, which leads to lower costs and rational use of resources	Fatigue or uniform wear failures are not easily detected by CBM measurements
Maintenance 4.0 Predictive	Predicting a problem before it occurs and being able to estimate the remaining useful life (RUL) of the machine and/or its assemblies	Optimizes resources and lowers costs by predicting the service life of machine assemblies to avoid premature replacement and reduce unnecessary maintenance	Given that the prediction is probabilistic rather than deterministic, there is a risk of false positives that could lead to unnecessary maintenance activities

3 From Experience - Driven to Data - Driven Maintenance

Standard EN 13306:2017 defines maintenance as a combination of all the technical and associated administrative activities required to keep equipment, installations and other physical assets in the desired operating condition or to restore them to this condition. The implementation of these activities and their effectiveness requires appropriate data management, i.e. data collection, analysis and the use of adequate models to support decision-making. Data in maintenance is needed for various purposes: 1. Planning, monitoring, comparing and evaluating different maintenance activities (e.g. inspection, repair). 2. Building models supporting decision-making processes (e.g. defining service intervals, purchasing spare parts and consumables). 3. In order to constantly improve activities (e.g. reduce costs, service time, energy consumption). Over the years, both the possibilities of obtaining, collecting and processing data, as well as the scope of their use have changed [18].

The data related to the operation of machines was collected primarily in the operator's memory and was used to support decisions regarding the maintenance, repair and replacement of the machine. As there were no dedicated technical services in the companies, each maintenance decision was based on the empirical experience of the operator. Operators handled the data manually based on empirical experience. In the era of Maintenance 2.0 (Second Industrial Revolution), the method of obtaining and processing data has changed. In production companies, specialized departments were established to carry out maintenance and repair tasks, and the operators of machines and devices in practice did not participate in these activities and had no influence on their scope. Maintenance data was increasingly being obtained by maintenance technicians. Maintenance managers began to use more systematic methods of documenting (procedures, instructions, maintenance records, etc.) and analyzing maintenance data. Raw data was saved in written documents rather than stored in the machine operator's memory. Machine related data was used to support decisions about machine maintenance, repair and replacement. In particular, statistical models were introduced to analyze the failure rate, determine the time intervals for performing maintenance activities and their scope, and wear spare parts. Nevertheless, although scientific methods were increasingly used to analyze data, data were still acquired and analyzed by hand. As a result, the maintenance data utilization rate remained relatively low.

With the next industrial revolution, the role of information technologies and the scope of their application in production processes have significantly increased. As a consequence, enterprises could obtain more data from various functional areas, including production and maintenance. Many factors influenced these new opportunities. First of all, in order to improve the maintenance management (e.g. spare parts management, maintenance planning), companies started to implement IT systems, e.g. CMMS (Computer Maintenance Management System) was used. Next, OEM started to use IT systems such as CAD and CAM to modify and optimize new machines. Thirdly, sensors and computers began to be used to automatically control machines. Machine data from operational and maintenance processes began to be collected in computer systems and managed by IT systems.

With the appearance of Industry 4.0, condition monitoring techniques have evolved from visual inspection and manual analysis of data sets to real-time data-generating sensors and advanced data analysis methods. The data source (i.e. engineering systems versus human systems) is one of the major differences between traditional and recent data applications. Due to the intense development of ICT and its applications, and the development of sensor technology maintenance managers have gained a lot new data sources (e.g. sensors enable continuous measurement, monitoring and reporting of the current operating status of production devices and their working environment). These new data sources are rapidly generating large amounts of different types of data, known as big data. As such, data-driven approaches have become an important solution for intelligent maintenance [19–22].To explore data, advanced data analysis is required. Karim et al. [23] developed the concept for Maintenance Analytics, which consists of four phases: descriptive analytics, diagnostic analytics, predictive analytics and prescriptive analytics. This approach was used by Ansari et al. [24] to build another concept known as Knowledge-Based Maintenance (KBM). According to [25] "KBM collects machine (systems), process, and products data, which are then transmitted to three areas that provide overall strategies of maintenance, namely i) risk-based maintenance, ii) condition- or time-based maintenance, and iii) Total Productive Maintenance (TPM) and lean maintenance." Ansari et al. [24] define KBM as "a functional unit responsible to i) continuously support value generation and ii) facilitate developing and protecting maintenance collective knowledge across smart factories, which is enhanced by need- or opportunity-driven knowledge detection, discovery, modelling and representation approaches". KBM employs a variety of methods [24] including advanced statistics, stochastics, real-time computing and analytics, machine learning algorithms, static rule-based to "knowledge discovery, prediction, optimisation, adaptation, (self-)learning and ideally self-direction" (Fig. 2).

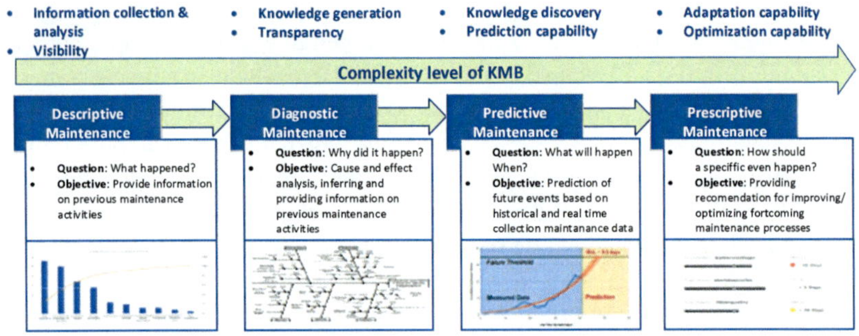

Fig. 2. Knowledge-Based Maintenance evolution; own elaboration on the base of [18, 24]

Taking into account the continuous development of methods of data acquisition and analysis, new concepts of data-based maintenance management will be created, enabling maintenance managers to improve the efficiency and availability of production equipment, reduce costs, and flexibility in approach to customer requirements. In addition, in the case of OEM, it will be possible to improve products to better meet the requirements of users and stakeholders throughout the life cycle.

To sum up; from the perspective of the data lifecycle, defined as "the process of data collection, transmission, storage, analysis, transfer, visualization, and application can be referred to the "data lifecycle"" [26] changes in the method of obtaining, storing, analyzing and managing data in different Maintenance ages are presented in Table 2.

Table 2. Comparison of maintenance data in different maintenance ages; source [18]

	Maintenance 1.0	Maintenance 2.0	Maintenance 3.0	Maintenance 4.0
Data Source	Operator experience	Maintainer and machines	Operator, maintainer, machines, information and computer systems	Operator, maintainer, information systems, suppliers data, OEM
Data Collection	Manual collection	Manual collection	Semi-automated collection	Automated collection via sensors and IoT
Data Storage	Operator memory	Written documents	Databases	Cloud
Data Analysis	Arbitrary	Reliability theory based on Bathtub curve assumption	Conventional algorithms	Fuzzy logic, Neutral networks, Machine learning Evolutionary algorithms,
Date Transfer	Verbal communication	Written documents	Digital files	Digital files
Data Management	N/A	Maintainer	Information systems	Cloud and Artificial Intelligence

4 Maintenance Scope from Time Perspective

Providing value to stakeholders and a balanced approach to their requirements is a key factor in the competitiveness of enterprises. Therefore, the production paradigms and the changing requirements of the stakeholders were adopted as a reference point to show the evolution of the scope of the maintenance functions. Each change in the production paradigm, and thus a change in how production is "performed", affects all functional areas of the enterprise. Therefore, along with the change in production paradigms, the approach to maintaining production equipment has changed, from the reactive approach corresponding to artisanal production, through preventive, lean (Lean Maintenance), green (Green Maintenance), to the contemporary approach in which it is considered a process, which should be managed in a sustainable perspective. The Lean TPM concept, also known as Lean Maintenance, is the answer to the Lean Manufacturing demand for reliable and stable machine operation. Smith and Hawkins [27] defined Lean Maintenance as a "proactive maintenance operation employing planned and scheduled maintenance activities through total productive maintenance (TPM) practices, using maintenance strategies developed through application of reliability centered maintenance (RCM) decision logic and practiced by empowered (self-directed) action teams using the 5S process, weekly Kaizen improvement events, and autonomous maintenance together with multi-skilled, maintenance technician-performed maintenance through the committed use of their work order system and their computer managed maintenance system (CMMS) or enterprise asset management (EAM) system". The concept of Lean Maintenance is aimed at identifying losses in maintenance processes and taking actions to eliminate or reduce them through proper management and continuous improvement. Losses can be identified in various maintenance processes, both those related to the direct implementation of maintenance activities (e.g. repairs, renovations) and those related to the flow of information [28–31]. In addition, losses in maintenance affect the effectiveness of customer service processes, production, logistics, etc. The scope of the impact of losses indicates, therefore, many groups of stakeholders inside the company that are affected by these losses, as well as those stakeholders with whom cooperation should be undertaken to reduce or eliminate these losses.

In the early nineties of the last century, the concept of Green Maintenance (GMn) was proposed as a support for the implementation of the Green Manufacturing paradigm. The concept of Green Maintenance is focused on the effective use of resources (consumables, spare parts, etc. according to Skoogh et al. [32], about 30% of energy consumption in industrial enterprises is wasted on machines in repair, idle and ready condition)), minimizing waste and the negative impact of production equipment on the environment throughout the life cycle of a technical facility, i.e. from design, through production, operation to decommissioning.

As described by [33] "proper maintenance is necessary to achieve optimal energy performance, while energy performance data is needed for effective maintenance management". Including the environmental dimension in the definition of maintenance obliges the managers to take into account the impact of maintenance activities on the natural environment in their decisions, identify the possibilities of reducing this impact and define, optimize and link environmental costs with maintenance costs [33, 34].

The practical implementation of the Green Maintenance concept in the enterprise depends on many factors, both internal (e.g. decision criteria when purchasing new machines/modernizing existing ones, maintenance strategy, manner of planning maintenance works, used repair technologies and materials, employees' competences and awareness) and external (e.g. design features of machines and devices). Therefore, the inclusion of new requirements (environmental issues) meant that maintenance must take into account the potential environmental hazards related to the occurrence of equipment failure in its activities and introduce risk analysis tools to decision-making processes related to the selection of maintenance strategies and proactively integrate the stages of the product life cycle. Meeting environmental commitments to internal and external stakeholders requires that different departments such as manufacturing, manufacturing, engineering treat maintenance as a complementary, interdependent function [35, 36].

With the growing influence of maintenance on various areas inside and outside the company, the challenges of sustainability have attracted the attention of both practitioners and academia. Sustainable development is an extremely important challenge for enterprises, and the range of issues it concerns (social, economic, environmental, technical, ethical) contributes to its ubiquity.

The maintenance function plays a vital role in promoting sustainability, reducing impact and integrating sustainability aspects to guarantee specified performance, availability and quality with limited resources and energy consumption. On the one hand, sustainable maintenance guarantees the compliance of processes and products and reduces their industrial impact on the economy, society and the surrounding environment. On the other hand, it must be a sustainable business function and limit its own revenues generated during all maintenance activities. According to Franciosi et al. [37] "Poor maintenance management leads to non-negligible economic, environmental and social impacts and obstacles to the sustainable manufacturing paradigm".

The economic dimension of sustainable development is one of the main areas that are heavily influenced by maintenance [36]. Although the maintenance itself is an unavoidable cost (every organization has machines and devices that need to be kept in motion), the cost of maintenance activities will not be so high if they are performed at the right time and scope [38, 39].

On the other hand, environmental aspects can be noticed throughout the entire life cycle of technical facilities, in which maintenance plays an important role at every stage of this cycle [40]. From that perspective, improperly defined and executed maintenance processes and activities lead to several environmental impacts, such as hazardous emissions, inefficient energy usage, ineffective resource consumption etc. [41, 42].

Building sustainable maintenance requires also taking into account social aspects. The social dimension of sustainable maintenance concerns, among others: the safety and health of employees, working hours (compliance with legal requirements), financial resources allocated to employee training and investments in new hardware and software supporting employees in their work [43]. Taking into account the fact that in the near future maintenance workers will need new competences that will allow for the effective use of modern technologies in the implementation of the operation process of complex and automated technical systems, the requirements for appropriate education and training will increase and may constitute a critical success factor of enterprise [44]. This applies in particular to the challenges of the fourth industrial revolution (Industry 4.0), to which the maintenance 4.0 concept is the answer.

5 Conclusions

When analyzing the development of requirements for the maintenance functions in the company resulting from changes in the production environment, it can be noticed that the maintenance function will take on a wider responsibility and is increasingly integrated with other functions in the company. The likely development scenario, and therefore the next step beyond communication, collaboration and integration, is the reorganization of companies to transform the functional organization into a fully integrated service-oriented organization. Digitalization can empower maintenance services by making use of collected data and advance technologies to monitor the equipment health, diagnose faults, predict and troubleshoot failures well before it could happen and even optimize the performance.

References

1. Koochaki, J., Bokhorst, J., Wortmann, H., Klingenberg, W.: Evaluating condition based maintenance effectiveness for two processes in series. J. Qual. Maint. Eng. **17**(4), 398–414 (2011)
2. Zonta, T., da Costa, C.A., da Rosa Righi, R., de Lima, M.J., da Trindade, E.S., Li, G.P.: Predictive maintenance in the Industry 4.0: a systematic literature review. Comput. Ind. Eng. **150**, 106889 (2020)
3. Iung, B., Levrat, E., Marquez, A.C., Erbe, H.: Conceptual framework for e-Maintenance: illustration by e-Maintenance technologies and platforms. Annu. Rev. Control. **33**(2), 220–229 (2009)
4. Johansson, N., Roth, E., Reim, W.: Smart and sustainable emaintenance: capabilities for digitalization of maintenance. Sustainability **11**(13) (2019)
5. Navas, M. A., Sancho, C., Carpio, J.: Disruptive maintenance engineering 4.0. International Journal of Quality & Reliability Management 37 (6/7), 853–871 (2020)
6. Turner, C.J., Emmanouilidis, C., Tomiyama, T., Tiwari, A., Roy, R.: Intelligent decision support for maintenance: an overview and future trends. Int. J. Comput. Integr. Manuf. **32**(10), 936–959 (2019)
7. Kinz, A., Bernerstaetter, R., Biedermann, H.: Lean smart maintenance – efficient and effective asset management for smart factories. In: MOTSP 2016 - 8th International Scientific Conference (2016)
8. Kumar, U., Galar, D.: Maintenance in the era of industry 4.0, issues and challenges. In: Quality, IT and Business Operations: Modeling and Optimization, pp. 231–250 (2018)
9. Bokrantz, J., Skoogh, A., Berlin, C., Wuest, T., Stahre, J.: Smart maintenance: an empirically grounded conceptualization. Int. J. Prod. Econ. **223**, 107534 (2020)
10. Compare, M., Baraldi, P., Zio, E.: Challenges to IoT-enabled predictive maintenance for industry 4.0. IEEE Internet Things J. **7**(5), 4585–4597 (2019)
11. Yoon, J.T., Youn, B.D., Yoo, M., Kim, Y., Kim, S.: Life-cycle maintenance cost analysis framework considering time-dependent false and missed alarms for fault diagnosis. Reliab. Eng. Syst. Safety **184**, 181–192 (2019)
12. Antosz, K., Jasiulewicz-Kaczmarek, M., Paśko, Ł, Zhang, C., Wang, S.: Application of machine learning and rough set theory in lean maintenance decision support system development. Eksploatacja i Niezawodnosc-Mainten. Reliab. **23**(4), 695–708 (2021)
13. Matyas, K., Nemeth, T., Kovacs, K., Glawar, R.: A procedural approach for realizing prescriptive maintenance planning in manufacturing industries. CIRP Ann. Manuf. Technol. **66**, 461–464 (2017)
14. Silvestri, L., Forcina, A., Introna, V., Santolamazza, A., Cesarotti, V.: Maintenance transformation through Industry 4.0 technologies: a systematic literature review. Comput. Indust. **123**, 103335 (2020)
15. Grijalvo Martín, M., Pacios Álvarez, A., Ordieres-Meré, J., Villalba-Díez, J., Morales-Alonso, G.: New business models from prescriptive maintenance strategies aligned with sustainable development goals. Sustainability **13**(1), 216 (2021)
16. Bokrantz, J., Skoogh, A., Berlin, C., Stahre, J.: Maintenance in digitalised manufacturing: delphi-based scenarios for 2030. Int. J. Prod. Econ. **191**, 154–169 (2017)
17. Kwon, D., Hodkiewicz, M.R., Fan, J., Shibutani, T., Pecht, M.G.: IoT-based prognostics and systems health management for industrial applications. IEEE Access **4**, 3659–3670 (2016)
18. Jasiulewicz-Kaczmarek, M., Legutko, S., Kluk, P.: Maintenance 4.0 Technologies—new opportunities for sustainability driven maintenance. Manag. Prod. Eng Rev. **11**(2), 74–87 (2020)

19. Cao, Q., et al.: KSPMI: a knowledge-based system for predictive maintenance in Industry 4.0. Robot. Comput. Integr. Manufac. **74**, 102281 (2022)
20. Gopalakrishnan, M., Subramaniyan, M., Skoogh, A.: Data-driven machine criticality assessment–maintenance decision support for increased productivity. Product. Plan. Control **33**(1), 1–19 (2022)
21. Pawellek, G.: Integrierte Instandhaltung und Ersatzteillogistik: Vorgehensweisen, Methoden, Tools. Springer, Berlin (2013)
22. Chen, C., Wang, C., Lu, N., Jiang, B., Xing, Y.: A data-driven predictive maintenance strategy based on accurate failure prognostics. Eksploatacja i Niezawodnosc – Mainten. Reliab. **23**(2), 387–394 (2021). https://doi.org/10.17531/ein.2021.2.19
23. Karim, R., Westerberg, J., Galar, D., Kumar, U.: Maintenance analytics–the new know in maintenance. IFAC-PapersOnLine **49**(28), 214–219 (2016)
24. Ansari, F., Glawar, R.: Tanja, Nemeth: PriMa: a prescriptive maintenance model for cyber-physical production systems. Int. J. Comput. Integr. Manuf. **32**(4–5), 482–503 (2019)
25. Lee, E.A.: What is real time computing? a personal view. IEEE Des. Test **35**, 64–72 (2017)
26. Siddiqa, A., Hashem, I.A.T., Yaqoob, I., Marjani, M., Shamshirband, S., Gani, A., et al.: A survey of big data management: taxonomy and state-of-the-art. J Network Comput Appl **71**, 151–166 (2016)
27. Smith, R., Hawkins, B.: Lean Maintenance: Reduce Costs, Improve Quality, and Increase Market Share. Elsevier Butterworth-Heinemann, MA (2004)
28. Antosz, K., Ratnayake, R.C.: Machinery classification and prioritization: empirical models and AHP based approach for effective preventive maintenance. In: 2016 IEEE International Conference on Industrial Engineering and Engineering Management (IEEM), pp. 1380–1386. IEEE (2016)
29. Jasiulewicz-Kaczmarek, M., Saniuk, A.: How to make maintenance processes more efficient using lean tools. In: Goossens, Richard H.M. (ed.) Advances in Social & Occupational Ergonomics. Proceedings of the AHFE 2017 International Conference on Social & Occupational Ergonomics 17–21 July 2017, Advances in Intelligent Systems and Computing, vol. 605, pp. 9–20 (2018)
30. Ribeiro, D.R.S., Forcellini, F.A., Pereira, M., Xavier, F.A.: An overview about the implementing of lean maintenance in manufacturing processes. J. Lean Syst. **4**(3), 44–59 (2019)
31. Marttonen-Arola, S., Baglee, D.: Assessing the information waste in maintenance management processes. J. Qual. Maint. Eng. **26**(3), 383–398 (2019)
32. Skoogh, A., Johansson, B., Hansson, L.: Data requirements and representation for simulation of energy consumption in production systems. In: Proceedings of the 44th CIRP Conference on Manufacturing Systems, pp. 1–3 (2011)
33. Lewis, A., Elmualim, A., Riley, D.: Linking energy and maintenance management for sustainability through three American case studies. Facilities **29**(5/6), 243–254 (2011)
34. Costantino, F., Di Gravio, G., Tronci, M.: Integrating environmental assessment of failure modes in maintenance planning of production systems. Appl. Mech. Mater. **295–298**, 651–660 (2013)
35. Ajukumar, V.N., Gandhi, O.P.: Evaluation of green maintenance initiatives in design and development of mechanical systems using an integrated approach. J. Clean. Prod. **51**, 34–46 (2013)
36. Fraser, K., Hvolby, H.H.: Maintenance management models: a study of the published literature to identify empirical evidence: a greater practical focus is needed. Int. J. Qual. Reliab. Manage. **32**(6), 635–664 (2015)
37. Franciosi, C., Di Pasquale, V., Iannone, R., Miranda, S.: Multi-stakeholder perspectives on indicators for sustainable maintenance performance in production contexts: an exploratory study. J. Qual. Maint. Eng. **27**(2), 308–330 (2020)

38. Jasiulewicz-Kaczmarek, M., et al.: Application of MICMAC, Fuzzy AHP and fuzzy TOPSIS for evaluation of the maintenance factors affecting sustainable manufacturing. Energies **14**(5), 1436 (2021)
39. Zhang, C., Zhang, Y., Dui, H., Wang, S., Tomovic, M.M.: Importance measure-based maintenance strategy considering maintenance costs. Eksploatacja i Niezawodnosc – Mainten. Reliab. **24**(1), 15–24 (2022). https://doi.org/10.17531/ein.2022.1.3
40. Macchi, M., Farruku, K., Holgado, M., Negri, E., Panarese, D.: Economic and environmental impact assessment through system dynamics of technology-enhanced maintenance services. Int. J. Ind. Syst. Eng. **23**(1), 36–56 (2016)
41. Al-Turki, U.M., Ayar, T., Yilbas, B.S., Sahin, A.Z.: Health, safety and sustainability in maintenance. In: Integrated Maintenance Planning in Manufacturing Systems. SAST, pp. 59–69. Springer, Cham (2014). https://doi.org/10.1007/978-3-319-06290-7_4
42. Xia, T., Xi, L., Du, S., Xiao, L., Pan, E.: Energy-oriented maintenance decision-making for sustainable manufacturing based on energy saving window. J. Manuf. Sci. Eng. **140**(5), 051001 (2018)
43. Vrignat, P., Kratz, F., Avila, M.: Sustainable manufacturing, maintenance policies, prognostics and health management: a literature review. Reliab. Eng. Syst. Saf. **218**, 108140 (2022)
44. Jasiulewicz-Kaczmarek, M., Drożyner, P.: Social dimension of sustainable development – safety and ergonomics in maintenance activities. In: Stephanidis, C., Antona, M. (eds.) UAHCI 2013. LNCS, vol. 8009, pp. 175–184. Springer, Heidelberg (2013). https://doi.org/10.1007/978-3-642-39188-0_19

Redesign of the Internal Logistics System of a Textile Supplier for the Automotive Industry

Ana Rita Matos, José Pedro Dias, Rui M. Lima(✉), Rui M. Sousa, and Maria Sameiro Carvalho

Department of Production and Systems, Algoritmi Centre, School of Engineering, University of Minho, Campus de Azurém, 4800-058 Guimarães, Portugal
{rml,rms,sameiro}@dps.uminho.pt

Abstract. The aim of this work is the design, implementation and evaluation of an internal logistic system based on a combination of Mizusumashi and e-Kanban, within a company of textile reinforcements production for the automotive industry. A critical review of literature allowed the selection of a set of steps for designing the Mizusumashi and e-Kanban systems, namely: specification of the quantity of material to be handled, the required working force and the material buffers, as well as the time of replenishment. The results showed a 50% reduction in the quantity of stock in the production area. This paper provides a critical look at internal supply lean tools and their application, showing that it is possible to redesign an internal supply system without almost any cost. Additionally, it provides a practical approach to Mizusumashi, which is a type of Lean solution that is not usually seen in process industries.

Keywords: Mizusumashi · Lean manufacturing · Process industry · Internal logistics system · Textile automotive industry

1 Introduction

The factors contributing to the increasing competitiveness in the industrial environment include the numerous changes in technology, the globalization of production industries and the reduction of the products life cycle. As a result, organizations have been facing market uncertainty and continuous change. To overcome the difficulties, companies have begun to develop new methods and perspectives to meet the market needs. So, to answer efficiently to this dynamic environment it should be created flexible company with several strategic options [1].

Aiming to increase the levels of quality and flexibility in customer service, with fewer costs and reduced life cycles, has gained relief from the application of Lean Manufacturing [2]. This approach is focused on creating the value desired by the client (not just value interpreted by the company), seeking to produce more with fewer resources and eliminating, or at least reducing, the activities that add no value to the product. Thus, the application of Lean Manufacturing reduces the use of resources such as space and time and leads to a decrease in stocks. One of the tools that support these goals is the

J. Machado et al. (Eds.): icieng 2022, LNME, pp. 49–60, 2023.
https://doi.org/10.1007/978-3-031-09360-9_5

Just-In-Time (JIT), which allows the company to have the right materials and the right amount at the right time.

From a logistical point of view, the JIT philosophy has stimulated the concept of materials delivery in small batches to production areas. Currently, many companies adopt the supermarket's methodology to meet the fast and flexible challenges caused by internal supply [3]. The Kanban and the Mizusumashi are two tools associated with JIT. The Kanban system is activated by the placement of orders (Kanbans) and acts as a mechanism of limitation of stock levels and control of flow, to deliver the correct materials and exact amount at the proper time and place. To complement this material flow controlled by Kanbans, the Mizusumashi system ensures the materials handling necessary to meet the customer orders and the production needs [4].

The reduction of available resources in the productive system caused by the implementation of Lean solutions leads to higher requirements for quality and flexibility in internal logistics. Increasing product diversity and introducing new models into the complexity of logistics in the supply chain has shown greater applicability and importance in this field. This sets up internal logistics in a production environment as the activity responsible for the efficient delivery of materials carried out with great frequency and involving smaller amounts at a time [5].

Many articles present different approaches useful to the implementation of Mizusumashi systems in the industrial environment [4, 6–11], although it is not usual to find publications that describe this implementation in process industries such as the textile industry involved in this project. The work presented in this article was developed in a textile reinforcements production company for the automotive industry, which can be predominantly characterized as a process industry, as opposed to discrete manufacturing companies (e.g., the textile clothing industry). Moreover, few works were found describing the details about the integration of Mizusumashi with the Kanban system.

The present study was developed to reduce transport waste related to inventories on the shop floor and concentrate the internal supply activities in one logistics team. These results will be pursued through the implementation of a Mizusumashi internal supply system without resorting to additional implementation costs.

The main objective of this study is to demonstrate the advantages of concentrating activities that do not add value to the final product (raw material replenishment, scrapping, and storage), performed by different production teams, on a single team by merging all those tasks in a Mizusumashi system. For that purpose, the study is divided into six main sections representing the approach followed to attain it.

Hence, the first chapter is an introduction containing the motivation for the study and the main objectives, followed by a second chapter where the importance of Lean production methodologies in the automotive industry, focused on Mizusumashi and Kanban methodologies, is stated. In the third chapter, the industrial context and the design plan for the solution are described. The fourth and fifth chapters focus on the data gathering and analysis of the results obtained. Finally, the main conclusion about the results is presented in the sixth and last chapter of this article.

2 Lean Production

Lean Thinking was created inside the automotive industry in the last century and was quickly applied to most business activities. Lean is an antidote in times of crisis with a philosophy based on the constant elimination of waste and the creation of value for all stakeholders [12]. In this philosophy, only the necessary products, at the necessary time, in the necessary quantities are manufactured [13]. According to Pattanaik, Sharma [14], Lean Manufacturing has been increasingly applied by worldwide leaders in the automobile industry. In a recent survey, about 36% of production companies in the United States implemented or are implementing Lean Manufacturing [14]. Its central concept is pulling production in which the flow on the shop floor is driven by the demand for production. This concept is a means by which the general business processes are organized to offer products with greater variety and higher quality, using fewer resources and less time than could be achieved by mass production methods [15]. Thus, by applying the JIT concept, companies aim to create a uniform material flow through the production process, which can avoid the major queues and intermediate buffers. This aspect applies not only to the production units involved but also to a higher level between the different units in the production chain [16]. Karlsson, Norr [16], say that there is a general understanding in the literature of how JIT will influence the frequency of deliveries and the container size carried. Most authors seem to agree with the importance of transport integration with the material flow in production.

A well-adapted transport system should contribute to the total return for the customer, reducing total costs, capital tied up in stocks, work in process (WIP) and transport times [17]. Thus, it is possible to keep a high service level for their manufacturing facilities the company as well as for their customers.

To complement thc introduction of the JIT concept, Coimbra [18] propose Total Flow Management (TFM). TFM is defined as an integrated concept to increase the flow of processes and the entirety efficiency (pull) of a supply chain, aiming at the reduction of the total delivery time of the tasks associated with the supply chain. TFM is a great opportunity for improvement and is based on the concept of eliminating waste, such as inventory, material handling and overproduction. This plan also includes the elimination of the use of forklift trucks, replacing them with a higher frequency transport using the Milk Run or Mizusumashi systems [18].

2.1 Mizusumashi

The Mizusumashi system is a materials replacement strategy associated with the internal supply chain (between warehouse and production), with the possibility of creating transport routes and delivery schedules [19]. The material is transferred to the work area only when it is needed. The inventory level is reduced because the existing stock in buffers of each production area just has to be coincident with the safety stock of each product [20]. The Mizusumashi process begins with the production orders and material requirements sent to the warehouse by the head of the production. Then, the Mizusumashi prepares and delivers the materials to the manufacturing areas [21]. Different approaches to Mizusumashi implementation can be found in the literature. Hanson, Finnsgård [22] applied logistic trains instead of trucks, replacing the traditional supply of pallets with plastic containers. In this way, these authors obtained a reduction of human labour, inventory reduction at the work centres, elimination of the stacker capacity lack which caused unpredictable lead delivery and delays in deliveries. Rother [9] used logistical trains at an automotive company in Germany, with supply synchronization with the production needs. The main results were: increased productivity; reduction of empty transport; reduction of the occupation of logistics transportation equipment; reduction of space occupied by materials in the production area; greater security. Alnahhal et al. [6] made the introduction of more frequent deliveries, considering requirements of demand, supply and transport control. This approach, led to a reduction of inventory at the work centres and the implementation of a more efficient and ecological supply system. Droste, Deuse [8] defined four factors with influence the implementation decision of an internal supply system: control materials, warehouse features, means of transport and where the product is used (Point of Use).

A plan to implement a lean approach to materials flow may be developed according to the following steps [23]: 1. Develop a supply plan for all materials; 2. Set a place to do the packing and loading of materials; 3. Create transportation routes; 4. Implement a pull signal to transmit information; 5. Apply continuous improvement in the system.

All these works describe a materials delivery system with increased frequency, allowing the producer to reduce the overall inventory levels when performing a wide range of products. Moreover, this reduction in stock levels reduces the space occupied and this can be reason enough to adopt Mizusumashi. Additionally, a key role for the production responsible is levelling the workflow and creating a work plan in regular service for the same way in the warehouse [23].

2.2 Kanban and Pull Implementation

Among the various tools of production management of industrial systems, based on the Lean Manufacturing philosophy, the Kanban is widely used today in industrial environments. The word Kanban is of Japanese origin and its meaning is label/card. This technique was created in Japan after World War II, by Taiichi Ohno, at Toyota Motor Company in the 50s, and was inspired by the American supermarket's system [24]. Kanban was created to fight two of the seven major wastes: overproduction and excessive stocks. According to Courtois et al. [25], this tool aims to:

- Produce/deliver only the product desired by the customer.
- Produce/deliver at the time the product is ordered.
- Produce/deliver the quantity ordered.

This approach is related to information systems allowing fast upstream to downstream convey and following requirements by a pull production system with the JIT concept [25]. The main types of Kanban used in the industry are specific Kanban, generic Kanban and electronic Kanban.

A specific Kanban controls the priority of production and the level of stock of specific items. This Kanban usually contains the following information: part number manufactured or transported, container capacity, and identification of upstream and downstream stations [25].

The generic Kanban has been designed to fill a very specific limitation of specific Kanbans: a reduced variety of materials [25]. It is used to control the WIP of a system in a dynamic environment, by providing a fixed number of Kanban to every workstation. This type of Kanban can be used for any type of product in the system. A production order can only be executed if a Kanban is available on the workstation. Otherwise, the production has to wait for a Kanban. Thus, the level of WIP and lead time are dependent on the number of production Kanbans [26].

Electronic Kanban (or e-Kanban) benefits from computer technology and information to manage its operation. Improving information transmission and processing efficiency has become a simple and effective method for many businesses to achieve the JIT production method. However, only the production levelling may cause smooth orders. In this sense, the electronic Kanban is also called electronic request [27]. The benefits of implementing electronic Kanban are the reduction of general steps in the work process, lead times, Kanbans in circulation, amount of materials, intermediate storage area, and inventory costs [28].

3 Industrial Context and General Design Plan

This section aims to describe the industrial context in which this work was developed, characterise the problem and present the general steps used to design the proposed solution in the form of Mizusumashi and Kanban systems.

3.1 Industrial Context

This company has five production stages: twisting, weaving, dipping cable/fabric, knitting fabric and thermoset fabric. These processes contribute in different ways to the three types of final products: dipped fabric, dipped cable and thermoset fabric. The twisting process is the process where yarn spools will be used to create greige bobbins of cable through a twisting process. This process has cycle times that can reach 25 h and the survival time of the yarn spools stocked can reach 72 h. The process is very dependent on manual work, mostly for replenishment activities on the twisting machines, namely the spindles. The replenishment process is controlled by the production managers of each shift, always taking into consideration the availability of each shift. This availability (men-hours), will define how many machines can be replenished per shift. With 136 positions on each machine, on a total of 64 machines, a rough calculation shows that one shift can replenish around 1000 spindles. As there is a natural mismatch of all machines, this supply does not happen simultaneously, implying a need for raw materials on the shop floor spikes.

The weaving process is the longest process, with cycle times that can reach one and a half weeks. Here the bobbins of cable produced on the twisting process are used to produce the greige fabric rolls. Both dipping stages are processed with cycle times very short, compared to another process in this company; here dipped cable bobbins or dipped fabric rolls can be produced in minutes. The dipping process uses dipping solutions, tensions, and temperature to improve the mechanical and chemical characteristics of the greige cable/fabric rolls. The knitting process also uses yarn spools and produces the semi-finished material greige knitting rolls; they follow to the thermosetting stage, to acquire better mechanical properties. The knitting process has cycle times of around 6 h while the thermosetting process, like the dipping process, have a cycle close to 20/25 min. From all the processes, twisting is the one where the raw material flow is greater; it can involve 120 pallets of raw materials per day and this work intends to improve the replenishment system that can go up to around 1000 spindles in one shift.

3.2 General Design Plan

According to Harris et al. [23], the most appropriate implementation plan to obtain a lean materials flow can follow the following steps: 1. Develop a supply plan for all materials; 2. Set places to do the packing and loading of materials; 3. Create transportation routes; 4. Implement a pull signal to transmit information; 5. Apply continuous improvement in the system. The first step based on stocks control and materials supply can be done through a Mizusumashi system in the plant. A Mizusumashi is a cyclical transport system with manual operations that handles the delivery of raw materials and finished products within the factory, using settled routes and times. This approach also includes the second and third steps of the implementation plan. To apply the Mizusumashi system is necessary to reflect on the available information transmission techniques. The approach most commonly used by the automotive industry is the Kanban system. The 4th step is the application of a Kanban system which is an information system that controls the produced quantity in all processes. This technique aims to pull parts when necessary and control inventories. The Kanban contains information about points of picking and loading materials, transport and type of product [29]. Finally, the company should create a continuous improvement system including the analysis of the internal supply system. This article is based on the development of steps 1 and 4 of the previous plan: develop a supply plan for all materials; implement a pull signal to transmit information. For the first of these steps, it was necessary to specify the quantity of material to be handled, the working force, the material buffers and the time of replenishment. Finally, for step 4 a solution based on e-Kanban was developed.

4 Internal Supply System Implementation

Before this study, the raw material supply of the company was mostly controlled by some operators who were responsible for the raw material requisition tasks. The experience that they have about the process determines the precision of the number of pallets of raw material requested. Nevertheless, even very experienced operators faced problems like distribution schedules not defined; no routes for distribution; non-efficient and non-standard information flow; mismatch between distribution times and the needed effective time of raw material; lack of control of the number of pallets on the shop floor. A very important aspect of the raw material in the company is related to the fact that most of the raw materials are at consignment, so only at the time that these raw material are delivered to production do they start to be counted as company stock. Hence, control about the quantity of raw material on the shop floor is very important, because the lesser the quantity the less will be the amount of value on the stock. Additionally, this control is very important because of the space occupied on the shop floor. Taking into consideration that the company works seven days a week, 24 h per day, the implementation of the Mizusumashi methodology must ensure that the production would not be affected. For that reason, the following steps were defined: specification of the quantity of materials movements; specification of the man-hour; specification of the minimum replenish time; specification of production buffers and distribution periods; specification of the e-Kanban system.

4.1 Specification of the Quantity of Materials Movements

To define the number of material pallets that would be necessary to transport materials from the warehouse to the production, a simulation tool was developed using sales forecasts from the company. Thus, it was possible to determine the number of material movements to be done by the Mizusumashi operator; this was very important because the number of material movements determines the labour time needed (man-hours). The forecasted quantity of movements by type of material per shift is:

- Raw material Pallets - 33
- Trolleys - 36
- Greige rolls - 17
- Dipped rolls - 20
- Greige Knitting rolls - 4
- Thermoset rolls - 19
- Dipped cable pallets - 4

4.2 Specification of the Man Workload

After the calculation of the number of movements for the Mizusumashi, the next logical step would be the calculation of the number of man-hours needed to execute it.

As this implementation aims to reduce the waste of movements in the productive areas, eliminating the movement waste from the production and transferring it to a logistic team was necessary to determine the activities that used more workload and the handling activities that were causing entropy on the production process. Of all the studied activities of the production operator, four of them were related to movement and transportation and should be excluded from production activities: raw material distribution, collecting packing wastes, scrap collecting and storage of knitting finished rolls. The man workload of these activities is presented in Table 1.

Hence, one operator (0.98) will be needed per shift to execute the 3 tasks listed. The workload was calculated taking into consideration all the resting times needed. This operator will be transferred from production to the logistic team; by doing this, no additional costs with operators will be added to the company to reduce this movement waste from production.

Table 1. Analysis of tasks allocated to the Mizusumashi.

Task	Men/shift
Raw material distribution and waste collecting	0.82
Scrap collecting	0.03
Knitting storage	0.13
Total	0.98

4.3 Specification of the Minimum Replenish Time

The definition of a supply schedule for materials is vital to avoid unexpected delays that can represent a threat to the production flow. The smaller survival time in twisting stage will determine the interval between replenishments [15]. The survival time for all the cables (around 50) produced in twisting was calculated based on formula (1).

$$Replenish\ Time = \frac{yarn\ spool\ weight * 1000/yarn\_dtex}{(RPM/TPM)/60} \times 0.75 + setup\_time + lead\_time \quad (1)$$

The size of a full bobbin of yarn will "survive" on the production depending on the ratio of yarn consumption, this ratio varies from material to material, and it is calculated through the RPM (rotation per minute) and TPM (torsion per meter) of each material (0,75 represents the assumption that spindles are replenished when the bobbin reaches 25% of its original size). The smallest time calculated for materials was around 6 h.

4.4 Specification of Production Buffers and Distribution Periods

It is fundamental to reduce the movements of the production operators regarding the raw material transport activities. For that reason, a clear strategy of buffers definition along the shop floor was needed to optimize de transport activities.

As the buffers were not previously completely defined, new loading points were defined in all the production stages. For this specific type of production, the operator has the power to decide if his shift can replenish or not more machines than was expected. Therefore, no strict routes for the raw material supply from the warehouse to production were defined. However, the same did not apply to the distribution periods, here a specific time of raw material delivery was needed to allow planning of the machines to be replenished to be as accurate as possible.

The calculated survival time was of 6 h, so the intervals between deliveries are expected to be 6 h, but, as one of the priorities was to reduce the number of pallets on the shop floor, this interval was considered too big. With a 6 h interval, the production would lose some dynamism regarding the efficient use of men available, as it would require a level of planning that the operator cannot assure. Thus, the defined delivery times were the following: 00:00 h; 04:00 h; 08:00 h; 12:00 h; 16:00 h; 20:00 h.

5 Implementation Results

Using "initial" as the situation before implementing the Mizusumashi, "intermediate" as the initial phase of implementation and "final" as the actual situation after some adjustments, is possible to verify that during and after the "intermediate" phase the decrease of the number of pallets requested by day was clear (Fig. 1). This decrease is related to the increase in the delivery's frequency, and due to the defined schedules for distribution, operators are now able to plan better their work.

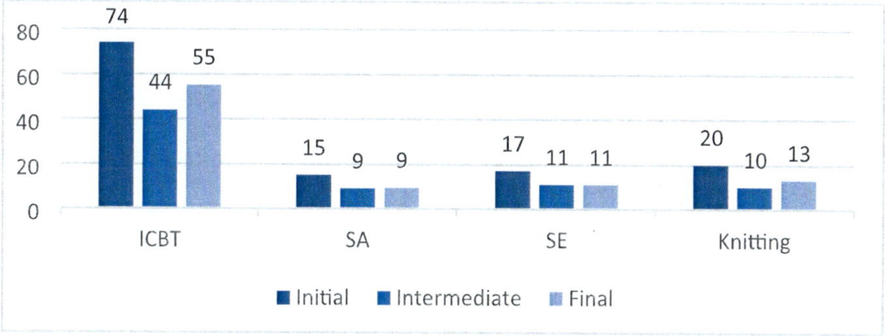

Fig. 1. The average number of requests for material delivery.

Regarding the average quantity of stock on the shop floor, it is possible to verify a 50% reduction (Fig. 2).

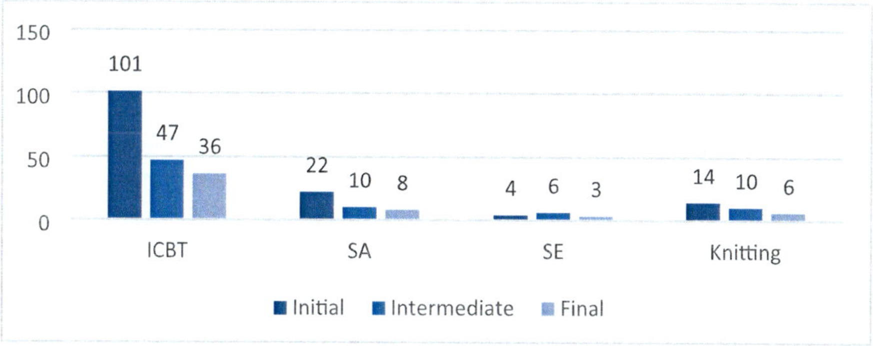

Fig. 2. The average stock found on the shop floor.

5.1 Discussion

The direct impact of the solution applied is the reduction of 50% of the occupied space with raw materials (Table 2). This reduction of 50% of stock on production has a very important financial impact, as it was explained before most of the raw materials are consignment materials.

Table 2. Results after implementation.

Occupied area (m^2)	ICBT (twisting)	SA (twisting)	SE (dipping)	Knitting
Initial	145	32	6	20
Intermediate	68	14	9	14
Final	52	11	5	9

Also, several other improvements were identified but not measured in this study, some due to the lack of time available, others because they are intangible improvements felt in the shopfloor but very hard to quantify, namely:

- Improvement of the efficiency of the production teams themselves; after the Mizusumashi implementation, those teams have more time to focus on their main activities instead of wasting time to plan and prepare the activities now in charge of the Mizusumashi;
- Movements inside the plant by the production teams were severely reduced. The production team has now more focus on their activities;
- Layout organization and space optimization;
- Information flow inside production areas about when and where the raw materials are needed.

Another interesting result is the awareness gained by the productions teams about the "rules" they now need to keep inside the productive areas. They used other lean techniques, such as 5Ss, despite they were never forced to do so; they naturally did that to preserve the correct functioning of the new internal supply system.

6 Final Remarks

Lean manufacturing offers a variety of tools that allow companies to improve their process. A result of this work is the effective reduction of the stock on the shop floor by the use of a Mizusumashi, showing that the non-added value activities of the production operators were substituted by a more successful internal logistic system with the same quantity of man workload. This first implementation of a Lean internal logistic system in the company brought confidence and big expectations during the next Lean improvements.

Systems like Mizusumashi are not so commonly used as they could, due to the afraid of incurring great costs, which generally are associated with this kind of implementation, mostly regarding the range of equipment associated with it. In this case, it was possible to show that the capacity to apply functional lean tools can be done by the improvement of the existing equipment, letting the greater investments in equipment to a medium-term period where the solutions are already firmly established in the culture of the company.

In future work, the company would like to understand the potential impact of Mizusumashi systems on the overall implementation and culture change necessary to implement Lean tools. This desire is related to the fact that during this implementation there was a change in the behaviour of the operators with a positive impact on their willingness and collaboration to implement and improve the Mizusumashi, once they felt and experienced the difference between the old and new distribution method.

Acknowledgements. This work has been supported by FCT – Fundação para a Ciência e Tecnologia within the R&D Units Project Scope UIDB/00319/2020.

References

1. Koste, L.L., Malhotra, M.K.: Theoretical framework for analyzing the dimensions of manufacturing flexibility. J. Oper. Manag. **18**(1), 75–93 (1999). https://doi.org/10.1016/S0272-6963(99)00010-8
2. Roos, D., Womack, J.P., Jones, D.: The Machine that Change the World-The Story of Lean Production. Rawson/Harper Perennial, New York (1990)
3. Maurizio Faccio, D.Y.C.D., Savino, M.M., Mazza, A.: Kanban-driven parts feeding within a semi-automated O-shaped assembly line: a case study in the automotive industry. Assem. Autom. **35**(1), 3–15 (2015). https://doi.org/10.1108/aa-07-2014-068
4. Nomura, J., Takakuwa, S.: Optimization of a number of containers for assembly lines: the fixed-course pick-up system. Int. J. Simul. Model. **5**(4), 155–166 (2006). https://doi.org/10.2507/IJSIMM05(4)3.066
5. Marodin, G.A., de Eckert, C.P., Saurin, T.A.: Avançando na implantação da logística interna lean: dificuldades e resultados alcançados no caso de uma empresa montadora de veículos. Revista Produção Online. **12**(2), 455–479 (2012)
6. Alnahhal, M., Ridwan, A., Noche, B.: In-plant milk run decision problems. In: 2014 International Conference on Logistics and Operations Management (GOL), pp. 85–92. IEEE (2014)
7. Kilic, H.S., Durmusoglu, M.B., Baskak, M.: Classification and modeling for in-plant milk-run distribution systems. Int. J. Adv. Manuf. Technol. **62**(9–12), 1135–1146 (2012). https://doi.org/10.1007/s00170-011-3875-4
8. Droste, M., Deuse, J.: A planning approach for in-plant milk Run processes to optimize material provision in assembly systems. Enabling Manufacturing Competitiveness and Economic Sustainability, p. 604–610. Springer, Heidelberg (2012). https://doi.org/10.1007/978-3-642-23860-4_99
9. Rother, T.: Less empty runs with milk run and tugger train systems. Maschinen Markt. (2011)
10. Ichikawa, H.: Simulating an applied model to optimize cell production and parts supply (Mizusumashi) for laptop assembly. Proceedings - Winter Simulation Conference, pp. 2272–2280 (2009)
11. Afonso, T., Alves, A.C., Carneiro, P., Vieira, A.: Simulation pulled by the need to reduce wastes and human effort in an intralogistics project. Int. J. Indust. Eng. Manage. **12**(4), 274–285 (2021). https://doi.org/10.24867/IJIEM-2021-4-294
12. Roldão, V., Ribeiro, J.S.: Gestão das Operações: Uma abordagem integrada. 1ª Edição, Lisboa: Monitor (2007)
13. Ohno, T.: Toyota Production System: Beyond Large-Scale Production. Productivity Press (1988)
14. Pattanaik, L.N., Sharma, B.P.: Implementing lean manufacturing with cellular layout: a case study. Int. J. Adv. Manuf. Technol.. **42**(7–8), 772–779 (2008). https://doi.org/10.1007/s00170-008-1629-8

15. Jina, J., Bhattacharya, A.K., Walton, D.A.: Applying lean principles for high product variety and low volumes: some issues and propositions. Logist. Inf. Manage. **10**(1), 5–13 (1997)
16. Karlsson, C., Norr, C.: Total effectiveness in a just-in-time system. Int. J. Oper. Prod. Manag. **14**(3), 46–65 (1994). https://doi.org/10.1108/01443579410058522
17. Li, F., Bao, S., Li, S.: Applications of lean logistics on engine manufacturing planning. SAE Int. J. Mater. Manuf. **7**(2), 308–312 (2014). https://doi.org/10.4271/2014-01-0772
18. Coimbra, E.A.: Total Management Flow: Achieving Excellence with Kaizen and Lean Supply Chains. Kaizen Institute (2009)
19. Reis, L., Varela, M.L.R., Machado, J.M., Trojanowska, J.: Application of lean approaches and techniques in an automotive company. Romanian Rev. Prec. Mech. Opt. Mechatron. **50**, 112–8 (2016)
20. Gyulai, D., Pfeiffer, A., Sobottka, T., Váncza, J.: Milkrun vehicle routing approach for shop-floor logistics. Procedia CIRP **7**, 127–32 (2013)
21. Galrao Ramos, A., Pereira Lopes, M., Silva Avila, P.: Development of a platform for lean manufacturing simulation games. IEEE Revista Iberoamericana de Tecnologias del Aprendizaje. **8**(4), 184–190 (2013). https://doi.org/10.1109/RITA.2013.2284960
22. Hanson, R., Finnsgård, C.: Impact of unit load size on in-plant materials supply efficiency. Int. J. Prod. Econ. **147**, 46–52 (2014). https://doi.org/10.1016/j.ijpe.2012.08.010
23. Green, J.C., Lee, J., Kozman, T.A.: Managing lean manufacturing in material handling operations. Int. J. Prod. Res. **48**(10), 2975–2993 (2010). https://doi.org/10.1080/00207540902791819
24. Liker, J.K.: The toyota way. Esensi (2005)
25. Courtois, A., Martin-Bonnefois, C., Pillet, M., Costa, H.: Ad-Verbum. Gestão da produção (1997)
26. Chang, T.M., Yih, Y.: Generic kanban systems for dynamic environments. Int. J. Prod. Res. **32**(4), 889 (1994). https://doi.org/10.1080/00207549408956977
27. Li-fang, W., Fan-Sen, K., Yu-chun, W., Shao-zhong, S., Jing-hua, C.: Planning and application research on JIT logistics information system for auto parts. In: World Automation Congress 2012, pp. 1–4. IEEE (2012)
28. Jarupathirun, S., Ciganek, A.P., Chotiwankaewmanee, T., Kerdpitak, C.: Supply Chain Efficiencies Through E-Kanban: A Case Study. International Conference on IT to Celebrate S Charmonman's 72nd Birthday (2009)
29. Matzka, J., Di Mascolo, M., Furmans, K.: Buffer sizing of a Heijunka Kanban system. J. Intell. Manuf. **23**(1), 49–60 (2012). https://doi.org/10.1007/s10845-009-0317-3

SAPEVO-H^2 a Multi-criteria Approach Based on Hierarchical Network: Analysis of Aircraft Systems for Brazilian Navy

Miguel Ângelo Lellis Moreira[1,2], Carlos Francisco Simões Gomes[1], Maria Teresa Pereira[3,4](✉), and Marcos dos Santos[2,5]

[1] Fluminense Federal University, Niterói, RJ 24210-346, Brazil
[2] Naval Systems Analysis Center, Rio de Janeiro, RJ 20091-000, Brazil
[3] School of Engineering, Polytechnic of Porto, 4200-072 Porto, Portugal
mtp@isep.ipp.pt
[4] Instituto de Ciência e Inovação em Engenharia Mecânica e Engenharia Industrial, 4200-465 Porto, Portugal
[5] Military Institute of Engineering, Urca, RJ 22290-270, Brazil

Abstract. The present study explores an extensive approach to the SAPEVO-M method, based on a hierarchical and hybrid analysis, dealing with variables of quantitative and qualitative natures. Named SAPEVO-H^2, the modeling, based on the multicriteria approach, enables the evaluation of alternatives in the light of multiple criteria, providing the integration of objectives of a problem, transcribed in attributes, and structured in a hierarchical model. Integrating multiple decision-makers, group and individual evaluations is possible, designating the evaluator in his area of expertise and decision-making responsibility. In search to explore the implementation feasibility of the model, a case study is carried out based on the hypothetical acquisition of Remote Piloted Aircraft Systems for the Brazilian Navy. The analysis clarifies the indications of preferences at each level of the hierarchy, exposing the construction of local and global results and performance. In the end, the main points of the modeling and the proposal for future works are presented.

Keywords: Multi-criteria decision aid · Hierarchical analysis · SAPEVO-M

1 Introduction

Based on the National Defense Policy of Brazil [1], The country plays a prominent role in the world environment, privileging peace and defending dialogue and peace for the solution of disputes between States. However, it is noteworthy that the country directs its decisions towards the continuous development of its defense strategies, expressed through political, economic, social, military, and scientific-technological development [2].

Decision-making in political and military environments involves different levels and areas, interconnecting strategic, tactical, and operational analyzes in favor of a direction

J. Machado et al. (Eds.): icieng 2022, LNME, pp. 61–74, 2023.
https://doi.org/10.1007/978-3-031-09360-9_6

aligned with the objectives in a given problematic situation [3]. Complementarily, it is necessary to consider that decision-making in these spheres is complex and, in many cases, must be strategic by the moment that generates influences not only in the military scope but impacts on other areas of society [4].

In the scenario of high-level decision-making, it is common the involvement of multiple decision-makers to determine and analyze aspects relevant to the problem in search to provide, from multiple perspectives, a consensus in the decision [5]. As presented in [6], with the involvement of multiple scenarios and circumstances, the increase in complexity in a given analysis becomes noticeable, with different points of view as to the importance of a decision variable. However, it is necessary to consider it in favor of a substantial evaluation and greater assertiveness in the final decision.

In decision-making, the Operational Research (OR) provides, through its algorithms and methodologies, the analysis of complex problems with a technical and scientific basis, providing its structuring, understanding, and in specific cases, the feasibility of a favorable solution for a given problematic [7]. Considering a set of different alternatives evaluated on multiple attributes, the Multi-Criteria Decision Aid models (MCDA), from OR, enable to support the decision-making process, contemplating techniques that allow the decision-maker (person/group/organization) to structure the variables and preferences of the evaluation, clarifying their respective degrees of importance in an interactive process with other actors [8].

Regarding real decision-making problems, uncertainty variables are intrinsic to the assessment, especially in situations involving multiple decision-makers, where members generally disagree on thresholds values and preference assignments [9]. However, as exposed in [10], even though group decisions are common to analyze real situations, in a particular way, few models consider the formation of subgroups to deal with problems at strategic, tactical, and operational levels.

In this context, the study explores an extensive approach of the SAPEVO-M method (Simple Aggregation of Preferences Expressed by Ordinal Vectors Group Decision Making) [11]. Given modeling, previously named SAPEVO-H^2 (SAPEVO Hybrid and Hierarchical) approaches an analysis with multiple decision-makers, however, providing the creation of subgroups of the set for analysis of specific parts of the problem built in a hierarchy format, not restricting itself to the evaluation of alternatives under multiple criteria, but also considering other attributes relevant. Additionally, to the ordinal evaluation model present in the traditional method, the modeling also provides the treatment of quantitative attributions by cardinal inputs.

As a way of validating the approach, is explored a case study based on strategic decision-making for a hypothetical acquisition of RPAS (Remotely Piloted Aircraft System) for the Brazilian Navy (BN). Given the decision-making process, the integration of multiple decision-makers at different hierarchical levels and knowledge concerning the variables relevant to the problematic situation is performed.

The article is structured into five sections. After the introduction, are explored the theoretical foundations of MCDA and other basic methodologies of the related model. In Sect. 3, the axiomatic structure of the SAPEVO-H^2 modeling is presented, exposing its particularities and extensive points concerning the previous models. Section 4 details the numerical application of the given model through a case study. Section 5 presents final considerations and proposals for future work.

2 Multi-criteria Decision Aid

The decision-making process, integrated into human activity, is characterized as the process of analysis for a set of actions in search of obtaining a favorable solution to a given problem [8]. In this scenario, MCDA provides the structuring and understanding of a problem in complex environments, considering risk and uncertainty, helping to clarify viable solutions for problems of a varied nature, made possible through an interactive and transparent axiomatic process [12].

MCDA methods operate in search to provide the preference establishment between alternatives, under the analysis of criteria with a tradeoff between them [13]. According to [14], commonly, there are three main types of issues dealt with in models belonging to MCDA, they are: Choice, exposing the most favorable alternative in a global context; Ranking, ordering from the most favorable alternatives to the least favorable ones with a form of the solution; and Classification, allocating the alternatives into dominance clusters.

2.1 SAPEVO-M Method

The SAPEVO-M method [11] can be understood as an axiomatic evolution of the SAPEVO method [15], introducing an evaluation format for multiple decision-makers in a decision-making process, in addition to the axiomatic improvement of the previously developed model, thus bringing an increase in its consistency. As presented in [16], the main feature of the model is related to the ordinal transformation process of preferences used to obtain the degrees of preference relations between the alternatives in each criterion and the generating of the importance degree of the variables, thus generating their respective weights.

The axiomatic process of the model is realized in two parts. Preliminarily, the transformation of the ordinal preference between criteria must be performed, expressed by a vector representing the weights of the criteria. Then, the ordinal transformation of the preference between alternatives within a given set of criteria, expressed by a matrix, is performed. A series of pairwise comparisons, between criteria or between alternatives in a criterion, denote the individual preference information of each decision-maker.

Based on an ordinal scale of seven points (−3, −2, −1, 0, 1, 2, 3), ranging from absolutely worse to absolutely better, are carried out a set of pairwise comparisons, enabling the indication of the intensity of preferences between two variables. A given model supports the two parts of the model analysis, serving as a basis for evaluating the criteria and evaluating the alternatives for each criterion. The normalization process of the variables (alternatives or criteria) is given by Eq. (1), where a_{ij} is the preference set between two variables on the matrix, allowing a transformation of ordinal scores into cardinal quantities, providing the aggregation of preferences and ordering of alternatives [11].

$$\upsilon = \frac{\sum a_{ij} - \min a_{ij}}{\max a_{ij} - \min a_{ij}} \tag{1}$$

3 Multi-criteria Decision Aid

The SAPEVO-H^2 approach consists of enabling the assessment of a given problem by multiple decision-makers, allowing each decision-maker to evaluate the hierarchy of attributes and criteria, in a partial or total way, that is, a decision-maker or a group of decision-makers evaluates all levels of the hierarchy or just part of these levels. Emphasizing that the type of analysis performed concerns the structuring of the problem. The process is analysis process is divided into four steps, they are:

1. Construction of the decision problem and hierarchical model;
2. Analysis of each hierarchical level;
3. Aggregation of preferences defined for hierarchy elements; and
4. Ranking of alternatives for local and global levels.

3.1 Construction of Evaluation Structure

The modeling integrates a set of decision-makers *D*, where $d_m \in D, m = 1, \ldots n$, enabling the decision-maker to evaluate a given problem based on a set of hierarchical levels *N*, where $n_h \in N, h = 1, \ldots l$.

Related to the achievement of a given strategic objective, representing the top of the hierarchy under evaluation, for each hierarchical level a set of elements is determined, which can be:

- Attributes set *T*, where $t_r \in T, r = 1, \ldots s$.
- Criteria set *C*, where $c_j \in C, j = 1, \ldots k$.
- Sub Criteria set *S*, where $s_u \in S, u = 1, \ldots v$.
- Alternatives set *A*, where $a_i \in A, i = 1, \ldots b$.

Considering the possible assessment sets, Fig. 1 provides an example of a configuration for a given approach, highlighting that for each set there is a minimum need of two elements, making it possible to compare the elements belonging to a set at each level of analysis.

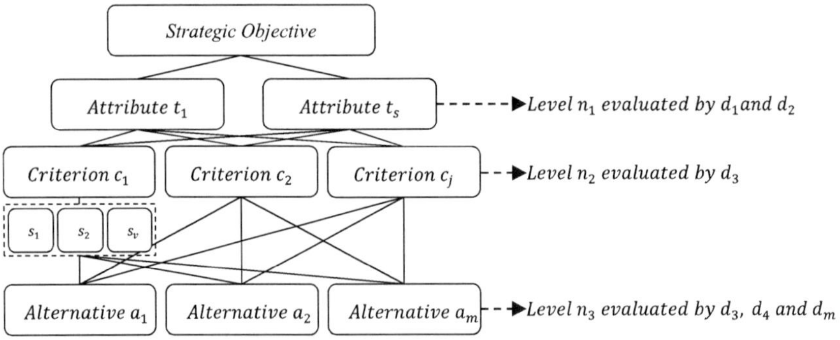

Fig. 1. Example of hierarchical structure for evaluation

3.2 Element Analysis

For all sets of evaluation in its respective levels, it makes possible the evaluation by multiple decision-makers or by only one decision-maker. Concerning the analysis of the respective levels of the sets of attributes and criteria, is used an ordinal scale based on seven points, as shown in Table 1.

Table 1. Ordinal scale

Verbal descriptions	Punctuation
Absolutely worse/less important	−3
Much worse/less important	−2
Worst/less important	−1
Equivalent	0
Best/more important	1
Much better/more important	2
Absolutely better/more important	3

3.2.1 Ordinal Evaluation with Multiples Decision-Makers

For an analysis of a given set of elements with multiple decision-makers, is used the ordinal scale (Table 1) for the pairwise assessment between the elements of the set. For each decision-maker d_m, is established a matrix with the indications of preference between the elements of the set. Through Eq. (2), in a linear way, a degree of preference is stablished for each element of the comparison matrix, where the normalization of the sums defines the degree of importance, where:

$$v = \frac{\sum a_i - min\ a_i}{max\ a_i - min\ a_i} \tag{2}$$

With the degrees obtained for each element, is performed the sum (3) for each decision-maker. Following the procedure, is obtained the normalization of the sums (4), indicating the respective importance of the elements for the attribute of the higher level. In this model, keeping the technique indicated in [11], if any criterion or attribute presents zero importance, is assigned for this element, 1% of the smallest value greater than zero.

$$e_i = \frac{\sum_{m=1}^{n} e_{im}}{m} \tag{3}$$

$$\upsilon_i = \frac{e_i}{\sum e_i} \tag{4}$$

3.2.2 Ordinal Evaluation with One Decision-Maker

For analysis by only one decision-maker at a given level, is used an axiomatic variation proposed in [9]. In this context, a scale based on an upper and lower limit is generated, defined as *maximum sum* (5) and *minimum sum* (6), respectively.

$$\text{maximum}\, sum = (n-1)3 \tag{5}$$

$$minimum\, sum = (n-1)(-3) \tag{6}$$

Using the ordinal scale (Table 1), are obtained the scores of the elements of a given set under evaluation, where the normalization of scores is made possible by Eq. (7), generating the degrees of importance of the elements. Subsequently, the Eq. (4) is used to normalize the obtained degrees.

$$\upsilon_i = \frac{\sum a_i - minimum\, sum}{maximum\, sum - minimum\, sum} \tag{7}$$

3.2.3 Cardinal Evaluation

Considering as a hybrid model, the modeling enables not only evaluating elements based on qualitative attributions but also quantitative analysis, which is the evaluation of alternatives in quantitative criteria, where the alternatives already have cardinal numerical attributions for each criterion.

For the evaluation, are considered two types of thresholds, representing in each quantitative criterion, a minimum limit L_{min}, and a maximum limit L_{max}. For each alternative, is calculated the degrees of importance a_{ij} through the Eq. (8).

$$a_{ij} = \begin{cases} 0 & if\ a_{ij} \le L_{\min} \\ \frac{a_{ij}-L_{\min}}{L_{\max}-L_{\min}} & if\ L_{\min} < a_{ij} < L_{\max} \\ 1 & if\ a_{ij} \ge L_{\max} \end{cases} \tag{8}$$

All decision concerning the value of thresholds need to be set by the analyst and the decision-makers of the process. In the end of quantitative analysis, for each a_{ij}, the Eq. (3) and (4) need to be processed.

3.2.4 Aggregation Process

Once assigned the degrees of importance of all elements of the hierarchy, the values are aggregated, indicating the relative importance of the element e_i regarding the element of the upper level s_j by $e_{ij} = e_i s_j$. For every two levels above a given element k_l, is obtained its importance through the additive aggregation model indicated in Eq. (9).

$$e_{ijl} = \sum_{l=1}^{n} e_{ij} k_l \tag{9}$$

With the procedure of additive aggregation, it is possible to obtain the performances of the alternatives, leading to their ranking in a global and local format. As it is a hierarchical model, it is possible to analyze the performance of alternatives in each criterion or attribute, clarifying the global result based on the attributions of decision-makers at different levels of the hierarchy. It is noteworthy that in addition to obtaining the performance of the alternatives, are clarified the criteria and attributes under analysis by indicating their respective importance in the decision context, representing the preference relations of a group of decision-makers, as well as of each individual decision-maker.

4 Case Study

The RPAS, commonly referred to as drones, are remote aircraft systems that have aroused the interest of civil and military organizations, through their multiple forms of implementation in tactical and strategic operations related to the defense and security of a country [17]. Currently, SARPs present the integration of high-level technologies, including sensors, radars, cameras, and radars, enabling their use in logistics, surveillance, and combat operations [9].

However, it is necessary to emphasize that the acquisition or development of a given technology does not restrict to the analysis of only operational attributes, where it is often necessary to align with the policies and strategic objectives of a given problematic situation [9]. In this context, the case study presents a hypothetical analysis for the acquisition of RPAS for the BN, enabling the alignment of this acquisition with the strategies of the problem, transcribed into four levels:

- Level 1: Representation of the BN strategic objectives;
- Level 2: Determination of areas of importance for the use of technology;
- Level 3: Set of criteria and sub-criteria for measuring system performance;
- Level 4: Set of alternatives for analysis.

The analysis was carried out with the support of a group of five stakeholders, both career officers from the BN, with vast experience in the study and implementation of RPAS. In the given context, two decision-makers assessed levels 1 and 2 of the hierarchy, representing the strategic approach to the problem. Related to the operational assessment of the problem, levels 3 and 4, we used the support of the other three decision-makers. Figure 2 exposes the problem structuring.

The first part of the evaluation is based on analyzing the elements of the first level set, carried out by decision-makers D_1 and D_2. As exposed in Table 2, based on the ordinal scale of attribution (Table 1), decision-makers assess *Technological Development* (TD), *Safeguarding the Life of the Military* (SL), *Mastery of Technologies for Combat* (MCT), and *Efficiency in Operations Tactics and Strategics* (EO).

As the assessment proceeds, it is necessary to assess level 2. For the given level, the areas of employment in Science and Technology (ST), Logistics (LO), and Combat Operations (CO) are assessed, where the analysis of the elements concerns each situation (element) of the upper level, previously evaluated. Tables 3, 4, 5, and 6 expose these evaluations.

After the evaluations of levels 2 and 3, the analysis of the most operational attributes of the problem is started, thus evaluating the influence criteria, which are: *Speed*

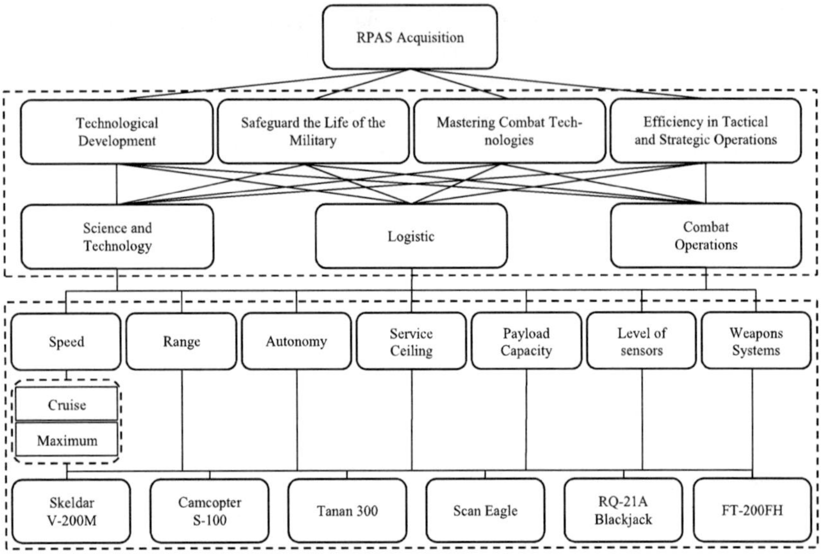

Fig. 2. Structuring of problematic situation

Table 2. Evaluation of elements at level 1 by D_1 and D_2

D_1	TD	SL	MCT	EO	*Punctuation*	*Pt. Normalized*	*Final Utility*	
TD	0	0	1	1	2	1		
SL	0	0	1	1	2	1		
MCT	-1	-1	0	-1	-3	0	TD	0.265
EO	-1	-1	1	0	-1	0.4	SL	0.478
D_2	TD	SL	MCT	EO	*Punctuation*	*Pt. Normalized*	MCT	0.003
TD	0	-2	0	-1	-3	0,111	EO	0.255
SL	2	0	2	1	5	1		
MCT	0	-2	0	-2	-4	0		
EO	1	-1	2	0	2	0.667		

Table 3. Evaluation of level 2 for *Technological Development* by D_1 and D_2

D_1	ST	LO	CO	*Punctuation*	*Pt. Normalized*	*Final Utility*	
ST	0	3	2	5	1		
LO	-3	0	-1	-4	0		
CO	-2	1	0	-1	0.333	ST	0.764
D_2	ST	LO	CO	*Punctuation*	*Pt. Normalized*	LO	0.109
ST	0	2	2	4	1	CO	0.127
LO	-2	0	1	-1	0.286		
CO	-2	-1	0	-3	0		

Table 4. Evaluation of level 2 for *Safeguarding the Life of the Military* by D_1 and D_2

D_1	ST	LO	CO	*Punctuation*	*Pt. Normalized*	*Final Utility*	
ST	0	-1	-3	-4	0		
LO	1	0	-2	-1	0.333		
CO	3	2	0	5	1	ST	0.125
D_2	ST	LO	CO	*Punctuation*	*Pt. Normalized*	LO	0.125
ST	0	1	-2	-1	0.333	CO	0.750
LO	-1	0	-3	-4	0		
CO	2	3	0	5	1		

Table 5. Evaluation of level 2 for *Mastery of Technologies for Combat* by D_1 and D_2

D_1	ST	LO	CO	*Punctuation*	*Pt. Normalized*	*Final Utility*	
ST	0	2	-1	1	0.667		
LO	-2	0	-3	-5	0		
CO	1	3	0	4	1	ST	0.452
D_2	ST	LO	CO	*Punctuation*	*Pt. Normalized*	LO	0.005
ST	0	3	0	3	1	CO	0.543
LO	-3	0	-3	-6	0		
CO	0	3	0	3	1		

Table 6. Evaluation of level 2 for *Efficiency in Op. Tactics and Strategics* by D_1 e D_2

D_1	ST	LO	CO	*Punctuation*	*Pt. Normalized*	*Final Utility*	
ST	0	-3	-3	-6	0		
LO	3	0	0	3	1		
CO	3	0	0	3	1	ST	0.005
D_2	ST	LO	CO	*Punctuation*	*Pt. Normalized*	LO	0.543
ST	0	-3	-2	-5	0	CO	0.452
LO	3	0	1	4	1		
CO	2	-1	0	1	0.667		

(*Spd*), *Range* (*Rg*), *Autonomy* (*Aut*), *Service Ceiling* (*SC*), *Payload Capacity* (*PC*), *Technological Level of Sensors* (*TLS*), and *Weapons Systems* (*WS*).

The evaluation of the criteria will be carried out uniquely for each implementation area established at level 2 of the hierarchy. However, instead of group analysis, will be treated an individual assessment for each decision-maker in their expertise. In this context, as explored in Sect. 3.2.2., the variables *maximum sum* = 18 and *minimum sum* = −18 will be considered. Tables 7, 8, and 9 expose the assessments.

Obtaining the weights of the criteria in each evaluated scenario, it is necessary to evaluate the alternatives. To analyze the qualitative criteria, are used the ordinal scale, and normalization of scores, performing this procedure for the *Technological Level of Sensors* and *Weapons Systems*. For the quantitative criteria are performed the cardinal assessment procedure, defining the cardinal values of the alternatives in each criterion,

Table 7. Evaluation of criteria at level 3 for *Science and Technology* by D_3

D_3	*Spd*	*Rg*	*Aut*	*SC*	*PC*	*TLS*	*WS*	Punctuation	Pt. Normalized	Weights
Spd	0	0	−1	0	−1	−3	−2	−7	0.305	0.087
Rg	0	0	−1	0	−1	−3	−2	−7	0.305	0.087
Aut	1	1	0	1	0	−2	−1	0	0.500	0.143
SC	0	0	−1	0	−1	−3	−2	−7	0.305	0.087
PC	1	1	0	1	0	−2	−1	0	0.500	0.143
TLS	3	3	2	3	2	0	1	14	0.889	0.254
WS	2	2	1	2	1	−1	0	7	0.694	0.198

Table 8. Evaluation of criteria at level 3 for *Logistic* by D_4

D_4	*Spd*	*Rg*	*Aut*	*SC*	*PC*	*TLS*	*WS*	Punctuation	Pt. Normalized	Weights
Spd	0	1	−3	1	−3	−3	2	−5	0.361	0.098
Rg	−1	0	−3	0	−3	−3	1	−9	0.250	0.068
Aut	3	3	0	3	0	0	3	12	0.833	0.227
SC	−1	0	−3	0	−3	−3	1	−9	0.250	0068
PC	3	3	0	3	0	0	3	12	0.833	0.227
TLS	3	3	0	3	0	0	3	12	0.833	0.227
WS	2	1	−3	−1	−3	−3	0	−7	0.305	0.083

Table 9. Evaluation of criteria at level 3 for *Combat Operations* by D_5

D_5	*Spd*	*Rg*	*Aut*	*SC*	*PC*	*TLS*	*WS*	Punctuation	Pt. Normalized	Weights
Spd	0	0	0	0	1	−1	−2	−2	0.444	0.127
Rg	0	0	0	0	1	−1	−2	−2	0444	0.127
Aut	0	0	0	0	1	−1	−2	−2	0,444	0.127
SC	0	0	0	0	1	−1	−2	−2	0.444	0.127
PC	−1	−1	-1	−1	0	−2	−3	−9	0.250	0.071
TLS	1	1	1	1	2	0	−1	5	0.639	0.183
WS	2	2	2	2	3	1	0	12	0.833	0.238

and each decision-maker, in their expertise, defines a minimum and maximum limit, L_{min} and L_{max} for each criterion. Table 10 exposes the cardinal data.

Table 10. Quantitative attributions to alternatives in each evaluation criterion

		Cruise sp.	*Maximum sp.*	*Range*	*Autonomy*	*Service ceiling*	*Payload Cp.*
Skeldar		130	150	200	5	9842	40
Camcopter		185	222	200	6	18000	50
Tanan 300		100	150	180	12	13000	80
Scan Eagle		110	148	100	24	19500	3.4
RQ-21A		112	167	93	16	20000	17.7
FT-200FH		80	110	100	12	12000	50
D_3	L_{min}	100	100	0	0	10000	10
	L_{max}	180	200	150	24	20000	70
D_4	L_{min}	120	120	50	6	5000	0
	L_{max}	200	250	150	18	12000	80
D_5	L_{min}	140	140	60	6	12000	0
	L_{max}	220	250	120	18	20000	30

Once established the cardinal data and preferences are carried out the aggregation process. One point of interest in the worked modeling concerns the feasibility of analyzing the performance of alternatives globally and locally, that is, at each level of hierarchy and attribute. In this scenario, the utility of the alternatives in all attributes of the hierarchy is obtained, at levels 1 and 2. Table 11 shows the utilities of the respective alternatives to the attributes.

Table 11. Utilities of alternatives at levels 1 and 2 of hierarchy

	Science and Technology	Logistic	Combat operations	Technological development	Safeguarding The Life of the Military	Mastery of Technologies for Combat	Efficiency in Operations
Skeldar	0.481	0.496	0.443	0.478	0.454	0.460	0.472
Camcopter	0.696	0.588	0.801	0.697	0.761	0.752	0.685
Tanan 300	0.452	0.533	0.469	0.463	0.475	0.462	0.504
Scan Eagle	0.443	0.384	0.345	0.424	0.362	0.389	0.367
RQ-21A	0.496	0.495	0.602	0.509	0.575	0.554	0.543
FT-200FH	0.699	0.593	0.495	0.661	0.533	0.588	0.549

Considering the attributes of level 2 in Table 11, it is understood that the alternative FT-200FH proved to be more favorable in the areas of *Science and Technology* and

Logistics, preceded by the Camcopter and RQ-21A aircraft in the first area and Camcopter and Tanan 300 in the second area. However, with a greater distance between the scores, SARP Camcopter was more favorable to the context of *Combat Operations*, preceded by RQ-21A and FT-200FH.

Evaluating level 1, still in Table 11, the alternative Camcopter proved to be the most favorable in the four attributes of the level in question. The alternative RQ-21A remained in the second position in three attributes and the third in one attribute. On the other hand, the alternative FT-200FH obtained the third position in three attributes and the second position in one attribute, showing a reduction in its scores when compared to its performance at level 2.

Complementing the analysis of the hierarchy, Table 12 shows the result in its global form, that is, the scores obtained after the aggregation process, along with the final ranking of the alternatives, considering all the attributes and criteria of the problem.

Table 12. Utilities of alternatives in the global context of decision-making

Alternatives	*Global ùtility*	*Ranking*
Skeldar	0.465	5°
Camcopter	0.725	1°
Tanan 300	0.479	4°
Scan Eagle	0.380	6°
RQ-21A	0.550	3°
FT-200FH	0.571	2°

At the end of the analysis, the Camcopter alternative becomes the most favorable in the main points of the decision. It is noteworthy that even given the alternative exhibiting the best performance, it is feasible to consider other systems at home for a future unfeasibility of the main alternative. In his way, were also identified Unfavorable alternatives as a solution to the problem, for example, the RPAS Scan Eagle, establishing the lowest score at all evaluated levels.

In addition to the final ranking obtained, it was also possible to construct rankings for each level and context of assessment, thus being able to indicate different types of solutions for each scenario to be implemented, also analyzing the feasibility of combined use of the solutions clarified in the proposed decision model.

5 Conclusion

The proposed study presents an extensive approach to the SAPEVO-M method, named SAPEVO-H^2, supporting a hierarchical analysis and analysis of variables of different natures. In addition to enabling the analysis and evaluation of criteria and alternatives in a ranking problematic, it enables the integration of attributes at different hierarchical levels, making it possible to build and evaluate high-level decision problems, integrating multiple decision-makers in different parts of the problematic situation.

As a practical application of the modeling, an evaluation was carried out based on the hypothetical acquisition of RPAS for the BN. In the evaluation, a four-level hierarchy was structured, with attributes (objectives/application areas), criteria and alternatives, aligned with the strategic objective of acquiring the given technology. With five decision-makers, were used group and individual assessments.

The approach presented to be favorable and aligned to other multi-criteria decision-making problems with multiple decision-makers in different types of problems. For future works, we search to evaluate other problems identifying improvement of the model, also the construction of consistency analysis for the pairwise evaluations, and the development of a computational web platform, exploring numerical and graphical resources, integrating the decision-makers in an online environment.

References

1. Brasil: Política Nacional de Defesa - Estratégia Nacional de Defesa. Ministério da Defesa, Brasília (2020)
2. Cepik, M., Licks Bertol, F.: Defense policy in Brazil: bridging the gap between ends and means? Def. Stud. **16**, 229–247 (2016)
3. Shortland, N., Alison, L., Barrett-Pink, C.: Military (in)decision-making process: a psychological framework to examine decision inertia in military operations. Theor. Issues Ergon. Sci. **19**, 752–772 (2018)
4. Haerem, T., Kuvaas, B.R., Bakken, B.T., Karlsen, T.: Do military decision makers behave as predicted by prospect theory? J. Behav. Decis. Mak. **24**, 482–497 (2011)
5. Zhang, Y., Xu, Z., Liao, H.: A consensus process for group decision making with probabilistic linguistic preference relations. Inf. Sci. (Ny) **414**, 260–275 (2017)
6. de Araújo Costa, I.P., et al.: Strategic study for managing the portfolio of IT courses offered by a corporate training company: an approach in the light of the ELECTRE-MOr multicriteria hybrid method. Int. J. Inf. Technol. Decis. Mak. 1–29 (2021). https://doi.org/10.1142/S0219622021500565
7. de Araújo Costa, I.P., et al.: Bibliometric studies on multi-criteria decision analysis (MCDA) applied in personnel selection. Front. Artif. Intell. Appl. **341** (2021). https://doi.org/10.3233/faia210239
8. dos Santos M., de Araujo Costa, I.P., Gomes, C.F.S.: Multicriteria decision-making in the selection of warships: a new approach to the AHP method. Int. J. Anal. Hierarchy Process. **13** (2021). https://doi.org/10.13033/ijahp.v13i1.833
9. Moreira, M.Â.L., Gomes, C.F.S., dos Santos, M., do Carmo Silva, M., Araujo, J.V.G.A.: PROMETHEE-SAPEVO-M1 a hybrid modeling proposal: multicriteria evaluation of drones for use in naval warfare. In: Thomé, A.M.T., Barbastefano, R.G., Scavarda, L.F., dos Reis, J.C.G., Amorim, M.P.C. (eds.) IJCIEOM 2020. SPMS, vol. 337, pp. 381–393. Springer, Cham (2020). https://doi.org/10.1007/978-3-030-56920-4_31
10. Zanazzi, J.L., Gomes, L.F.A.M., Dimitroff, M.: Group decision making applied to preventive maintenance systems. Pesqui. Operacional. **34**, 91–105 (2014)
11. Gomes, C.F.S., Santos, M.D., Teixeira, L.F., Sanseverino, A.M., Barcelos, M.R.: SAPEVO-M a group multicriteria ordinal ranking method. Pesqui. Operacional. **40**, 1–20 (2020)
12. Roy, B.: Paradigms and challenges. In: Greco, S., Ehrgott, M., Figueira, J.R. (eds.) Multiple criteria decision analysis. ISORMS, vol. 233, pp. 19–39. Springer, New York (2016). https://doi.org/10.1007/978-1-4939-3094-4_2

13. de Almeida, I.D.P., et al.: Study of the Location of a second fleet for the Brazilian navy: structuring and mathematical modeling using SAPEVO-M and VIKOR methods. In: Rossit, D.A., Tohmé, F., Mejía Delgadillo, G. (eds.) ICPR-Americas 2020. CCIS, vol. 1408, pp. 113–124. Springer, Cham (2021). https://doi.org/10.1007/978-3-030-76310-7_9
14. Ishizaka, A., Nemery, P.: Multi-criteria Decision Analysis. Wiley, Chichester (2013)
15. Gomes, L.F.A.M., Mury, A.R., Gomes, C.F.S.: Multicriteria ranking with ordinal data. Syst. Anal. Model. Simul. **27**, 139–145 (1997)
16. Maêda, S.M., Basílio, M.P., Costa, I.P., Moreira, M.Â., dos Santos, M., Gomes, C.F.: The SAPEVO-M-NC method. Front. Artif. Intell. Appl. **341**, 89–95 (2021). https://doi.org/10.3233/faia210235
17. Giones, F., Brem, A.: From toys to tools: the co-evolution of technological and entrepreneurial developments in the drone industry. Bus. Horiz. **60**, 875–884 (2017)

Firefly and Cuckoo Search Algorithm for Scheduling Problems: A Performance Analysis

Carlota Moreira[1], Catarina Costa[1], André S. Santos[2](✉), João A. Bastos[3], Leonilde R. Varela[4], and Marlene F. Brito[5]

[1] Institute of Engineering, Polytechnic Institute of Porto - ISEP/IPP, Porto, Portugal
{1180775,1180788}@isep.ipp.pt
[2] Interdisciplinary Studies Research Center (ISRC), Institute of Engineering, Polytechnic Institute of Porto - ISEP/IPP, Porto, Portugal
abg@isep.ipp.pt
[3] INESC TEC, Institute of Engineering, Polytechnic Institute of Porto - ISEP/IPP, Porto, Portugal
jab@isep.ipp.pt
[4] Algoritmi Research Centre, University of Minho, Guimarães, Portugal
leonilde@dps.uminho.pt
[5] Instituto de Ciência e Inovação em Engenharia Mecânica e Engenharia Industrial - INEGI, Porto, Portugal
mab@isep.ipp.pt

Abstract. Meta-heuristics are some of the best-known techniques to approach hard optimization problems, however, there are still questions about what makes some meta-heuristics better than others in a specific problem. This paper presents an analysis of the Firefly and Cuckoo Search Algorithm, such as others meta-heuristics. In order to assess the performance of the Firefly Algorithm and the Cuckoo Search Algorithm, they were compared with other well-known optimization techniques, such as Simulated Annealing and Local Search. Both meta-heuristics analysed in an in-depth computational study, reaching the conclusion that both techniques could be useful in Scheduling Problems and lead to satisfactory solutions quickly and efficiently. Moreover, the results of the analysis show that the Firefly Algorithm, despite having a high runtime, performs better than the other techniques.

Keywords: Meta-heuristics · Firefly algorithm · Cuckoo search algorithm · Simulated annealing

1 Introduction

Metaheuristics, along with other optimization approaches aim at achieving optimal or near optimal solutions to a given problem under study, namely in the context hard problems, such as the ones occurring in manufacturing management [1–6]. They employ

J. Machado et al. (Eds.): icieng 2022, LNME, pp. 75–88, 2023.
https://doi.org/10.1007/978-3-031-09360-9_7

search strategies in a given space and focus on areas that will allow the most effective achievement of the desired solution. Thus, they start from a set of initial solutions and throughout the application of the algorithm chosen to solve the problem, a progressive logic of the search for the optimal solution is followed [7].

The success of this optimization encounter depends on several factors such as the characteristic of the objective function, the type of constraint and the selection of the optimization algorithm [8].

More and more optimization algorithms are used by machine learning techniques, applied in engineering and industry when the goal is to optimization something, whether it is to minimize cost or maximize production efficiency, for example [7].

Modern meta-heuristic algorithms that have become popular are increasingly inspired by nature, such as the Firefly Algorithm and the Cuckoo Search Algorithm. Firefly Algorithm is based by the luminous attractiveness of fireflies, and, on the other hand, Cuckoo Search Algorithm is inspired by the breeding of behaviour of cuckoos, more specifically by the parasitism of the litter [9].

In fact, a peculiarity of these types of meta-heuristics is their capability to adapt to any problem. They can be used to solve Flow Shop Scheduling Problems (FSSP), which is used in this article to analyse the performance of both metaheuristics, as well as any other hard optimization problem.

The aim of this article is to compare the performance of the Firefly Algorithm and the Cuckoo Search Algorithm with some of the most widely used performance methods that act as a basis for the evaluation, in order to assess the efficiency, effectiveness and robustness of each of them.

2 Literature Review

This literature review will emphasize heuristics and meta-heuristics used throughout the paper to solve the problem and further analyse the solutions obtained.

2.1 Firefly Algorithm

The Firefly Algorithm (FA) is inspired by the nature of fireflies, being introduced by Xin-She-Yang [10] in 2008, which follows three idealized rules, namely: (1) fireflies are attracted to others regardless of gender, (2) attractiveness is inversely proportional to distance and proportional to brightness, with the less bright insects moving towards the brighter ones, however, in the event that they are equally bright, the firefly will move randomly, and (3) brightness is affected or determined by the value of the objective function, and is proportional to the quality of the solution [11].

In order to improve the solution, the firefly needs to move towards the fireflies that have a brighter light emission than its own, however in the event that the luminosity of both fireflies are equal, then a random move is generated [12], as shown in Table 1.

Table 1. Firefly algorithm pseudocode [13]

```
Begin
Generate initial population of fireflies x_i (i=1, 2,..., n)
 While t < MaxGeneration
  For i=1 to n fireflies
   For j=1 to n fireflies
    If (I_j > I_i) Movement from x_i towards x_j
    ElseIf (I_j = I_i) Random movement of x_i
    End if
   End for
  End for
  Rank the fireflies and find the best one
 Loop
 Post-Processing of the results
End
```

The variables β, r_{ij} and Movement are defined from the following equations, where Eq. 1 represents the attractiveness of fireflies, Eq. 2 indicates the distance between firefly *i* and firefly *j* and Eq. 3 reveals the movement of the firefly *i* [12].

$$\underline{\text{Attractiveness } \beta}: \beta = \beta_0 e^{-\gamma r^2} \tag{1}$$

$$\underline{\text{Distance } r_{ij}}: r_{ij} = \sqrt{\sum (x_i - x_j)^2} \tag{2}$$

$$\underline{\text{Movement:}} \, x_i = x_i + \beta(x_j - x_i) + \alpha\left(random - \frac{1}{2}\right) \tag{3}$$

The third parcel of the Movement equation refers to the randomness parameter, where random consists of a uniformly distributed random number [12].

This algorithm was developed to solve continuous problems, but it can be discretized so that it can be used to solve discrete optimization problems.

2.2 Cuckoo Search Algorithm

Cuckoo Search Algorithm is based on the brood parasitic behaviour of some cuckoo species in combination with the Lévy Flight behaviour of some birds and fruit flies [14]. Cuckoo lays their eggs in nests of other host birds, which may or may not be of the same species. If the host bird discovers the egg in their nest, it may destroy it or abandon the nest and rebuild it elsewhere. The analogy made by this algorithm consists in identifying the nests as possible solutions to the optimization problem, as shown in Table 2 [15]. Although this new optimization approach discovered by Yang and Deb [16] is generally used in solving continuous problems, in this article a discrete approach is applied to a flow shop scheduling problem.

- Lévy Flights

Foraging and the route to obtain food is in general random, since the next step is based on the current location and the probability of moving to the next location [14]. This probability can be modelled mathematically, where the model is represented by small random steps followed, in the long run, by large steps [17].

The new solution X_i will be generated by Eq. 4, for the ith nest through the equation of Lévy Flights [18]:

$$X_i^{n+1} = X_i^n + \alpha \oplus Lévy(\beta) \tag{4}$$

where α represents the step size and must be greater than zero and the random step length follows the following Lévy Flight distribution as shown in Eq. 5:

$$Lévy(\beta) \sim s^{-\beta} \tag{5}$$

In order to randomly generate flights through the symmetric Lévy distribution, the simplified Mantegna algorithm is applied [17]. Thus, the next nests are obtained using Eq. 6 and the Eq. 7 represents the stepsize:

$$X_i^{n+1} = X_i^n + stepsize \times random \tag{6}$$

$$stepsize = \alpha \times step \times (X_i - X_{best}) \tag{7}$$

Additionally, the step can be calculated using two Gaussian distributions U and V, as shown in Eq. 8.

$$step = \frac{U}{|V|^{\frac{1}{\beta}}} \tag{8}$$

with U and V shown in Eq. 9:

$$U \sim N\left(0, \sigma^2\right), V \sim N(0, 1) \tag{9}$$

where $U \sim N\left(0, \sigma^2\right)$ means that the samples are generated from a distribution normal Gaussian distribution with mean zero and variance σ^2 calculated by Eq. 10 [19]:

$$\sigma^2 = \left(\frac{\Gamma(1+\beta) \times \sin\frac{\pi \times \beta}{2}}{\Gamma\left(\left(\frac{1+\beta}{2}\right) \times \beta \times 2^{\frac{\beta-1}{2}}\right)}\right)^{\frac{1}{\beta}} \tag{10}$$

Table 2. Cuckoo search algorithm pseudocode

```
Begin
 Makespan Objective Function
 Generate initial population of n nests X_i = (i=1, 2, ..., n)
 While t < Interruption Criteria
   Get a random cuckoo through Lévy Flights
   Evaluate Fitness/Makespan F_i
   Choose a random nest j
   If (Fi > Fj),
      Replace j with the new solution
   End if
  Abandon a fraction p_a of the worst nests and build new ones
  Keeping the best solutions
  Rate the solutions and find the best one
 Loop
  Post-Processing of the results
End
```

Next, Local Seach and Simulated Aneealing will also be covered in a literature review, since they are well known techniques that will be used as a baseline for the analysis of the other techniques.

2.3 Local Search

The Local Search Algorithm [20] is a heuristic that seeks to solve optimization problems, with the main objective being to reach the "best" solution among several possible candidates in the search space.

The algorithm implies that, starting from an initial solution, pointwise changes are applied in each iteration that lead to the greatest reduction in the fitness value [21].

The algorithm terminates when there is no way to improve the current solution, in the neighbourhood structure.

The main disadvantage is that it settles for the local optimum, without guaranteeing its quality in the totality of the solution space, since it lacks diversity [21].

2.4 Simulated Annealing

Simulated Annealing is a stochastic approach that aims to overcome the local optimum by accepting worse solutions with a probability [22]. This probability is based on the annealing process and is obtained as a function of the temperature system [23].

This iterative method is one of the most used techniques in solving combinatorial optimization problems, being a local random search technique based on the principles of physics, namely the annealing process [24]. This process, in physics, consists of heating a solid at high temperatures and then gradually cooling it to reach the minimum energy state. This state can only be achieved if the temperature used for heating is high enough and if cooling is done slowly enough [25].

In the practical application of meta-heuristics, the temperature should gradually decrease as the algorithm moves forward, in order to decrease the probability of accepting solutions. In this way, worse solutions are rejected more often as the algorithm progresses. The probability of accepting a worse solution at each iteration is given by [26]:

$$p = e^{-\frac{\Delta}{T}} \tag{11}$$

where Δ is the difference between new and current solution and T is a parameter.

3 Flow Shop Scheduling Problems

Effectively, Flow Shop sequencing problems require that each job follows a specific sequence of operations, where each job does not pass through the same machine more than once, and each machine can only process one job at a time.

On the other hand, this project includes the analysis and implementation of several meta-heuristics for the case, where some of them (Cuckoo Search Algorithm and Firefly Algorithm) are intended to be applied to continuous problems [27]. The solution adopted for this problem was based on discretization appeared in Table 3:

Table 3. Solution discretization

Randomly generated values	0,368284404	0,091932714	0,636843383	0,845850885	0,751828754
Job sequence	1	2	3	4	5
	⬇		Sorting in ascending order	⬇	
Randomly generated values	0,091932714	0,368284404	0,636843383	0,751828754	0,845850885
Job sequence	2	1	3	5	4

3.1 Implementation and Parameterization of Meta-heuristics

In order to improve the performance of the meta-heuristics must the tuned to the problem. To understand the effect of the combination of the various parameters of the meta-heuristics on their performance, Design of Experiments (DOE) was applied to analyse each meta-heuristic. This allows the control factors to be manipulated, determining their effect on the value of the objective function, according to H. Shim, N. Shin, A. Stern et al. [28].

Table 4. Results of parameter tuning - SA

Parameters	Value	Range
T	9601	{6400; 8001; 9601}
L	150	{150; 2250; 4500}
α	0,8	{0,8; 0,87; 0,95}
IC	10000	{10000; 20000; 30000}

For Simulated Annealing, parameter tuning was applied to T, L, α and IC, and the following ranges of values were considered, as shown in Table 4:

The assigned levels were based on studies by E. Aarts and J. Korst [29], R. Kolisch and S. Hartmann [30], F. Martinez-Rios and J. Frausto-Solis [31].

For the Firefly Algorithm, the parameters γ, β_0, α_0 and number of fireflies (n), as evidenced in Table 5. The combination factor IC × n was set to 10000 iterations in order to obtain a good relation between solution quality and computational effort.

Table 5. Results of parameter tuning - FA

Parameters	Value	Range
γ	0,1	{0,1; 0,5; 1}
β_0	1	{0,1; 1}
α_0	0,1	{0,1; 0,15; 0,2}
n	40	{20; 30; 40}

The defined levels of variation the parameters were based on the studies by Udaiyakumara and Chandrasekaran [11], Khadwilard et al. [13].

For the Cuckoo Search Algorithm meta-heuristic, the parameters β, $P(a)$ and the number of solutions (n) were varied, as demonstrated in Table 6. The combination factor IC × n was set to 10000 iterations.

Table 6. Results of parameter tuning - CS

Parameters	Value	Range
B	0,4	{0,4; 1,1; 1,5}
α	1	{0,1; 1}
P(a)	0,7	{0,3; 0,5; 0,7}
n	30	{20; 30; 40}

The range of values experienced were grounded in the studies of S. Roy [18], H. Wang [16], J. García and B. Crawford [32] and D. Belbachir and F. Boumediene [33].

4 Analysis of Results

By analysing the computational results obtained from multiple runs of each metaheuristic, it is immediately identifiable the stochastic nature of the metaheuristic.

From the results obtained, the difference between heuristics and meta-heuristics is well established. Local Search quickly converges to the local optimum, which highlights its lack of diversity and excessive intensity. On the other hand, the Simulated Annealing, Firefly Algorithm and Cuckoo Search Algorithm, allow for in-depth exploration of the solutions space, thanks to the higher level of diversity, and thus obtain a better solution for the desired objective.

4.1 Results and Runtime for a Specific Instance

In the interest of evaluating the performance of the different meta-heuristics presented for the given instance, the following graphs are presented in Fig. 1 and Fig. 2, show the obtained values of makespan and runtime 40 runs of the same instance.

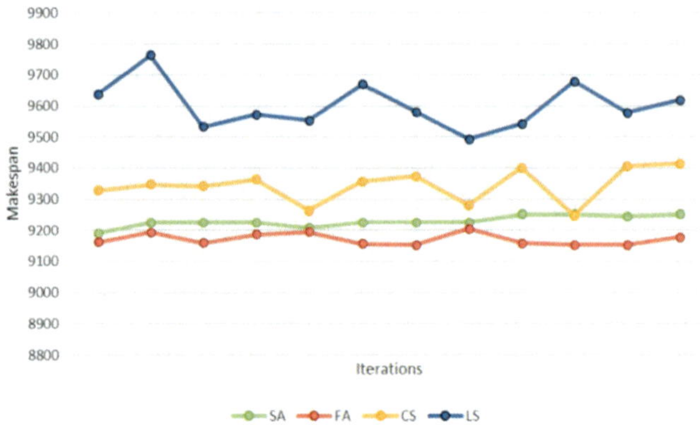

Fig. 1. Makespan of LS, SA, FA and CS

In Table 7, the statistical data for the above sample are shown in comparison with the best solution found for the problem.

The data in Table 7 is purely statistical data that tries to evaluate the quality of the solutions and the variation of the final solutions obtained by the different techniques.

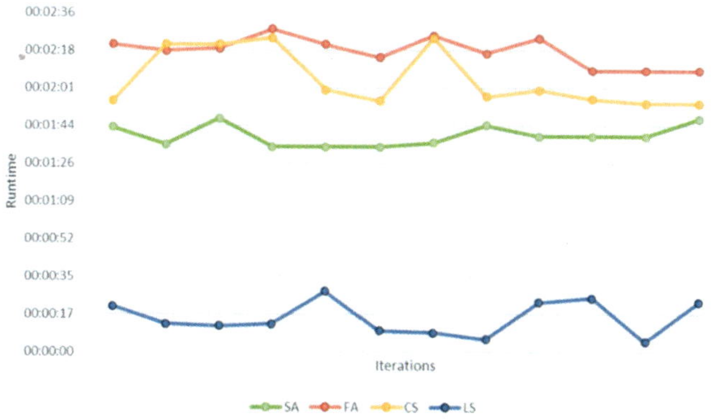

Fig. 2. Runtime of LS, SA, FA and CS

Table 7. Evaluation of the quality of the solutions

	SA	FA	CS	LS
Mean	9221	9175	9372	9599
SD	25,88	20,28	59,43	92,44
Runtime	00:01:45	00:02:23	00:02:02	00:00:19
Best	9190	9151	9248	9387
Worst	9284	9212	9469	9816
Makespan difference	0,42%	0,00%	1,05%	2,51%

From the graphics presented, it can be seen that the Firefly Algorithm are generally better than the others, however, regarding the runtime, it is highest. In contrast, the Simulated Annealing presents a solution, on average with runtime 54.20% lower, although 0.42% worse than the best solution found. The Cuckoo Search Algorithm presents 1.32% worse results, with a similar runtime to that of Simulated Annealing. Local Search, on the other hand, despite the reduced runtime, offers a solution 2,51% worse, on average, than the best-found solution.

From the simulation, it can be concluded that Simulated Annealing presents a very satisfactory solution in a more efficient way, among the presented meta-heuristics, since it requires less runtime. However, it should be noted that the Firefly Algorithm is more effective, in that it provides better solutions in comparable situations. Although the Cuckoo Search Algorithm is not as effective in terms of minimizing makespan, it manages to be more efficient than the Firefly Algorithm.

As for the robustness of the algorithms, it can be gauged that the four presented offer fairly consistent and similar solutions, which is evidenced by the Fig. 1 and Fig. 2. In addition, from Table 7, the solutions obtained by the Firefly Algorithm present a lower standard deviation in relation to the other algorithms, this means that there is not a large discrepancy between the various solutions achieved from this meta-heuristic.

4.2 Results and Runtime for Different Instances

An extended computational study was carried out with 30 different instances [34], so that, despite the more random nature of the meta-heuristics, the results were credible. In these instances, the processing time of the various jobs on each of the five machines was generated uniformly, in the interval $\{1, 99\}$.

For the 30 instances, exactly the same conclusion was reached, as the one presented above. The solution quality/runtime relationship only became more evident in Fig. 3. Obversing the two extremes, the Local search presents the worst solution with a lower runtime, while the Firefly algorithm presents a better solution, requiring more runtime.

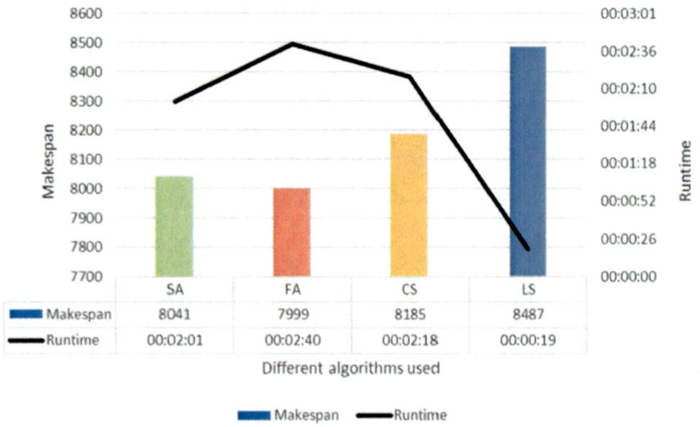

Fig. 3. Comparison between LS, SA, FA and CS

4.3 Influence of the Number of Jobs on the Robustness of the Algorithms

Furthermore, in Fig. 4 can be analized the discrepancy in solutions obtained by each technique in instances with a larger number of jobs. For problems with a larger number of jobs, in general, the solutions become less robust, as the value of the standard deviation increases with the number of jobs in the problems. The lower values of the standard deviation for problems with a smaller number of jobs are consistent due to the fact that the number of possible solutions is smaller and the technique is able to evaluate a higher percentage of solutions. By varying the number of jobs the exact the same conclusion was reached, that is, the Firefly Algorithm provides better solutions, but requires more runtime, while the Cuckoo Search Algorithm and, principally, the Simulated Annealing reach slightly worse solutions with much better runtime. It can also be seen that the algorithms require progressively more execution time to converge to a solution as the problem size increases.

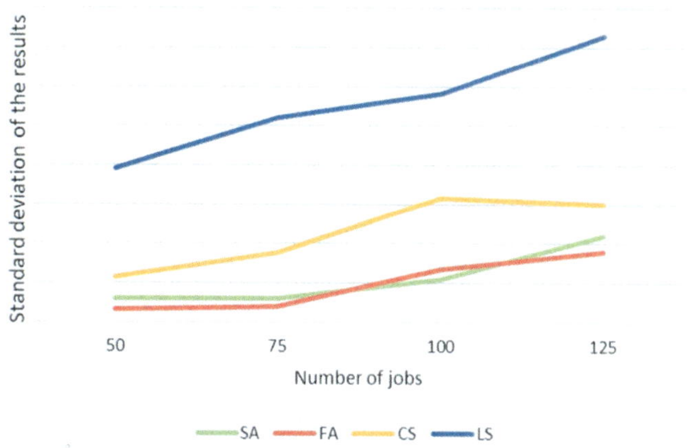

Fig. 4. Variation of the standard deviation with increasing number of jobs

To sum, it was observed that the Firefly Algorithm achieves better solutions, regarding the objective function, but in a less efficient way than Simulated Annealing and the Cuckoo Search Algorithm, since it consumes more execution time.

Regarding the robustness of the solution, it varies depending on the number of jobs. However, in general, it can be seen that the Firefly Algorithm and the Simulated Annealing both present a high robustness, noting that for a significant number of jobs, the Firefly Algorithm presents more constant results.

5 Conclusions

In this study, an attempt has been made to adapt the Firefly Algorithm and Cuckoo Search Algorithm to solve the scheduling of a flow shop problem by discretizing the methods and performing sensitivity analysis and parameterization.

Once developed, computationally implemented, and tested, these algorithms have proven that they can be very useful in problem solving in that they generally lead to satisfactory solutions quickly and efficiently.

The experimental analysis developed shows that the Firefly Algorithm provides better results than Local Search, Simulated Annealing and Cuckoo Search Algorithm, regardless of its dimension of the instance, despite requiring more execution time.

Regarding the efficiency and effectiveness of the metaheuristics, it can be concluded that Simulated Annealing is more efficient; however, Firefly is more effective since it was able to find solutions with lower makespan values. In concern of Cuckoo Search Algorithm, although is a good technic, it is not as effective in terms of makespan minimization, but it is more efficient than the Firefly Algorithm.

From the graphs presented throughout the article, the Firefly Algorithm presents better solutions than the others do, since it is 0.43%, 1.06% and 2.58% better than Simulated Annealing, Cuckoo Search Algorithm and Local Search, respectively. However, it has the highest runtime compared to the other methods developed, with a difference of 54.20% compared to the runtime of Simulated Annealing.

As future work, we consider improving the population or initial solution by integrating a heuristic or basic priority rules and developing a new hybrid meta-heuristic that combines the best aspects of the algorithms presented, in order to understand the influence on the objective function. On the other hand, it would also be interesting to expand the use of these algorithms in other decision-making contexts.

Acknowledgement. This work was supported by national funds through the FCT-Fundação para a Ciência e Tecnologia through the R&D Units Project Scopes: UIDB/00319/2020, and EXPL/EME-SIS/1224/2021.

References

1. Arrais-Castro, A., Varela, M.L.R., Putnik, G.D., Ribeiro, R.A., Machado, J., Ferreira, L.: Collaborative framework for virtual organisation synthesis based on a dynamic multi-criteria decision model. Int. J. Comput. Integr. Manuf. **31**(9), 857–868 (2018). https://doi.org/10.1080/0951192X.2018.1447146
2. Sousa, R.A., Varela, M.L.R., Alves, C., Machado, J.: Job shop schedules analysis in the context of industry 4.0. In: 2017 International Conference on Engineering, Technology and Innovation: Engineering, Technology and Innovation Management Beyond 2020: New Challenges, New Approaches, ICE/ITMC 2017 - Proceedings, 2018-January, pp. 711–717 (2018). https://doi.org/10.1109/ICE.2017.8279955
3. Gangala, C., Modi, M., Manupati, V.K., Varela, M.L.R., Machado, J., Trojanowska, J.: Cycle time reduction in deck roller assembly production unit with value stream mapping analysis. In: Rocha, Á., Correia, A.M., Adeli, H., Reis, L.P., Costanzo, S. (eds.) WorldCIST 2017. AISC, vol. 571, pp. 509–518. Springer, Cham (2017). https://doi.org/10.1007/978-3-319-56541-5_52
4. Trojanowska, J., Żywicki, K., Varela, M.L.R., Machado, J.M.: Shortening changeover time - an industrial study. In: The Proceedings of the 2015 10th Iberian Conference on Information Systems and Technologies (CISTI), pp. 1–6 (2015). https://doi.org/10.1109/CISTI.2015.7170373
5. Vieira, G.G., Varela, M.L.R., Putnik, G.D., Machado, J.M., Trojanowska, J.: Integrated platform for real-time control and production and productivity monitoring and analysis. Romanian Rev. Precision Mech. Opt. Mechatron. **2016**(50), 119–127 (2016)
6. Reddy, M.S., Ratnam, C.H., Agrawal, R., Varela, M.L.R., Sharma, I., Manupati, V.K.: Investigation of reconfiguration effect on makespan with social network method for flexible job shop scheduling problem. Comput. Ind. Eng. **110**, 231–241 (2017)
7. Bulatović, R.R., Dordević, S.R., Dordević, V.S.: Cuckoo search algorithm: a metaheuristic approach to solving the problem of optimum synthesis of a six-bar double dwell linkage. Mech. Mach. Theory **61**, 1–13 (2013)
8. Khari, M., Kumar, P.: An effective meta-heuristic cuckoo search algorithm for test suite optimization. Informatica **41**(3), 363–377 (2017)
9. Yang, X.S., Deb, S.: Cuckoo search via Lévy flights. In: 2009 World Congress on Nature & Biologically Inspired Computing, NABIC 2009 - Proceedings, pp. 210–214 (2009)
10. Yang, X.-S.S., Karamanoglu, M.: Nature-Inspired Metaheuristic Algorithms Second Edition, vol. 4, no. C (2013)
11. Udaiyakumar, K.C., Chandrasekaran, M.: Application of firefly algorithm in job shop scheduling problem for minimization of Makespan. Procedia Eng. **97**, 1798–1807 (2014)

12. Sureja, N.M.: Solving random travelling salesman problem using firefly algorithm. Int. J. Innov. Technol. Explor. Eng. **9**(4), 1037–1041 (2020)
13. Khadwilard, A., Chansombat, S., Thepphakorn, T., Thapatsuwan, P., Chainate, W., Pongcharoen, P.: Application of firefly algorithm and its parameter setting for job shop scheduling. In: First Symposium on Hands-on Research and Development, vol. 1, no. 1 (2011)
14. Marichelvam, M.K., Prabaharan, T., Yang, X.S.: Improved cuckoo search algorithm for hybrid flow shop scheduling problems to minimize makespan. Appl. Soft Comput. J. **19**, 93–101 (2014)
15. Ribeiro Barros de Alencar, M., Alencar de Souza, B., Luiz Araújo Neves, W., Santos Freire Ferraz, R.: Aplicação de Algoritmo de Busca Cuco no Dimensionamento Ótimo de Gerador Fotovoltaico para Redução de Custos, pp. 731–736 (2019)
16. Wang, H., Wang, W., Sun, H., Li, C., Rahnamayan, S., Liu, Y.: A modified cuckoo search algorithm for flow shop scheduling problem with blocking. In: 2015 IEEE Congress on Evolutionary Computation, CEC 2015 - Proceedings, no. May 2019, pp. 456–463 (2015)
17. Nadjemi, O., Nacer, T., Hamidat, A., Salhi, H.: Optimal hybrid PV/wind energy system sizing: application of cuckoo search algorithm for Algerian dairy farms. Renew. Sustain. Energy Rev. **70**, 1352–1365 (2017)
18. Roy, S., Sinha Chaudhuri, S.: Cuckoo search algorithm using Lèvy flight: a review. Int. J. Mod. Educ. Comput. Sci. **5**(12), 10–15 (2013)
19. Nascimento, G.M., Carvalho, P.V.M.: Uma Variante Melhorada do Algoritmo Busca Cuco usando uma Estratégia de Quasi Opposition – Based Learning, pp. 60–69 (2017)
20. Desale, S., Rasool, A., Andhale, S., Rane, P.: Heuristic and meta-heuristic algorithms and their relevance to the real world: a survey. Int. J. Comput. Eng. Res. Trends **351**(5), 2349–7084 (2015)
21. Stutzle, T., Ruiz, R.: Iterated Local Search: A Concise Review T. St u Technical report, Univ. Libr. Bruxelles, April 2018
22. Melouk, S., Damodaran, P., Chang, P.Y.: Minimizing makespan for single machine batch processing with non-identical job sizes using simulated annealing. Int. J. Prod. Econ. **87**(2), 141–147 (2004)
23. Roa-Sepulveda, C.A., Pavez-Lazo, B.J.: A solution to the optimal power flow using simulated annealing. Int. J. Electr. Power Energy Syst. **25**(1), 47–57 (2003)
24. Azizi, N., Zolfaghari, S.: Adaptive temperature control for simulated annealing: a comparative study. Comput. Oper. Res. **31**(14), 2439–2451 (2004)
25. De Vicente, J., Lanchares, J., Hermida, R.: Placement by thermodynamic simulated annealing. Phys. Lett. Sect. A Gen. At. Solid State Phys. **317**(5–6), 415–423 (2003)
26. D'Amico, S.J., Wang, S.J., Batta, R., Rump, C.M.: A simulated annealing approach to police district design. Comput. Oper. Res. **29**(6), 667–684 (2002)
27. Kota, L., Jármai, K.: Discretization of the firefly algorithm for the travelling salesman problem. DAAAM International Scientific Book, pp. 495–512 (2013). ISSN 1726-9687, ISBN 978-3-901509-84-1
28. Shim, H., et al.: Estudo paramétrico em dutos rígidos com vãos livres por meio de técnicas de doe. Adv. Opt. Mater. **10**(1), 1–9 (2018)
29. Aarts, E., Korst, J.: Simulated Annealing and Boltzmann Machines: A Stochastic Approach to Combinatorial Optimization and Neural Computing, pp. 19–22 (1975)
30. Kolisch, R., Hartmann, S.: Heuristic algorithms for the resource-constrained project scheduling problem: classification and computational analysis. In: Węglarz, J. (ed.) Project Scheduling, pp. 147–178. Springer, Boston (1999). https://doi.org/10.1007/978-1-4615-5533-9_7

31. Martinez-Rios, F., Frausto-Solis, J.: A simulated annealing algorithm for the satisfiability problem using dynamic Markov chains with linear regression equilibrium. Simulated Annealing Adv. Appl. Hybridizations (2012)
32. García, J., Crawford, B., Soto, R., Castro, C., Paredes, F.: A k-means binarization framework applied to multidimensional knapsack problem. Appl. Intell. **48**(2), 357–380 (2017). https://doi.org/10.1007/s10489-017-0972-6
33. Belbachir, D., Boumediene, F., Hassam, A., Ghomri, L.: Adaptation and parameters studies of CS algorithm for flow shop scheduling problem. Int. J. Electr. Comput. Eng. **11**(3), 2266–2274 (2021)
34. Kwak, S.G., Kim, J.H.: Introduction basic concepts of central limit theorem central limit theorem: the cornerstone of modern statistics KJA. Korean J. Anesth. **70**(2), 144–156 (2017)

Power Series and Inequalities on the Parameters of a Strongly Regular Graph

Luís Almeida Vieira[1,2(✉)] and Vasco Moço Mano[3]

[1] Faculty of Engineering of Porto, Section of Mathematics, Porto, Portugal
lvieira@fe.up.pt

[2] University of Porto, Center of Mathematics of University of Porto, Porto, Portugal

[3] University of Porto, Porto, Portugal

Abstract. In the following, we deal with the problem of the existence of primitive strongly regular graphs with 1 in its spectrum and such that its regularity is between certain values with respect to the order of the graph. We consider the natural correspondence between the adjacency matrix of a strongly regular graph and the Euclidean Jordan algebra, $\mathcal{B}$, generated by the natural powers of that matrix. Considering the spectrum of a constructed series of Hadamard powers an element of that Euclidean Jordan algebra $\mathcal{B}$, we establish a necessary condition for the existence of this kind of primitive strongly regular graphs.

Keywords: Euclidean Jordan algebra · Graph theory · Strongly regular graph

1 Introduction

The aim of this paper is to establish an existence condition over the parameters and the elements of the spectrum of a primitive strongly regular graph where the positive eigenvalue is one and its regularity k and its order n verify the equality $\frac{n}{5} + 1 < k < \frac{2n}{5}$.

The foundations of our research lie with the association of a particular Euclidean Jordan algebra to the adjacency matrix of a primitive strongly regular graph. This allows us to deduce an admissibility condition on the parameters and over the elements of the spectrum of such a graph.

This paper is divided in four sections. Some properties of strongly regular graphs are exposed in Sect. 2 and the main results about Euclidean Jordan algebras are described in Sect. 3. In Sect. 4 we associate a three dimensional real Euclidean Jordan algebra $\mathcal{A}$ to a strongly regular graph via its adjacency matrix. This enables us to establish an admissibility condition for the existence of this graph involving the elements of its spectrum and its parameters.

J. Machado et al. (Eds.): icieng 2022, LNME, pp. 89–99, 2023.
https://doi.org/10.1007/978-3-031-09360-9_8

2 Some Results About Strongly Regular Graphs

In this preliminary section we will introduce some relevant definitions and results on Graph Theory. For further reading, please refer to [1].

The graphs that we deal with in our work are simple, that is, without edges with the same extreme vertices or edges with the two extreme points in the same vertex. On a simple graph G an edge is defined by two extreme vertices, therefore when needed we can use the notation $G = (V(G), E(G))$ where $V(G)$ represents the vertices of G and $E(G)$ represents the set of edges of G. From now on we only consider simple graphs. Let $G = (V(G), E(G))$ be a simple graph. Then one defines the order of G has being the cardinal of $V(G)$ and one defines the dimension of G as being the cardinal of $E(G)$.

Given a vertex v of G one defines the degree of v, as being the number of incident edges on v. A graph G is called k-regular if the degree of all its vertices is k.

Given a graph $G = (V(G), E(G))$ of order n one defines the adjacency matrix of G as being the real matrix $A = [a_{ij}]$ of order n such as defined in (1):

$$a_{ij} = \begin{cases} 1 & \text{if} \quad v_i v_j \in E(G) \\ 0 & \text{if} \quad v_i v_j \notin E(G) \end{cases}. \tag{1}$$

The complement graph of a simple graph G, denoted by $\overline{G}$, is a simple graph that has the same set of vertices of G and such that two distinct vertices of $\overline{G}$ are adjacent on $\overline{G}$ if and only if they are not adjacent on G.

We observe that if G is a k-regular graph of order n then $\overline{G}$ is a $n-k-1$-regular graph.

A simple graph of order n is complete if all pair of distinct vertices are adjacent vertices. In this case, the degree of each of its vertices is $n-1$.

A simple graph $G = (V(G), E(G))$ is the null graph if its set of edges is the empty set and the set of vertices is not empty.

Given a graph $G = (V(G), E(G))$ a path of G is an non-empty sequence of vertices and edges of G, $v_1 e_1 v_2 e_2 v_3 \cdots v_{k-1} e_{k-1} v_k$ such that $v_1, v_2, \cdots, v_k \in V(G)$ and $e_1, e_2, \cdots, e_{j-1} \in E(G)$ and each edge e_i has extremes v_i and v_{i+1} and v_k for $i = 1, \cdots, k-1$, such that all vertices are distinct with the exception of the vertex v_1 and the vertex v_k and such that all edges are distinct. A simple graph is connected if for any two distinct vertices there exists a path that joins them. A graph that is not connected is called disconnected or unconnected.

Given a simple and regular graph, G, if every pair of vertices in $V(G)$ has a fixed number of common neighbors, then G is called strongly regular. Specifically, G can be identified with the so-called parameter set $(n, k; \lambda, \mu)$ where n is the order of the graph, k is the standard regularity, λ is the number of common neighbors to each pair of adjacent vertices and μ is the number of common neighbors to each pair of non-adjacent vertices in G.

Let's consider the $(n, k; \lambda, \mu)$-strongly regular graph, G, and A its adjacency matrix. It is well known that A satisfies the following equality:

$$\lambda A - \mu(A + I_n - J_n) + k I_n = A^2.$$

Note that I_n and J_n denote the identity matrix of order n and the matrix of ones also of order n, respectively. The eigenvalues of A, that are k, θ and τ, where θ and τ are defined by the equalities (2) and (3), are called the eigenvalues of G:

$$\theta = \frac{\lambda - \mu + \sqrt{(\lambda - \mu)^2 + 4(k - \mu)}}{2}, \tag{2}$$

$$\tau = \frac{\lambda - \mu - \sqrt{(\lambda - \mu)^2 + 4(k - \mu)}}{2}. \tag{3}$$

It is known that, if G is a $(n, k; \lambda, \mu)$-strongly regular graph, then its complement is a $(n, n-k-1; n-2-2k+\mu, n-2k+\lambda)$-strongly regular graph. The complement of G is usually denoted by $\overline{G}$.

A strongly regular graph G is *primitive* if and only if G and $\overline{G}$ are connected. A $(n, k; \lambda, \mu)$-strongly regular graph is not primitive or imprimitive if and only if $\mu = k$ or $\mu = 0$. In the following we only consider $(n, k; \lambda, \mu)$-strongly regular graphs such that $0 < \mu < k < n - 1$.

3 Euclidean Jordan Algebras: An Introduction

A *Jordan algebra* $\mathcal{B}$ with finite dimension is an algebra over a field $\mathbb{K}$ equipped with a product of vectors $\Box : (u, v) \mapsto u\Box v$ that is a bilinear mapping, from $\mathcal{B} \times \mathcal{B}$ to $\mathcal{B}$, such that we have $u\Box v = v\Box u$ and $u\Box(u^{2\Box}\Box v) = u^{2\Box}\Box(u\Box v)$, for any vectors u and v of $\mathcal{B}$ where $u^{2\Box} = u\Box u$. For instance $\mathcal{B} = \mathbb{R}^n$ with the operation $\Box$ such that for any $w = (w_1, w_2, \cdots, w_n)$ and $t = (t_1, t_2, \cdots, t_n)$ of $\mathbb{R}^n$ we have

$$w\Box t = (w_1 t_1, w_2 t_2, \cdots, w_n t_n)$$

is a Jordan algebra. Indeed, for any $u = (u_1, u_2, \cdots, u_n)$, $v = (v_1, v_2, \cdots, v_n)$ and $w = (w_1, w_2, \cdots, w_n)$ of $\mathcal{B}$ we have the following calculations.

$$\begin{aligned} u\Box v &= (u_1, u_2, \cdots, u_n)\Box(v_1, v_2, \cdots, v_n) \\ &= (u_1 v_1, u_2 v_2, \cdots, u_n v_n) \\ &= (v_1 u_1, v_2 u_2, \cdots, v_n u_n) \\ &= v\Box u \end{aligned}$$

and

$$\begin{aligned} u\Box(u^{2\Box}\Box v) &= (u_1, u_2, \cdots, u_n)\Box\left((u_1^2, u_2^2, \cdots, u_n^2)\Box(v_1, v_2, \cdots, v_n)\right) \\ &= (u_1, u_2, \cdots, u_n)\Box(u_1^2 v_1, u_2^2 v_2, \cdots, u_n^2 v_n) \\ &= (u_1(u_1^2 v_1), u_2(u_2^2 v_2), \cdots, u_n(u_n^2 v_n)) \\ &= (u_1^2(u_1 v_1), u_2^2(u_2 v_2), \cdots, u_n^2(u_n v_n)) \\ &= (u_1^2, u_2^2, \cdots, u_n^2)\Box(u_1 v_1, u_2 v_2, \cdots, u_n v_n) \\ &= u^{2\Box}\Box(u\Box v). \end{aligned}$$

If $\mathcal{B}$ contains an element, $\mathbf{e}$, which is non-zero, such that $\mathbf{e}\Box u = u\Box\mathbf{e} = u$, for all u in $\mathcal{B}$, then $\mathbf{e}$ is denominated the *unit* of $\mathcal{B}$. In the remaining of this work, we only deal with Jordan algebras that are real, with finite dimension and containing a unit $\mathbf{e}$.

Example 1. The algebra over the field $\mathbb{R}$ of real symmetric matrices of order m, $\mathcal{B} = \mathrm{Sym}(m, \mathbb{R})$ with the operation $u\Box v = \frac{uv+vu}{2}$ for u and v elements of $\mathcal{B}$. $\mathcal{B}$ is a Jordan algebra. Let's consider u an element of $\mathcal{B}$. Now, we define the natural powers of u as being defined in the following manner: $u^{0\Box} = \mathbf{e}, u^{1\Box} = u$ and $u^{2\Box} = u\Box u$. And, therefore we introduce the natural power of order n of u in the following way $u^{n\Box} = u\Box u^{(n-1)\Box}$ for any natural number n. Noting that $u^{2\Box} = u\Box u = \frac{uu+uu}{2} = \frac{u^2+u^2}{2} = u^2$ then recurring to induction over n we deduce that $u^{n\Box} = u^n$ for any natural number n.

Now we show that $\mathcal{B}$ is a Jordan algebra. Since $u\Box v = \frac{uv+vu}{2} = \frac{vu+uv}{2} = v\Box u$, for all u and v in $\mathcal{B}$, then we have the following:

$$\begin{aligned}
u\Box(u^{2\Box}\Box v) &= u\Box(u^2\Box v) = u\Box\left(\frac{u^2v+vu^2}{2}\right)\\
&= \frac{u\left(\frac{u^2v+vu^2}{2}\right) + \left(\frac{u^2v+vu^2}{2}\right)u}{2} = \frac{\frac{u^3v+uvu^2+u^2vu+vu^3}{2}}{2}\\
&= \frac{\frac{u^3v+u^2vu+vu^3+uvu^2}{2}}{2} = \frac{\frac{u^2(uv+vu)+(vu+uv)u^2}{2}}{2}\\
&= \frac{\frac{u^2(uv+vu)+(uv+vu)u^2}{2}}{2} = \frac{u^2(u\Box v) + (u\Diamond v)u^2}{2}\\
&= \frac{u^{2\Box}(u\Box v) + (u\Box v)u^{2\Box}}{2} = u^{2\Box}\Box(u\Box v).
\end{aligned}$$

An *Euclidean Jordan algebra*, $\mathcal{B}$, is a Jordan algebra equipped with an scalar product $\cdot|\cdot$ such that for any u, v and $w \in \mathcal{B}$ we have $(u\Box v)|w = v|(u\Box w)$.

Remark 1. Let $\mathcal{B} = \mathrm{Sym}(n, \mathbb{R})$. Considering the scalar product $\cdot|\cdot$ such that for any u and $v \in \mathcal{B}$ we have $u|v = \mathrm{trace}(u\Box v)$, then $\mathcal{B}$ is an Euclidean Jordan algebra. Indeed, we have that for u, v and w in $\mathcal{B}$ the following calculations

$$
\begin{aligned}
(u\Box v)|w &= \operatorname{trace}((u\Box v)\Box w)\\
&= \operatorname{trace}\left(\frac{\frac{uv+vu}{2}w + w\frac{uv+vu}{2}}{2}\right)\\
&= \left(\frac{uv+vu}{2}w\right)\\
&= \operatorname{trace}\left(\frac{uvw+vuw}{2}\right)\\
&= \operatorname{trace}\left(\frac{uvw}{2}\right) + \operatorname{trace}\left(\frac{vuw}{2}\right)\\
&= \operatorname{trace}\left(\frac{vuw}{2} + \frac{vwu}{2}\right)\\
&= \operatorname{trace}\left(v\left(\frac{uw}{2} + \frac{wu}{2}\right)\right)\\
&= \operatorname{trace}(v\,(u\Box w))\\
&= \operatorname{trace}\left(\frac{v\,(u\Box w) + (u\Box w)\,v}{2}\right)\\
&= \operatorname{trace}(v\Box(u\Box w))\\
&= v|(u\Box w).
\end{aligned}
$$

Extended surveys about Jordan algebras and Euclidean Jordan algebras can be found in [2–5]. Several application of Euclidean Jordan algebras in mathematics are developed in the articles [6–13] and [14]. We must also refer the thesis [15–17] that recur to Euclidean Jordan algebras as a tool to develop some results on mathematics.

If u in $\mathcal{B}$ is such that $u^{2\Box} = u$ then we call it an idempotent $\mathcal{B}$. Let u and v be two idempotents of $\mathcal{B}$. They are called *orthogonal* if $u\Box v = 0$. A set $\{f_1, f_2, \ldots, f_l\}$, such that its elements are idempotents, orthogonal with respect to $\Box$ and satisfying $f_1+f_2+\cdots+f_l = \mathbf{e}$, is called a complete system of orthogonal idempotents.

Example 2. Let's consider the symmetric matrices F_{jj} for $j = 1, \cdots, n$ of the Euclidean Jordan algebra $\mathcal{B} = \mathrm{Sym}(n, \mathbb{R})$ such that each matrix F_{jj} is such that the entry jj of F_{jj} is equal to 1 and such the other entries distinct from jj are zero. Let k be a natural number such that $1 < k < n-2$ then the set, $\mathcal{S} = \{\sum_{j=1}^{k} F_{jj}, \sum_{j=k+1}^{n} F_{jj}\}$ is a complete system of non zero orthogonal idempotents of $\mathcal{B}$.

Example 3. Let's consider the Euclidean Jordan algebra $\mathcal{B} = \mathbb{R}^n$ equipped with the operation of multiplication of vectors $u\Box v = (u_1v_1, u_2v_2, \cdots, u_nv_n)$, the canonical basis $<f_1, f_2, \cdots, f_n> = <(1, 0, \cdots, 0), (0, 1, \cdots, 0), \cdots, (0, 0, \cdots, 1)>$ of $\mathcal{B}$ and let k be a natural number such that $2 < k \leq n-2$. Then, we conclude that $\mathcal{S} = \{c_1, c_2\} = \{\sum_{j=1}^{k} f_j, \sum_{j=k+1}^{m} f_j\}$ is a complete system of non zero idempotents of $\mathcal{B}$.

From now on we suppose that $\mathcal{B}$ is a n-finite dimensional Euclidean Jordan algebra.

The rank of an element u in $\mathcal{B}$ is the smallest natural number l, such that $\{\mathbf{e}, u^{1\Box}, u^{2\Box}, \ldots, u^{l\Box}\}$ is a linearly dependent set. In this case, the notation adopted is $\operatorname{rank}(u) = l$. Furthermore, the number $\max\{\operatorname{rank}(u) : u \in \mathcal{B}\}$ is called the rank of the algebra $\mathcal{B}$ and we denote it by $\operatorname{rank}(\mathcal{B})$. Vectors of $\mathcal{B}$ whose rank is equal to the rank of the algebra are called regular. These elements form a dense and open set in $\mathcal{B}$.

We now take a regular element, u, of $\mathcal{B}$ with rank equal to $r \in \mathbb{N}$. It is easy to verify that $\{\mathbf{e}, u^{1\Box}, u^{2\Box}, \cdots, u^{r\Box}\}$ is not a free set of $\mathcal{B}$. Removing the last element of the previous set, one can straightforwardly conclude that we obtain a free set of $\mathcal{B}$. Therefore, we can affirm the existence of $\gamma_1(u), \gamma_2(u), \ldots, \gamma_{r-1}(u)$ and $\gamma_r(u)$ in $\mathbb{R}$, such that $u^{r\Box} - \gamma_1(u)u^{r-1\Box} + \cdots + (-1)^r\gamma_r(u)\mathbf{e} = 0$, where 0 represents the zero vector of $\mathcal{B}$. The characteristic polynomial of u is then defined as the polynomial in α

$$p_u(\alpha) = \alpha^r - \gamma_1(u)\alpha^{r-1} + \cdots + (-1)^r\gamma_r(u).$$

Given a fixed basis of $\mathcal{B}$, each $\gamma_i(u)$ can be observed as a homogeneous polynomial of degree i with respect to the coordinates u in that basis. The definition of p_u can be extended to all elements of $\mathcal{B}$ by density and continuity. We call the zeros of p_u the eigenvalues of u.

For each element u in an Euclidean Jordan algebra $\mathcal{B}$, there are unique real numbers $\gamma_1(u), \gamma_2(u), \ldots, \gamma_{k-1}(u)$ and $\gamma_k(u)$, all distinct, and a unique complete system of orthogonal idempotents $\{f_1, f_2, \ldots, f_k\}$, such that

$$u = \gamma_1(u)f_1 + \gamma_2(v)f_2 + \cdots + \gamma_k(v)f_k. \tag{4}$$

The decomposition of u described in (4) is usually addressed as the *first spectral decomposition* of u and the numbers $\gamma_j(u)$'s appearing in (4) are precisely the eigenvalues of u.

An idempotent u is primitive if it is a non-zero idempotent of $\mathcal{B}$ and if it is impossible to write it as the sum of two non-zero orthogonal idempotents. A complete system of orthogonal idempotents where each of its elements is primitive is usually addressed as a *Jordan frame.*

Example 4. Consider the Euclidean Jordan algebra $\mathcal{B} = \mathbb{R}^n$ equipped with componentwise vector multiplication and the usual scalar product of $\mathbb{R}^n$. Then, the set $\mathcal{S} = \{f_1, f_2, \cdots, f_n\}$, where $<f_1, f_2, \cdots, f_n>$ is the canonical basis of $\mathbb{R}^n$, is a Jordan frame of $\mathcal{B}$.

For each u in an Euclidean Jordan algebra $\mathcal{B}$ such that $\operatorname{rank}(\mathcal{B}) = r$, there is a Jordan frame $\{f_1, f_2, \cdots, f_r\}$ and real numbers $\gamma_1, \gamma_2, \cdots, \gamma_{r-1}$ and γ_r such that

$$u = \gamma_1(u)f_1 + \gamma_2(u)f_2 + \cdots + \gamma_r(u)u_r. \tag{5}$$

The decomposition (5) is called the second spectral decomposition of u.

4 Power Series and Some Conditions Involving the Parameters of a Strongly Regular Graph

In this section, we are supposing that G is a $(n, k; \lambda, \mu)$-strongly regular graph, such that $0 < \mu < k < n-1$ and $\frac{n}{5} + 1 < k < \frac{2n}{5}$, whose adjacency matrix A has the eigenvalues, k, θ and τ. We will clarify later the reason behind imposing this limitations on k. From now on we suppose that the eigenvalues k, θ and τ of A are all distinct.

Next, we will associate to A a real 3-dimensional Euclidean Jordan algebra with rank equal to 3. Let $\mathcal{B} = \mathrm{Sym}(n, \mathbb{R})$ be the Euclidean Jordan algebra of the symmetric matrices of order n endowed with the operation of multiplication of vectors $\Box$ such that for any u and v in $\mathcal{B}$ we have $u \Box v = \frac{uv+vu}{2}$, where uv and vu represents the usual product of square matrices of u by v and of v by u, and with the scalar product $u|v = \mathrm{trace}(u \Box v)$. Let's consider the subalgebra $\mathcal{B}'$ of $\mathcal{B}$, that is spanned by $\{I_n, A^k | k \in \mathbb{N}\}$. Since A has three distinct eigenvalues, we conclude that $\mathcal{B}'$ is an Euclidean Jordan algebra with rank and dimension 3. Next we consider the set $\mathcal{S} = \{F_1, F_2, F_3\}$ that is the unique Jordan frame of $\mathcal{B}'$ with respect to A. Its elements are:

$$
\begin{aligned}
F_1 &= \frac{1}{n} J_n, \\
F_2 &= \frac{|\tau| n + \tau - k}{(\theta - \tau) n} I_n + \frac{n + \tau - k}{(\theta - \tau) n} A + \frac{\tau - k}{(\theta - \tau) n} (J_n - A - I_n), \\
F_3 &= \frac{k - \theta + \theta n}{(\theta - \tau) n} I_n + \frac{k - n - \theta}{(\theta - \tau) n} A + \frac{k - \theta}{(\theta - \tau) n} (J_n - A - I_n).
\end{aligned}
$$

Next, we introduce some notation. The set $M_n(\mathbb{R})$ denotes set of real square matrices of order n. Given the matrices $C = [c_{ij}]$, $D = [d_{ij}]$ of $M_n(\mathbb{R})$, the matrix $C \circ D$ is the matrix such that $C \circ D = [c_{ij} d_{ij}]$ the componentwise product of the matrices C and D. And the Kronecker product of the matrices C and D is the matrix $C \otimes D$ such that $C \otimes D = [c_{ij} D]$. For any natural number k and for any $B \in M_n(\mathbb{R})$ we define the powers $B^{k\circ}$ with the equalities: $B^{0\circ} = J_n$, $B^{1\circ} = B$ and for $k \geq 2$ we define $B^{k\circ} = B^{(k-1)\circ} \circ B$.

Let's consider the power series, S_3, defined as

$$
S_3 = \sum_{k=1}^{+\infty} k((F_3)^{2\circ})^{(k-1)\circ}.
$$

We then have

$$
\begin{aligned}
S_3 = \sum_{k=1}^{+\infty} k \Bigg(& \frac{(k - \theta + \theta n)^2}{((\theta - \tau) n)^2} I_n + \frac{(k - n - \theta)^2}{((\theta - \tau) n)^2} A \\
& + \frac{(k - \theta)^2}{((\theta - \tau) n)^2} (J_n - A - I_n) \Bigg)^{(k-1)\circ}.
\end{aligned}
$$

Therefore we can conclude that

$$S_3 = \sum_{k=1}^{+\infty} k\left[\left(\frac{(k-\theta+\theta n)^2}{((\theta-\tau)n)^2}\right)^{k-1} I_n + \left(\frac{(k-n-\theta)^2}{((\theta-\tau)n)^2}\right)^{k-1} A + \left(\frac{(k-\theta)^2}{((\theta-\tau)n)^2}\right)^{k-1} (J_n - A - I_n)\right]$$

and we can deduce that

$$S_3 = \frac{1}{(1-(\frac{k-\theta+\theta n}{(\theta-\tau)n})^2)^2} I_n + \frac{1}{(1-(\frac{k-n-\theta}{(\theta-\tau)n})^2)^2} A + \frac{1}{(1-(\frac{k-\theta}{(\theta-\tau)n})^2)^2}(J_n - A - I_n)).$$

Next, we consider the element $F_3 \circ S_3$ that relatively to the Jordan frame $\mathcal{S}$ of $\mathcal{B}'$ has the spectral decomposition $F_3 \circ S_3 = \sum_{i=1}^{3} q_{i33} F_i$. Since

$$\begin{aligned} S_3 &= \lim_{n\to+\infty} S_n \\ &= \lim_{n\to+\infty} \sum_{k=1}^{n} k((\frac{(k-\theta+\theta n)^2}{((\theta-\tau)n)^2})^{k-1} I_n + (\frac{(k-n-\theta)^2}{((\theta-\tau)n)^2})^{k-1} A \\ &\quad + (\frac{(k-\theta)^2}{((\theta-\tau)n)^2})^{k-1}(J_n - A - I_n)) \end{aligned}$$

and considering the spectral decomposition $S_n = q_{1n}F_1 + q_{2n}F_2 + q_{3n}F_3$, then we conclude that the eigenvalues q_{in} are non-negative. This conclusion arises from the fact that the coefficients of the Hadamard powers of $(F_3)^{\circ 2}$ are positive and the property that holds for any square matrices A and B of order n that states that the smallest eigenvalue of $A \circ B$ is greater or equal that the product between the smallest eigenvalue of A and the smallest eigenvalue of B. Therefore, the eigenvalues of S_n are non-negative. Now since we have $S_3 = \sum_{i=1}^{3} q_{i3} F_i$ and $q_{i3} = \lim_{n\to+\infty} q_{in}$, then we conclude that the eigenvalues q_{i3} are non-negative. Finally, we have that the eigenvalues q_{i33} are non negative since we have mineigenvalue$(A \circ B) \geq$ mineigenvalue(A)mineigenvalue(B) for any real square symmetric matrices A and B of order n. Now let's analyse the eigenvalue q_{133}. Since we have $q_{133} \geq 0$ then we conclude that:

$$\frac{k-\theta+\theta n}{(\theta-\tau)n}\frac{1}{(1-(\frac{k-\theta+\theta n}{(\theta-\tau)n})^2)^2} + \frac{k-n-\theta}{(\theta-\tau)n}\frac{1}{(1-(\frac{k-n-\theta}{(\theta-\tau)n})^2)^2}k + \frac{k-\theta}{(\theta-\tau)n}\frac{1}{(1-(\frac{k-\theta}{(\theta-\tau)n})^2)^2}(n-k-1) \geq 0. \quad (6)$$

Next, since we have

$$\frac{k-\theta+\theta n}{(\theta-\tau)n} + \frac{k-n-\theta}{(\theta-\tau)n}k + \frac{k-\theta}{(\theta-\tau)n}(n-k-1) = 0,$$

and rewriting the inequality (6), we deduce the inequality (7):

$$\frac{k-\theta+\theta n}{(\theta-\tau)n}\left(\frac{1}{\left(1-\left(\frac{k-\theta+\theta n}{(\theta-\tau)n}\right)^2\right)^2}-\frac{1}{\left(1-\left(\frac{k-\theta}{(\theta-\tau)n}\right)^2\right)^2}\right)$$
$$+\frac{k-n-\theta}{(\theta-\tau)n}\left(\frac{1}{\left(1-\left(\frac{k-n-\theta}{(\theta-\tau)n}\right)^2\right)^2}-\frac{1}{\left(1-\left(\frac{k-\theta}{(\theta-\tau)n}\right)^2\right)^2}\right)k. \tag{7}$$

Since we are supposing that $k<\frac{2}{5}n$ and applying the Lagrange Theorem to the function f, such that $f(x)=\frac{1}{(1-x)^2}$, to the interval $\left[\left(\frac{k-\theta}{(\theta-\tau)n}\right)^2,\left(\frac{k-\theta+\theta n}{(\theta-\tau)n}\right)^2\right]$ and to the interval $\left[\left(\frac{k-\theta}{(\theta-\tau)n}\right)^2,\left(\frac{n+\theta-k}{(\theta-\tau)n}\right)^2\right]$, and noting that the derivative of f is an increasing function on these intervals, we obtain the inequality (8):

$$\frac{k-\theta+\theta n}{(\theta-\tau)n}\left(\left(\frac{k-\theta+\theta n}{(\theta-\tau)n}\right)^2-\left(\frac{k-\theta}{(\theta-\tau)n}\right)^2\right)\frac{2}{\left(1-\left(\frac{k-\theta+\theta n}{(\theta-\tau)n}\right)^2\right)^3}$$
$$+\frac{k-n-\theta}{(\theta-\tau)n}\left(\left(\frac{k-n-\theta}{(\theta-\tau)n}\right)^2-\left(\frac{k-\theta}{(\theta-\tau)n}\right)^2\right)\frac{2}{\left(1-\left(\frac{k-\theta}{(\theta-\tau)n}\right)^2\right)^3}k\geq 0. \tag{8}$$

We can rewrite inequality (8) as inequality (9):

$$\frac{n+\theta-k}{(\theta-\tau)n}\left(\left(\frac{n+\theta-k}{(\theta-\tau)n}\right)^2-\left(\frac{k-\theta}{(\theta-\tau)n}\right)^2\right)\frac{1}{\left(1-\left(\frac{k-\theta}{(\theta-\tau)n}\right)^2\right)^3}k$$
$$<\frac{k-\theta+\theta n}{(\theta-\tau)n}\left(\left(\frac{k-\theta+\theta n}{(\theta-\tau)n}\right)^2-\left(\frac{k-\theta}{(\theta-\tau)n}\right)^2\right)\frac{1}{\left(1-\left(\frac{k-\theta+\theta n}{(\theta-\tau)n}\right)^2\right)^3}. \tag{9}$$

Finally, after some algebraic manipulation, we deduce the inequality (10):

$$\frac{n+\theta-k}{(\theta-\tau)n}\left(\frac{1}{\theta-\tau}\frac{n-2k+2\theta}{(\theta-\tau)n}\right)\frac{1}{\left(1-\left(\frac{k-\theta}{(\theta-\tau)n}\right)^2\right)^3}k$$
$$<\frac{k-\theta+\theta n}{(\theta-\tau)n}\left(\frac{\theta n+2k-2\theta}{(\theta-\tau)n}\frac{\theta}{\theta-\tau}\right)\frac{1}{\left(1-\left(\frac{k-\theta+\theta n}{(\theta-\tau)n}\right)^2\right)^3}. \tag{10}$$

Next, taking into account that we are supposing that $\frac{1}{5}n+1 < k < \frac{2}{5}n$ and also that $\theta = 1$, after some algebraic manipulation of (10) we obtain the inequality (11):

$$k < 21\left(\frac{25|\tau|^2 + 50|\tau| + 25}{25|\tau|^2 + 50|\tau| - 24}\right)^3. \tag{11}$$

Now since $\tau = k - \mu$, we can deduce:

$$k < 21\left(\frac{25|k-\mu|^2 + 50|k-\mu| + 25}{25|k-\mu|^2 + 50|k-\mu| - 24}\right)^3$$

and we can establish the following result.

Theorem 1. *Let G be a $(n,k;\lambda,\mu)$-strongly regular graph such that $0 < \mu < k < n-1$, $\frac{n}{5}+1 < k < \frac{2n}{5}$ and with the distinct eigenvalues $k, \theta = 1$ and τ. Then, we have*

$$k < 21\left(\frac{25|k-\mu|^2 + 50|k-\mu| + 25}{25|k-\mu|^2 + 50|k-\mu| - 24}\right)^3 \tag{12}$$

The result presented in Theorem 1 furnish us a feasibility condition for the existence of a $(n,k;\lambda,\mu)$-strongly regular graph, G, such that $0 < \mu < k < n-1$, $\frac{n}{5}+1 < k < \frac{2}{5}n$ and that 1 is the positive eigenvalue of G. From the inequality (12) we can easily see that, if G is a $(n,k;\lambda,\mu)$-strongly regular graph whose parameters satisfy the conditions required, then if the regularity is sufficiently big then we can conclude immediately that such a strongly regular graph G with $\theta = 1$ cannot exist. In other words, our result shows that the regularity of these strongly regular graphs complying the conditions in consideration can never be too large.

Acknowledgments. In this work Luís Vieira was partially supported by the Center of Research of Mathematics of University of Porto (UID/MAT/00144/2013), which is funded by FCT (Portugal) with national (MEC) and European structural funds through the programs FEDER, under the partnership agreement PT2020.

References

1. Diestel, R.: Graph Theory. Springer, Berlin (2000)
2. Jacobson, N.: Structure and Representation of Jordan Algebras, volume XXXIX of Colloquium Publications American Mathematical Society, Providence, Rhode Island (1968)
3. Koecher, M.: The Minnesota Notes on Jordan Algebras and Their Applications. Springer, Heidelberg (1999). https://doi.org/10.1007/BFb0096285
4. McCrimmon, K.: A Taste of Jordan Algebras. Springer, New York (2003). https://doi.org/10.1007/b97489

5. Faraut, J., Korányi, A.: Analysis on Symmetric Cones. Oxford Mathematical Monographs, Clarendon Press, Oxford (1994)
6. Faybusovich, L.: Euclidean Jordan algebras and interior point algorithms. Positivity **1**(4), 331–357 (1997)
7. Faybusovich, L.: Linear systems in Jordan algebras and primal dual interior point algorithms. J. Comput. Appl. Math. **86**, 149–175 (1997)
8. Faybusovich, L.: A Jordan algebraic approach to potential redution algorithms. Mathematitische Zeitschrift **239**, 117–129 (2002)
9. Gowda, M.S., Sznajder, R.S.: Schur complements, Schur determinenatal and Haynsworth inertia formulas in Euclidean Jordan algebras. Linear Algebra Appl. **432**, 1553–1559 (2010)
10. Gowda, M.S., Sznadjer, R.S., Tao, J.: Some P properties for linear transformations on Euclidean Jordan algebras. Linear Algebra Appl. **393**, 203–232 (2004)
11. Vieira, L.A.: Euclidean Jordan algebras and inequalities over the spectrum of a strongly regular graph. J. Phys.: Conf. Ser. **1564**, 1–10 (2020)
12. Vieira, L.A.: Generalized inequalities associated to the regularity of a strongly regular graph. J. Comput. Methods Sci. Eng. **19**(3), 673–680 (2019)
13. Cardoso, D.M., Vieira, L.A.: On the optimal parameter of a self-concordant barrier over a symmetric cone. Eur. J. Oper. Res. **169**, 1148–1157 (2006)
14. Helmke, U., Ricardo, S., Yoshizawa, S.: Newton's algorithm in Euclidean Jordan algebras with applications to robotics. Commun. Inf. Syst. **2**(3), 283–298 (2002)
15. Baes, M.: Spectral functions and smoothing techniques on Jordan algebras. Ph.D. Thesis, University Catholique of Louvain (2006)
16. Schmieta, S.H.: Application of Jordan algebras to the design and analysis of interior point algorithms for linear, quadratically constrained quadratic, and semi definite programming. Ph.D. thesis, Rutgers University (1999)
17. Vieira, M.V.C.: Jordan algebraic approach to symmetric optimization. Ph.D. thesis, Technische Universiteit Delft (2007)

Check for updates

New Trends in the Management of Optometry Offices

Ioana Udrea[1](✉), Cristian Gabriel Alionte[1], Viorel Ionut Gheorghe[1], Tudor Cătălin Apostolescu[2], and Corina-Ioana Cobzac[1]

[1] Faculty of Mechanical Engineering and Mechatronics, University POLITEHNICA of Bucharest, Bucharest, Romania
ioana.udrea@upb.ro

[2] Faculty of Informatics, "Titu Maiorescu" University, Bucharest, Romania

Abstract. Lately, there has been a very rapid development of IoT systems and cloud platforms that host the data sent by them. However, storing and manipulating data in the cloud involves some limitations, in order to have full control over them and a very large number of available records, it is preferable to store in a local database. Also very useful is the integration of data received from IoT systems in existing databases, used for management or accounting operations.

In this context the purpose of the paper is to design an MS Access database specific to an optometry office and to record the indoor climate parameters and calculated data of these in it. An existing sensor system will be used for this purpose. A Python application to retrieve data from the cloud on the ThingSpeak platform, where the sensor system sends them, will be developed. Moreover, the Python application will also make a chart in which the real-time value of the Predicted Mean Vote (PMV) thermal comfort index will be displayed, the employees of the optometry cabinets, being able to see this index whenever they want. Comparisons with applications from other fields of activity, which display thermal comfort indices, will be made. The novelty of the paper is that it is the first time that thermal comfort is analyzed in optometry offices, the measured values being recorded in a database specific to their activity.

Keywords: Optometry office · IoT · MS Access · ThingSpeak · PMV · Human thermal comfort

1 Introduction

Smart connected things, seen as a constantly evolving field, is known as the Internet of Things (IoT) [1]. Significant advances in wireless communication and technology have led to an increase in both the number of IoT devices and a decrease in the power required for them, they are getting smaller and smaller, and the costs are declining. The development of IoT systems is in a continuous expansion, the increase of their use being exponential [2]. There are many IoT systems that can even be integrated into BMS building management systems. Most of the data taken by sensors in BMS systems are used in applications designed by the BMS system provider or at most related to the

J. Machado et al. (Eds.): icieng 2022, LNME, pp. 100–113, 2023.
https://doi.org/10.1007/978-3-031-09360-9_9

building information model (BIM) [3]. The field where wireless sensor networks have been first used is environment monitoring [4].

The study of human thermal comfort is a concern today. The satisfaction of employees at work but also of customers can be measured by thermal comfort indices, they are also related to employee productivity. Work environments have various peculiarities, there may be certain temperature and humidity conditions imposed by professional standards. Thus, the study of human thermal comfort is well done individually, in various work environments, such as schools, office buildings, residential buildings, hospitals or clinics [5].

The study of human thermal comfort was also performed using IoT systems integrated in BIM and even the correlation of data with the CAD plans of the building [6, 7]. The quality of the indoor work environment is an increasingly important aspect in many office buildings and medical centers [8], so we intend to study thermal comfort in an optometry office.

We want direct access to the data read by the sensors and their integration in a database specific to an optometry office. From what we have seen in the literature, this has not been done so far.

This paper aims to study human thermal comfort in optometry offices, by measuring certain parameters that enter the PMV equation and registering them in a management database of optometry offices, owned by a certain company, for example.

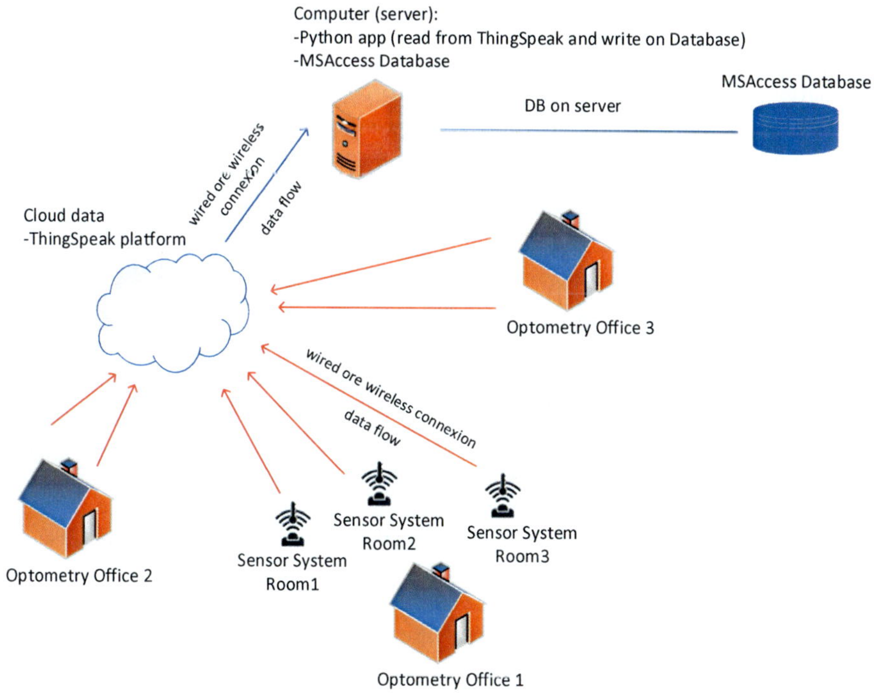

Fig. 1. The architecture of the proposed solution

As seen in Fig. 1, suppose a certain owner has three optometry offices, Optometry Office 1, 2 and 3. Each office can have one or more rooms. The management of these offices is performed in an MS Access database, see Fig. 1. This paper aims to design a simplified database for the management of any optometry offices and in which data can be integrated read by the sensors that can be mounted in all the rooms of optometry offices. Figure 1 shows three such offices that symbolically send data to "a cloud" (the cloud).

From the open platform for cloud computing such as ThingSpeak, ThingsBoard, Kaa, SiteWhere [9] was chosen ThingSpeak, an open IoT platform with MATLAB analytics. In this paper, the data in the cloud, located on the ThingSpeak platform, is read, and then sent to the database using a code written in Python. The system that sends data to the cloud must read the parameters with the help of which it can be determined in the thermal comfort index. Such a system was developed in [10] and will be used in this paper.

The study of thermal comfort in an optometry office does not imply anything special compared to the study of thermal comfort in a medical office in general. But this paper integrates the parameters read by the sensors and the PMV thermal comfort index in a database specific to optometry cabinets. Thermal comfort assessment is useful in general [11, 12] but especially in spaces where people live or work [13, 14]. Thermal comfort in buildings is important from the point of view of increasing the energy efficiency of the building as well as from the point of view of increasing the satisfaction of employees and implicitly their productivity. These are the main aspects of optimizing human thermal comfort in building.

In hospitals, the parameters of indoor thermal climate are specific to the various activities that take place here. Thus, for operating rooms, certain values must be met, according to the standards, for dry bulb temperature, relative humidity, room air change, outdoor air change, outdoor air intake [15]. Optometry cabinet does not have special requirements in terms of indoor thermal parameters, it can be assimilated with an office, category "Offices and spaces with similar activity (single offices, open plan offices, conference rooms, auditorium, cafeteria, restaurants, classrooms)", in terms of recommended indoor temperatures [16].

Digitization is growing in all fields of activity. Thus, web-type applications have been created that usually have in their structure specific databases for various fields of activity or that cumulate several domains, as is the case of enterprise-type applications. All of these have a database, called the database layer [17, 18]. This paper does not aim to create a complex application, with more users, which also integrates the database, this would be too complex task but aims to design a relational database, specific to the activities carried out in an optometry office. It can be part of the layers of a future web application. It is designed to be scalable so that several optometry cabinets can be connected to it.

Considering the presented ones, it is considered useful, being a novelty, the integration of the parameters related to thermal comfort with a database designed for keeping records of the optometry cabinets. This paper aims, in addition to the integration of sensors, even the design of the database.

When the solution proposed by this paper will be integrated in a complex application, alerts can be sent if the PMV thermal comfort index exceeds certain values. The thermal comfort index read in several in-rooms from various optometry practices can be compared with the values declared by the subjects through questionnaires, in order to be validated. In this situation it is possible to see in what intervals the thermal sensation of the employees and clients of an optometry office falls and if certain specific needs appear.

In the following, the paper will present the sensor system used for measurements, the designed database and the Python application which records the parameters read by the sensors in the database. After that, the results of the application are related to standards and compared to similar applications. Finally, based on the results obtained, conclusions are drawn and future directions for study and development are proposed.

2 Method and Procedure

The sensor system used is an existing one, it measures the internal thermal comfort parameters, it will be presented in the following. The database specific to the optometry cabinet is then designed, the structure of the tables and the relationships between them is described. Then follows the description of the method by which the data read by the sensors are taken in real time and recorded in the database.

2.1 Sensor System

The sensor system used in this study is composed of Raspberry Pi board, a breadboard with two sensors module, BMP180 [19] (compose of a temperature and a barometric pressure sensor) and DHT22 (composed of a temperature and a humidity sensor) [20] and a multiplexing connection board [10]. In Fig. 2a) a detailed image can be seen, with a box in which the two sensors are mounted on a circuit board, the design of the box is described in [21]. In Fig. 2b) is an overview of the entire system, on the left is Raspberry Pi and multiplexing board and on the right sensor box. The connection between them is made with the help of a GPIO cable.

It is a system that transmits data in the cloud, on the ThingSpeak platform [22]. ThingSpeak is an IoT platform that uses open-source software and allows live data streaming in the cloud. The solution proposed in [10] calculates, with certain approximations, which are presented in the respective paper, the PMV thermal comfort index. The ThingSpeak. platform offers the possibility to display the received data in real time, by configuring a "Channel", for each type of received data, temperature, humidity, air pressure and PMV. On the IoT system (more specific on the Raspberry Pi board), which can be seen in Fig. 2, runs both the software that sends the data to the cloud and the PMV index calculator.

The PMV index calculation formula, see Eq. 1, was determined by Fanger [23] and is implemented in software running on the Raspberry Pi board in [10]. The terms that go into this equation take into account thermal sensation, internal heat production of the body, heat loss through skin, heat loss by sweating, latent respiration heat loss, dry

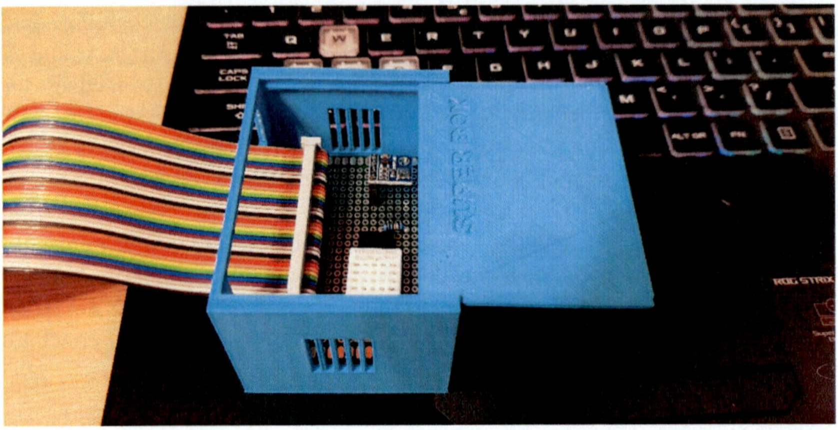

a)

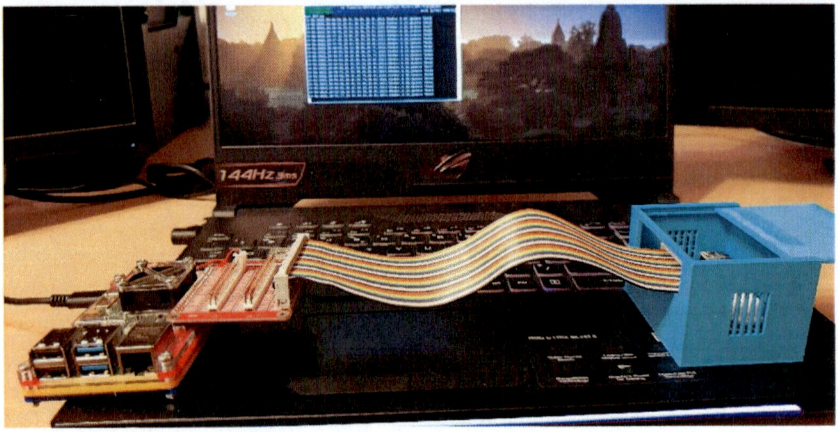

b)

Fig. 2. Sensor system used in this study: a) detail with sensors; b) sensor box connected to Raspberry Pi board

respiration heat loss, heat loss by radiation. In addition, there are the terms t_{cl}, h_c and f_{cl} which are described in Eqs. (2–4).

$$PMV = \left(0.352e^{-0.042\left(\frac{M}{A_{Du}}\right)} + 0.032\right) \cdot \Big[\frac{M}{A_{Du}}(1-\eta) - 0.35[43 - 0.061\frac{M}{A_{Du}}(1-\eta) - p_a\Big] - 0.42\Big[\frac{M}{A_{Du}}(1-\eta) - 50\Big] - 0.023\frac{M}{A_{Du}}(44 - p_a) - 0.0014\frac{M}{A_{Du}}(34 - t_a) - 3.4 \cdot 10^{-8} f_{cl}\left[(t_{cl}+273)^4 - (t_{mrt}+273)^4\right] + f_{cl}h_c(t_{cl} - t_a)\Big] \tag{1}$$

$$t_{cl} = 35.7 - 0.028 \cdot (M - W) - I_{cl} \cdot \{3.96 \cdot 10^{-8} \cdot f_{cl} \cdot [(t_{cl}+273)^4 - (\overline{t_r} + 273)^4] + f_{cl} \cdot h_c \cdot (t_{cl} - t_a)\} \tag{2}$$

$$h_c = \begin{cases} 2.38 \cdot |t_{cl} - t_a|^{0.25} \, for \, 2.38 \cdot |t_{cl} - t_a|^{0.25} > 12.1 \cdot \sqrt{v_{ar}} \\ 12.1 \cdot \sqrt{v_{ar}} \quad for \, 2.38 \cdot |t_{cl} - t_a|^{0.25} < 12.1 \cdot \sqrt{v_{ar}} \end{cases} \tag{3}$$

$$f_{cl} = \begin{cases} 1.00 + 1.290 \cdot I_{cl} \; for \; I_{cl} \leq 0.078 \text{ m}^2\text{K/W} \\ 1.05 + 0.645 \cdot I_{cl} \; for \; I_{cl} > 0.078 \text{ m}^2\text{K/W} \end{cases} \tag{4}$$

where,

W – effective mechanical power [W/m^2]
M – metabolic rate [W/m^2]
f_{cl} – clothing surface area factor
I_{cl} – clothing insulation [m^{2}K/W]
$\bar{t}_r$ – mean radiant temperature [°C]
t_a – air temperature [°C]
p_a – water vapour partial pressure [Pa]
v_{ar} – relative air velocity [m/s]
t_{cl} – clothing surface temperature [°C]
h_c – convective heat transfer coefficient [W/(m^{2}K].

2.2 MS Access Database

Below is a data base design, specific to an optometry office. In addition to the tables that describe the usual economic operation of an optometry cabinet, the database must be designed so that it can integrate the data received by sensors that measure thermal comfort parameters in various rooms of optometry cabinets.

With the help of MS Access, a relational database was designed, the most widely used database model today [24]. In general, in the design of databases we start from simpler models "Conceptual Data Model", after which the model is developed by adding fields and data types, reaching the "Logical Model" and finally we have described in- the whole structure of the tables with the data types specific to the respective database, with the relationships between the tables, with the specification of the primary and foreign keys and other details, which even represents the "Database Model" [25]. In our case, the database being relatively simple, we made only "Conceptual Data Model" and "Data-Model Database", no need for the intermediate help of "Logical Model" (Fig. 3).

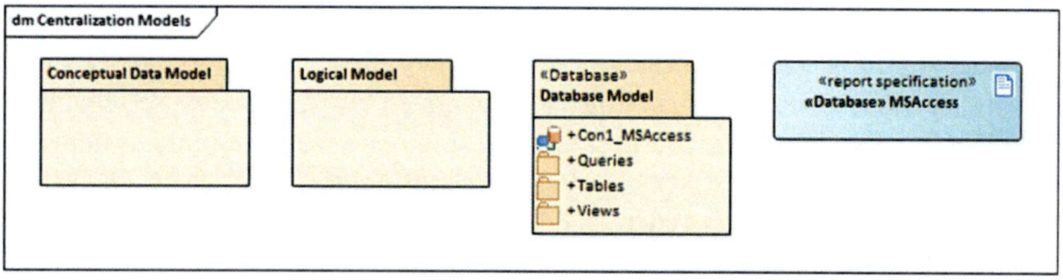

Fig. 3. Centralization of data models; diagram made with enterprise architect software [26]

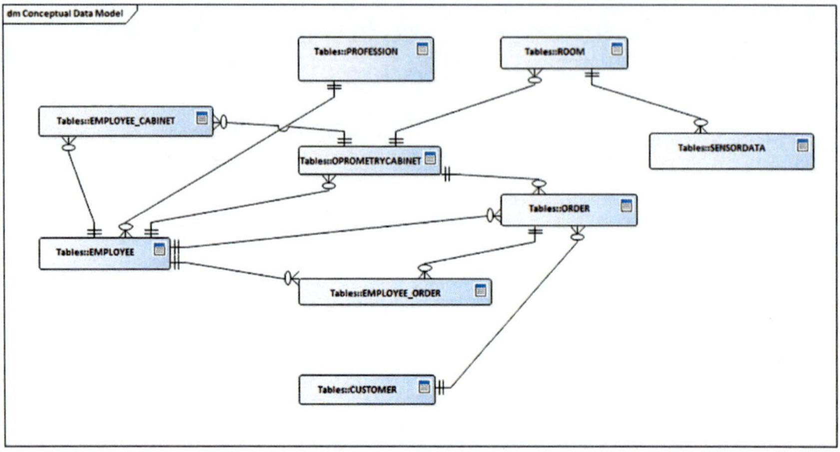

Fig. 4. Conceptual data model proposed for an optometry office; diagram made with enterprise architect software [26]

Conceptual data model shows how the business see its information and is independent of any technology or solution. In Fig. 4, Conceptual Data Model, for the proposed database, only the tables and the relationships between them are represented, using the ER (Entity-Relationship) notation. The actors involved are employees and customers. From the point of view of the optometry office, each employee has a profession, can work in an optometry office and can be the head of such an office and is responsible for a certain order. The many to many relationships that would have appeared between EMPLOYEE and ORDER and between EMPLOYEE and CABINET are solved by entering the intermediate tables EMPLOYE_ORDER, respectively EMPLOYEE_CABINET.

In Fig. 5, the final, complete model of the database is presented. All fields are specified here, the data types are MS Access specific. The space occupied by an optometry office in a building is quite small, it usually consists of several rooms. For this reason, a hierarchical division of space into offices and rooms was considered sufficient. In this sense there are the OPTOMETRYCABINET, parent table, and the ROOM, child table. Sensor systems can be placed in each room in each optometry office. In any of these rooms we can make measurements. The ROOM table has been introduced, which has the primary key composed of the office code (CodCO code field) and the room number (Room field).

The SENSORDED table contains the sensor recording data. The primary key in this table is Id_record (which is the ID of the json generated by ThingSpeak, see chapter 2.3 and Appendix A2). Since we can perform measurements with multiple systems, this field when it reaches the MS Access database is not unique. As a result, the primary key consisting of Id_record, CodCO and Room was made. At the database level, an SQL script was made to delete all the entries in the SENSORDATA table. Now it is run manually, the system is being tested but when it will be needed it will be accessed from the application code written in Pyton and will be run either at a certain time interval or when the number of records in the SENSORDATA table reaches a certain value.

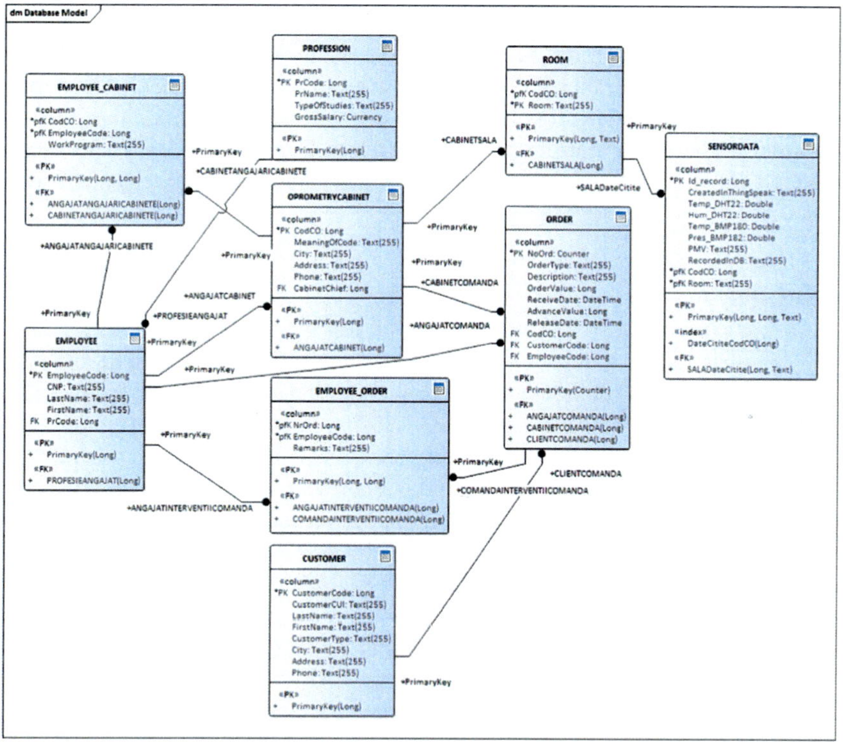

Fig. 5. Database model implemented in MS Access; diagram made with enterprise architect software [26]

2.3 Python App

A Python app has been written to take data from the ThingSpeak platform (from the JSON file provided by it). After the data has been read, it is written to an MS Access database. The Python code is on the same machine as the database (see Fig. 1).

The application checks if the record with the same ID (the one obtained from JSON file), the same office and the same room still exist in the database. If it still exists, the cycle repeats itself until it finds a new value that appeared in JSON, which no longer exists in the database. The time of about 5 s between recordings is quite short (in ThingSpeak the sensors send to 15 s), but it was chosen so for the testing stage, it can be easily changed.

In Appendix A2 it is just an interpreted and arranged image suggestively displayed hierarchically of attribute-value, easy to read, of a part of JSON file generated by ThingSpeak, the actual file, as it is known, is a raw data and it is not relevant to be put to work. The application is scalable, allows one to add as many offices and rooms as he wants. The application takes into account, of course, the structure of the database in which the records are inserted.

The pyplot module from the Matplotlib library of Python was used to create the graph in Fig. 6. The graph has been designed to be updated with each new record in the MS Access database, thus allowing the display of real-time data. On the horizontal axis is the time formatted as HH:MM:SS and on the vertical axis the calculated PMV values. The limit values between the different Categories of PMV, presented in Table 1, are highlighted with dashed lines of different colors, see the legend of the graphs in Fig. 6.

3 Results

For an optometry office, indoor climate parameters are read and with their help the PMV thermal comfort index is determined. The Fanger model [23] depends on six parameters (water vapor pressure, air temperature, air velocity, mean radiant temperature, water vapor pressure, metabolic rate and clothing level). The PMV calculation is presented in the paper [10]. The validation of the proper software that calculates PMV is done with the ISO7730 [27] standard. Regarding the data entered, certain parameters could not be measured and for this reason they were approximated. For the most accurate measurements, the system must be positioned in the middle of the room, where the mean radiant temperature is as close as possible to the air temperature. The sensor system does not accurately calculate PMV, nor is it wanted, it wants to be an indicator of places where problems can occur.

The standard EN 15251 [16] proposes and explains four categories of human thermal comfort, both for the Fanger model [23] and for the adaptive model [28]. In this paper we focus on classical comfort, which follows Fanger's theory. For this purpose, Table 1 is presented. In addition to the PMV values from the EN 15251 standard, we can also remember the recommended design values of the indoor temperature for small offices. Thus, for Category A it is the range 21–25.5C, for Category B it is the range 20–26 and for Category C it is the range 19–27. In the MS Access database, in addition to the PMV index, the indoor temperature measured by the sensors is also transmitted. Although this is only a parameter that influences the thermal comfort, the internal temperature values in various rooms of the optometry offices can also be increased.

Python graphs give us real-time data for PMV values recorded in different rooms of optometry offices. In Fig. 6a) the values have a slight ascending tendency, being all in Category I of comfort. In Fig. 6b) the values have a strong decreasing tendency, passing in an interval of a few minutes through all the comfort categories, from Category I to Category III and even exceeding it a little. This could happen, for example, during the winter, when the outside temperature is very low, and a window would be open in a room. In such situations, if the building is equipped with a HVAC system, with fresh air inlet, it could be recommended to the employees of the optometry office not to open the windows when the outside temperatures are low. Another utility of the solution realized in this paper is the use of data recorded in the MS Access database for statistical calculations. Graphs could be made at room or office level, over longer periods of time, to show the thermal comfort of both employees and customers. Where problems arise, field measurements could be made with more powerful and expensive systems, such as ConfortSense [29].

Table 1. Description of applicability for the categories proposed by EN 15215 and the corresponding comfort intervals

Category	Example	PMV values
I	High level of expectation and is recommended for spaces occupied by very sensitive and fragile people with special needs such as: disabled people, the sick, very young children and the elderly	$-0{,}2 < PMV < 0{,}2$
II	Normal waiting level and can be used for new buildings and renovations	$-0{,}5 < PMV < 0{,}5$
III	Acceptable, moderate standby level that can be used for existing buildings	$-0{,}7 < PMV < 0{,}7$
IV	This category can be accepted for a limited period of the year	$PMV < -0{,}7$ and $PMV > 0{,}7$

In [7], the authors present the integration of thermal comfort parameters in a facility management application called ARCHIBUS [30]. For this, in the existing application a screen is developed in which the current value is displayed, as well as graphs of variation in time for the parameters: temperature, humidity and air quality. Also the values calculated based on the temperature and humidity for a heat index are displayed in the respective screen. The calculation formula for the heat index can easily be changed, thus choosing from the equations developed by the National Weather Service (NWS) [31], Rothfusz model [32] and Temperature-Humidity Index (THI), developed by Kibler [33]. Because the existing application, ARCHIBUS allows this, the values of the indoor climate parameters were also associated with the CAD plans of the building, being able to even intervene on them in the sense of coloring a certain room when the temperature exceeds a certain threshold. In this paper, PMV was chosen as the index of thermal comfort and not a "simpler" index that takes into account only temperature and humidity because PMV is a more accurate index, although in the initial paper in which it was developed certain approximations have been made. In future studies, a sensor could be used to measure indoor air velocity, on which approximations were made, the PMV calculation module being already designed.

In terms of the data recorded in the MS Access database, if it existed sensor systems that record in multiple rooms, we can analyze which rooms have, for example, PMV or temperature issues, which time intervals the problems occur. For this, the data can be taken from the database and processed in other software or can be directly queried on the database. Not only the data recorded in the data base, but real time data is useful, employees can see at any time the current thermal comfort parameters and indices.

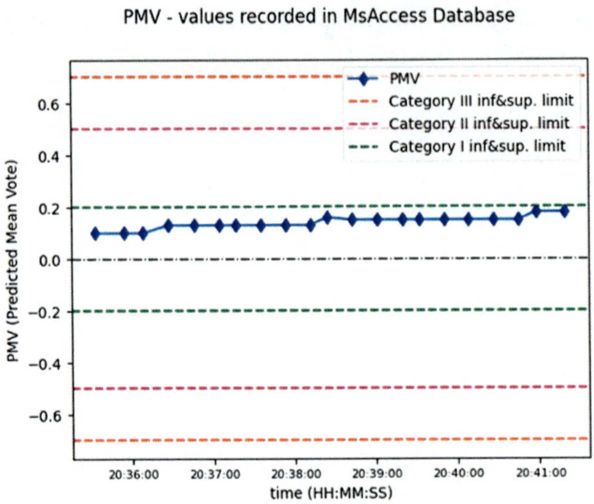

a)

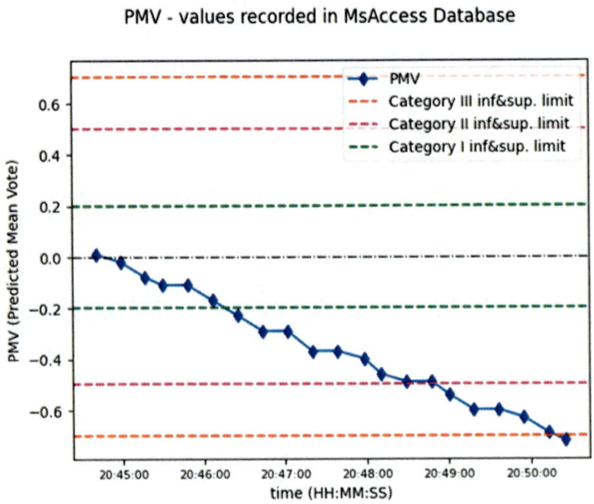

b)

Fig. 6. Real-time data for PMV values recorded in MS Access database: a) upward trend; b) strongly decreasing trend

4 Conclusion

Multilayer applications, which also have a database layer, are increasingly used today. IoT is also on the rise, more and more data sent by sensors being sent to the cloud from where it can be accessed at any time. Studying the general activities of an optometry practice, this paper aims to design a specific database for it and integrate in this database the indoor climate data read by sensors and transmitted in the cloud. In this regard, a database, not very complex but very useful for managing the activities of an optometry practice, has been developed. In addition to the classic tables related to activities related to employees, customers, orders, and others, it also includes tables of the hierarchical structure of the space, office-room, in which data read by real-time sensors are recorded, as well as a thermal comfort index, PMV.

At this moment, the system presented as a whole in this paper works for the transmission in the database of the information taken from any number of optometry offices. But the registration in the database of the other information referring to the optometry offices is done locally. MS Access can be accessed in the LAN by several users, but in order to be able to complete the data from the optometry offices located at different locations, a Web Interface for the database must be developed. In this situation it would be quite simple to display or send alerts if the indoor temperature or PMV index exceeds certain values.

From the point of view of thermal comfort, the optometry office is similar to any medical office, it can be said with any office in general, but in this paper are integrate thermal comfort data into a database with the specificity of an optometry office. If desired, other types of sensors can be easily integrated into the system.

For several rooms and in several optometry offices, complex dashboards can be made that would allow the simultaneous visualization of the thermal comfort parameters in several cabinets. It can be concluded that in this paper it is for the first time when in a database specific to the activities of an optometry office are integrated data read by sensors related to indoor human thermal comfort.

References

1. Saha, H.N., Saha, N., Ghosh, R., Roychoudhury, S.: Recent trends in implementation of Internet of Things—a review. In: 2016 IEEE 7th Annual Information Technology, Electronics and Mobile Communication Conference (IEMCON) (2016). https://doi.org/10.1109/IEMCON.2016.7746345
2. Nord, J.H., Koohang, K., Paliszkiewicz, J.: The Internet of Things: review and theoretical framework. Expert Syst. Appl. **133**, 97–108 (2019). https://doi.org/10.1016/j.eswa.2019.05.014
3. Tang, S., Shelden, D.R., Eastman, C.M., Pishdad-Bozorgi, P., Gao, X.: A review of building information modeling (BIM) and the Internet of Things (IoT) devices integration: present status and future trends. Autom. Constr. **101**, 127–139 (2019). https://doi.org/10.1016/j.autcon.2019.01.020
4. Shinde, S.R., Karode, A.H., Suralkar, S.R.: Review on-IOT based environment monitoring system. Int. J. Electron. Commun. Eng. Technol. **8**(2), 103–108 (2017). http://www.iaeme.com/IJECET/issues.asp?JType=IJECET&VType=8&IType=2
5. Yuan, F., et al.: Thermal comfort in hospital buildings – a literature review. J. Build. Eng. **45**, 103463 (2022)
6. Valinejadshoubi, M., Moselhi, O., Bagchi, A., Salem, A.: Development of an IoT and BIM-based automated alert system for thermal comfort monitoring in buildings. Sustain. Cities Soc. **66**, 102602 (2021). https://doi.org/10.1016/j.scs.2020.102602
7. Kraus, V.F., Udrea, I.: Advanced workplace management platform for monitoring and management of indoor climate parameters. In: 2021 IOP Conference Series: Earth and Environmental Science, vol. 664, p. 012047 (2021). https://doi.org/10.1088/1755-1315/664/1/012047

8. Leung, M., Chan, A.H.S.: Control and management of hospital indoor air quality. Med. Sci. Monet. **12**(3), SR17-23 (2006)
9. Best & Top Open Source IoT Platforms to Develop the IOT Projects, H2S Media Team. https://www.how2shout.com/tools/best-opensource-iot-platforms-develop-iot-projects.html. Accessed 30 Jan 2022
10. Udrea, I., et al.: IoT solution for monitoring indoor climate parameters in open space offices. In: E3S Web of Conferences, vol. 180, p. 02012 (2020). TE-RE-RD 2020. https://doi.org/10.1051/e3sconf/202018002012
11. Dogeanu, A., Tacutu, L., Iatan, E., Nicolae, A.M., Lungu, C.I.: Asphalt heat recovery application for sustainable green energy. Appl. Sci. **12**(3), 1196 (2022)
12. Elnaklah, R., Walker, I., Natarajan, S.: Moving to a green building: Indoor environment quality, thermal comfort and health. Build. Environ. **191** (2021). https://doi.org/10.1016/j.buildenv.2021.107592
13. Bode, F., Nastase, I., Croitoru, C.V., Sandu, M., Dogeanu, A.: A numerical analysis of the air distribution system for the ventilation of the crew quarters on board of the international space station. In: E3S Web of Conferences, vol. 32, p. 01006 (2018)
14. Alam, M.S., Salve, U.R.: Factors affecting on human thermal comfort inside the kitchen area of railway pantry car - a review. J. Therm. Eng. **7**(8), 2093 (2106)
15. Khodakarami, J., Nasrollahi, N.: Thermal comfort in hospitals – a literature review. Renew. Sustain. Energy Rev. **16**, 4071–4077 (2012)
16. EN 15251: Indoor environmental input parameters for design and assessment of energy performance of buildings addressing indoor air quality, thermal environment, lighting and acoustics, EN, Automation in Construction (2007)
17. Udrea, I., Kraus, V.F., Popescu-Cuta, A.: Facility management platform sensors monitoring for improving workplace services. In: IBIMA 2020, 36th IBIMA Conference Proceedings (2020)
18. Udrea, I., Kraus, V.: Construction log book – a web-based application that supports a building on its entire life-cycle. In: TE-RE-RD 2018, 7th International Conference of Thermal Equipment, Renewable Energy and Rural Development, Drobeta Turnu Severin, Romania, 31 May–2 June 2018, pp. 505–510. Ed. Politehnica Press (2018). ISSN: 2359-7941, ISSN-L: 2359-7941
19. BMP180: Barometric Pressure/Temperature/Altitude Sensor. https://www.adafruit.com/product/1603. Accessed 26 Feb 2022
20. DHT 22, temperature-humidity sensor. https://www.adafruit.com/product/385. Accessed 26 Feb 2022
21. Udrea, I., et al.: Optimizing the placement of indoor thermal climate sensors of an IoT solution. In: Proceedings of the 2020 12th International Conference on Electronics, Computers and Artificial Intelligence, ECAI (2020). https://doi.org/10.1109/ECAI50035.2020.9223120
22. ThingSpeak: IoT Analytics Platform. https://thingspeak.com/. Accessed 21 Jan 2022
23. Fanger, P.O.: Thermal comfort. Analysis and applications in environmental engineering. Published in the United States by McGraw-Hill Book Company 1972. Originally presented as the author's thesis, Danmarks Teknishe Højskole (1970)
24. Duminica, D., Apostolescu, T.C.: Sisteme de gestiune a bazelor de date in optometrie. Editura Universitara, Bucuresti (2016). ISBN: 978-606-28-0474-9
25. BABOK: A Guide to the Business Analysis Body of Knowledge. IIBA (International Institute of Business Analysis, Toronto (2015). Version 3.0, ISBN-13: 97978-1-927584-03-3
26. Enterprise Architect Version 13 - visual modelling and design tool, SPARK Systems. https://community.sparxsystems.com/news/1002-sparx-systems-releases-enterprise-architect-13. Accessed 21 Jan 2022

27. ISO-7330: International Standard ISO 7330: Ergonomic of the thermal environment - analytical determination and interpretation of thermal comfort using calculation of the PMV and PPD indices and local thermal comfort criteria (2005)
28. de Dear, R., Brager, G., Cooper, D.: ASHRAE RP-884. Developing and adaptive model of thermal comfort and preference. Technical report, The American Society of Heating, Refrigeration and Air-Conditioning Engineers, Inc., and Environmental Analytics, Atlanta (1997)
29. ComfortSense, measurement system, developed by Dantec Dynamics. https://www.dantecdynamics.com/solutions-applications/solutions/thermal-comfort/comfortsense/. Accessed 29 Jan 2022
30. ARCHIBUS Inc. https://archibus.com/products/. Accessed 26 Feb 2022
31. National Weather Service, Hagerstown Maryland - Weather Station. https://www.i4weather.net/hiwc.html. Accessed 26 Feb 2022
32. Rothfusz, L.P.: The heat index equation. National Weather Service Technical Attachment (SR 90–23) (1990)
33. Kibler, H.H.: Thermal effects of various temperature-humidity combinations on holstein cattle as measured by eight physiological responses. Missouri Agricultural Experiment, Exp. Stn Res. Bull. 862, Mt Vernon, ed: Environmental Physiology and Shelter Engineering, LXVII (1964)

The Role of Digitalization, Servitization and Innovation Ecosystem Actors in Boosting Business Model Innovation – A Literature Review

Mariza Mesquita[1](✉), Ana Correia Simões[2], and Vasco Teles[2]

[1] School of Economics and Management, University of Porto, Porto, Portugal
marizagabri@gmail.com
[2] INESC TEC, Porto, Portugal

Abstract. The Industry 4.0 technologies and servitization are requiring manufacturers to reinvent themselves to remain competitive. In this sense, companies are putting more emphasis on the customer experience, while associating services to their products with the support of emerging technologies. At the same time, actors in the innovation ecosystem such as universities, research institutes, and service providers are involved in the value co-creation process. Thus, this study aims to systemize and present the main findings of a literature review regarding the role of digitalization, servitization, and innovation ecosystem actors in boosting innovation in business models. The study involved 585 articles in international journals and conference proceedings published till 2021. A detailed selection process led 10 articles for further analysis. As a consequence of digitalization and servitization, there has to be an alignment among ecosystem actors to capture the co-created value generated by BMI. Moreover, the combination of BMI, servitization, and digitalization may also be structured into frameworks according to the different levels of each of those factors, thus allowing for companies to assess and position themselves in such frameworks and identify the path to follow.

However, any of these articles addressed the combination of the three topics proposed: digitalization, servitization, and the contribution of the innovation ecosystem actors, even if they showed a clear interdependence between these areas, leading to common findings impacting BMI.

Keywords: Digitalization · Industry 4.0 · Innovation ecosystem · Servitization · Business model innovation · Business model

1 Introduction

Servitization, the provision of services by manufacturers, associated with physical products [1] and the Industry 4.0 (I4.0), are two phenomena that are changing the current industry scenario. While the former creates value for the customer, the latter focuses on technologies and production methods to deliver the product/output to the customer. In

J. Machado et al. (Eds.): icieng 2022, LNME, pp. 114–127, 2023.
https://doi.org/10.1007/978-3-031-09360-9_10

this way, both phenomena are creating new challenges to the current Business Models (BMs) leading, in many cases, to their innovation [2].

It is in this context, where customers want to receive improved services, with higher quality and speed [3], along with new experiences, that servitization has gained increasing relevance in this fourth industrial revolution [4]. Industry starts to find in servitization a way to modify or even create new BMs [5]. Thus, there arises a concern to create value for the customer while creating an opportunity for the manufacturer [6].

Consequently, it is in the combination between servitization, and digital technologies used in the industry, that the theme digital servitization appears, responsible for the emergence of several BMs based on platforms and boosted by technology [6, 7]. Due to the high dynamics of the market, for companies to survive they need to experiment and explore several innovative BMs, otherwise focusing on only one BM may generate inertia [7]. Additionally, Sjödin et al. [8] highlighted the need to create new ways of working to experiment and explore new BMs and, consequently, co-create value based on digital servitization.

In this sense, several studies have been dedicated to investigating servitization in industry as mentioned by [8] and [9]. However, it is known that servitization and I4.0 have generated not only change in manufacturer's products and processes, but mainly a restructuring of BMs across the value chain, even generating impact on raw material suppliers [4]. This is because the process of technology adoption and servitization extends beyond the company's boundaries, going beyond the simple relationship between the manufacturer and the customer [7]. Value creation, as evidenced by [8], occurs through interdisciplinary teams, from different areas and companies, whose knowledge extends from suppliers to customers. Therefore, it is necessary to understand this dynamic from the point of view of the innovation ecosystem, where several actors are involved in the value co-creation process [7, 8, 10].

Although the literature seems to be unanimous that there is a need to understand the dynamics [4] and relationships [8] between the various actors of the innovation ecosystem, such as partners or even competitors, very little has been researched on these dynamics concerning servitization [7, 10]. Additionally, there is limited research relating business model innovation (BMI) and I4.0 solution providers [5].

In summary, there is still a gap in the literature that addresses, in a systematic way, the effects of digitalization and servitization [6] associated with the role of different innovation ecosystem actors, such as research institutes or solution providers, in business model innovation of industrial companies. Thus, this study aims to systemize and present the main findings of a literature review regarding the combined role of digitalization, servitization and innovation ecosystem actors in boosting innovation in business models.

After this introduction, a second section is presented with a description of the fundamental concepts used in this research, namely, business model, innovation in BM, servitization, and digitalization. Afterwards, the method used in this research is explained. The main results of the study are then presented in section four, summarizing the main findings collected and analyzed from the literature. Finally, the main conclusions of this study are presented.

2 Main Concepts

This section intends to introduce and define the basic concepts that are directly related to the topic of this research. In this section, the concepts are defined in the context and scope they are used in this research. The following topics are approached in this section: (i) Business model definition, (ii) Innovation in Business Model, (iii) Servitization, and (iv) digitalization.

2.1 Business Model

The BM determines the company's success [11]. As a result, studies related to this topic have helped researchers and managers in several areas [12]. However, this concept still generates discussions and divergences in the literature [12, 13], since many scholars have defined it subjectively [14].

On the one hand, certain researchers identify BM as a simpler concept, more specifically as a way of operating and profiting in a business [15–17], highlighting the value delivered as a determining factor for monetization [18].

On the other hand, researchers who define it as a more structured concept have increased, such as [19–21, 23], mainly influenced by companies with extraordinary profits that emerged with the internet boom, such as Google [12]. This has contributed to increasing its importance in the literature [12].

For this research, the widely used and accepted [25] concept of [20] was used through the tool called Business Model Canvas (BMC). Its modular perspective shows the operation of the business based on the creation, delivery, and capture of value. According to these authors, nine components are essential for value creation, delivery, and capture: value propositions, customer segments, customer relationships, channels, key partners, key activities, key resources, revenue streams and cost structure.

2.2 Innovation in Business Model

Authors as [26] understand that the terms BM and BMI are related, with the latter being more recent in the literature and positioning itself as an expansion of the former. Initially, the term BMI emerged with companies that did not produce products but created and delivered value through services [12]. This trend has gained more traction in digital businesses, driven by the internet and innovation [12, 14, 15, 22, 24].

Researchers [26] argue that BMI can be assumed as *'designed, new and non-trivial'* changes to its elements or the architecture that connects them. Beyond this, it is a strategy to position the company among other actors, such as customers, suppliers, competitors, partners, and institutions. These changes can be the connection with different actors, the alteration or addition of activities that increase value for the company and the network they belong [27].

In turn, the term can also be defined as the construction and implementation of a new BM, which can be the transformation from one model to another, the diversification of an existing model, the acquisition of new models, or even the construction of an entirely new model [28]. These options are similar to the changes in BMs proposed by [29]: creation, extension, revision, and discontinuation.

In a nutshell, companies must go beyond process and product changes and recognize new approaches to value creation to generate innovation in their BMs [30]. That is why nowadays we are witnessing a change in companies that move from physical reality, with offers of tangible products, to a digital reality-oriented towards services and customer experience [3], a strategy also known as servitization.

2.3 Servitization

Initially described as a process of change in BMs, servitization consists of offering services added to physical products [31]. Other authors [1, 32] present a more specific view by treating servitization as the union between the offer of products and services but focused on delivering value and user experience.

With the advance of technologies, companies have increasingly opted to use them to drive servitization [33]. Therefore, the servitization and digital technologies combination generate digital servitization (DS) [34]. DS is the creation of new services, or the implementation of improvements in existing services, through digital technologies [6].

It is responsible for offering new services [6], creating new service-oriented BMs [19, 35], and increasing companies' competitive advantage through value co-creation [1, 36]. It is also a process of change to integrate systems formed by product, service, and software, responsible for the creation and capture of value [7], whose definition is commonly found in the literature by the acronym PSS (Product Service-System) [1, 37, 40]. However, there are studies that [8] point out that this transformation takes place not only within manufacturers but also in the ecosystems that create, deliver, and capture value for the service.

In summary, manufacturers have been applying the DS strategy [6, 9]. In this way, it is evident that digitalization has accelerated the transformation of service-oriented BMs [6, 39], while ecosystem actors become decisive for the successful implementation of a PSS business model [19].

2.4 Digitalization

New technologies have implied significant changes in the processes and results of companies, especially in manufacturing [41]. These changes were driven by a complex system of related technologies [42] and appear in the literature under terms such as digital transformation, smart production, digitalization, or the fourth industrial revolution [2, 4, 43].

The implementation of I4.0 results in a digital maturation that allows the integration of products and organizational processes through technologies to provide more value to the company and customers [2].

The fourth industrial revolution consists of manufacturing operations systems [44] and base technologies [45]. Manufacturing operations systems have four dimensions [2]: *Smart Manufacturing*, responsible for technology-based production processes; *Smart Products*, refers to the added value to the final products; *Smart Supply Chain*, concerns the way raw materials and products will be delivered; and *Smart Working*, related to new ways of working supported by emerging technologies.

Authors as [2] point out four core technologies: (1) Big Data; (2) Cloud Computing; (3) IoT; and (4) Analytics, such as data mining and machine learning. However, other authors identify other elements that can also be considered basic technologies such as Cyber-Physical Systems (CPS), integrated software systems, 3D printing, cobots, artificial intelligence, augmented reality, and cyber security [41, 43, 46].

For [46], the adoption process of these technologies is called digitalization. Thus, in addition to bringing new opportunities, advantages, and challenges to companies [41, 47, 48], digitalization has also generated innovations in BMs of the industry [2, 48, 49].

3 Method

This paper presents a comprehensive literature review about the role of digitalization, servitization and innovation ecosystem actors in boosting innovation in business models, organizing the findings in two main topics: (1) digitalization and the participation of innovation ecosystem actors, and (2) the contribution of digitalization, servitization and innovation ecosystem to BMI. The choice of conducting a literature review is due to the fact that this research method creates a solid basis for gaining knowledge on the subject studied, facilitates the development of theory, and discovers gaps from previous research [50]. The literature review was conducted between October 2020 and June 2021.

The analysis was developed throughout a series of steps (described below) that allowed building a database containing the most important findings. From the database, it was possible to analyze the information, using various perspectives. This provided a complete overview of the characteristics of the literature about this topic. Figure 1 illustrates the entire process.

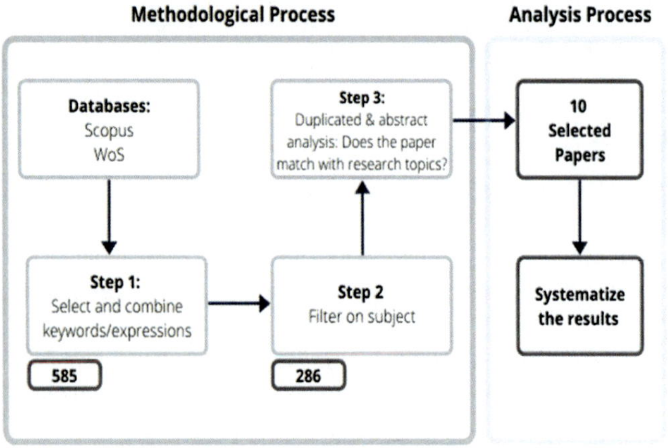

Fig. 1. Methodological and analysis processes

The research used Scopus and the Web of Science as sources for the research included in this study. Articles were extracted following a pre-established research logic of subsequent analysis (see above). Using the two selected databases considered, at the first step, 14 keywords/expressions were used to search in the titles, abstracts and keywords, and collect a broad number of papers. Only English literature was considered. After combining these refined keywords (Table 1), a total of 585 papers were analyzed. At step two, a filter was applied with the subject area chosen (Business and Management) where 286 papers were selected and considered to further analysis. Next, 24 papers were eliminated, since they were duplicates. Finally, after a careful examination, 252 papers did not match the topics studied in this work. This last step led to 10 papers for full and detailed analysis, which were aligned with the aim of this study and would contribute to answer to research question raised in this study.

Table 1. Search keyword/expression

Topic	Keywords
Digitalization	Industr* 4.0 Smart factor* I4.0 Fourth industrial revolution Digital transformation IIoT Digitalization Digitization
Servitization	Servitization Digital servitization Service provider Service supplier
Business model innovation	BMI Business model innovation

4 Findings

Ten articles were identified whose investigations showed a clear interdependence between the factors analysed, namely digitalization, servitization and innovation ecosystem. For this reason, they should be analysed together rather than isolated, as shown in Table 2. The innovation in BM is common to all of them.

4.1 Digitalization and the Participation of Innovation Ecosystem Actors

While improving the competitiveness of companies, digitalization enables the creation of value and revenues [38] while bringing changes in companies, namely about their relationships with suppliers and customers [4]. Data access becomes a valuable resource for differentiating manufacturers from other competitors. More than that, the data become an asset for manufacturers to understand their customers and devise specific strategies for the market niche where they operate [39].

Table 2. Main findings of the selected articles

Topic	Findings	References
Digitalization	Digitalization has increased and changed the relationship with actors of the ecosystem such as suppliers and customers. Once the relationship with key partners has solidified, changes in the key resources and value propositions emerge, directly impacting BMI	[2, 39]
Servitization	Digital servitization is the main strategy to digitalization. It has changed the customer relationship and the value proposition since there is a concern with the customer experience, resulting in BMI	[7, 39]
Innovation Ecosystem	The actors are responsible for providing assistance in technology transfer, methods, practices, techniques, partnerships, among others, but mainly they have an active participation in the cocreation value process which contributes to the BMI	[7, 39, 54, 58]
Business Model Innovation	It is a clear consequence of the use of digitalization and servitization by companies. The relations with actors of the innovation ecosystem boost changes in the value proposition of their business models	[2, 5, 7, 39, 54–59]

It is expected an increase in automation and connectivity in companies [4], with changes in human relationships, organizational methodologies, and processes [48]. These changes will also affect not only the internal environment of companies, but also their relationships in the external environment, such as the production chain, including suppliers, partners, and customers [4]. This may require a "reengineering" of their strategies and management practices [41, 48], as well as new ways of manufacturing products and their consumption, thus enabling or pushing the creation of new BMs.

In this sense, IoT and Big Data have become two of the main technologies for digitizing business and increasing interaction between companies, suppliers, and customers [39]. The use of these technologies allows, in addition to autonomous and more efficient management of resources and equipment, better monitoring and customer behaviour, which is why many manufacturers implement them to offer new services to their customers.

These technologies reach not only products, but also services [39]. Opportunities such as process redesign and digital servitization [7, 39, 47] can be seen as consequences of these strategies and are related to BMI [41, 48], since they help in the value creation and capture [39].

Unlike the old commercial relationships between companies and suppliers, nowadays, innovation ecosystems involve partners of different sizes and types [42]. To survive and grow, companies need to adapt to this reality and make changes to their production and management practices [48].

Despite the benefits mentioned, many companies still have difficulties in going through the digitalization process and this is due not only to the changes in their BMs, but also to the changes in their relationships with the actors in the innovation ecosystem they relate to, such as customers, suppliers, and solution providers [38].

The greater the complexity of technologies, the greater the need for support from partners such as solution providers research institutes [42, 48] while applying technology transfer (TT) from the latter to companies as part of their evolution process towards digitalization [43]. Therefore, the adoption of new technologies to foster digitalization must consider the capabilities and characteristics of the adopters and partners, such as solution providers and start-ups, that will be involved in this process [41]. The position of the manufacturer in the innovation ecosystem, sometimes as an adopter, sometimes as a solution provider to another partner, determines the investments that will be made and, consequently, the impacts on their BMs [41, 43]. Once in a position of adopter, the company enlarge its cost structure in exchange for new solutions to provide innovative value propositions. As a solution provider, the company shares knowledge, practices, and technologies to boost the key partners' revenue streams and expand its market share, validating its value proposition.

The TT process helps companies to enlarge their market, to increase competitiveness and economic growth [43].TT includes technologies, equipment, knowledge, or other resources, and the corresponding dissemination [41]. At the same time, it involves participants in the innovation ecosystem [42], such as industrial companies, their suppliers, solution providers, research institutes, universities, and government entities [43, 51, 53].

In short, I4.0 technologies and digital servitization have an impact on manufacturers' BMs [41] while encouraging new or improved relationships with different actors in the ecosystem, directly creating a positive impact in the development of knowledge, new technologies, BMIs, and competitiveness [48, 51].

4.2 Contribution of Digitalization, Servitization and the Innovation Ecosystem to BMI

A large amount of research explores servitization [7, 54, 58] and digitalization [5, 55–57, 59] with BMI separately, to understand their impacts. Only two studies gather these subjects [2, 39]. When it comes to the innovation ecosystem, two studies [5, 59] mentioned the researched companies as actors of the ecosystem, but only [7] and [59] indeed gather the subjects servitization and digitalization, respectively, with the industrial and innovation ecosystem. All these studies mentioned can be split into three groups:

1. The first focuses on understanding the impacts of digitalization on the BMI, mostly on SMEs from the manufacturer's point of view [5, 56, 57]. Beyond the changes in the value proposition, SMEs showed significant changes in the key resources of their BMs, especially new equipment, and production facilities to meet the requirements

of Industry 4.0. Additionally, the acquisition of new human resources or the training of the existing team was necessary to handle the new technologies [56].

Furthermore, digital transformation brings more benefits in the production of SMEs, although it still has little impact in areas such as the supply chain [5]. Also, [55] propose a microanalysis validating the importance of the leader in digital transformation. Of the identified studies, only [59] provide a macro analysis by measuring, for example, the level of digitalization in wine manufacturers, noting that there is a variety of BMIs according to the position the manufacturer occupies in the innovation ecosystem.

Moreover, the types of BMI arise according to the level of interactivity and relationship between the different actors, in the so-called divergence-convergence framework. For cases where there is cooperation between several actors at the same time, the authors classify BMIs as disruptively convergent. Cooperation between two groups of actors is called partially convergent BMI. Finally, relationships that focus only on core competencies without cooperation of actors are called divergent BMI.

2. The second line of investigation relates the concepts of servitization and manufacturer's BMIs, especially the relationship between innovations in products and services, and performance indicators [54, 58]. In a nutshell, servitization has a positive impact on long-term product innovation. Otherwise, servitization without product innovation only generates short-term results. In this line of investigation, only [7] explore the perspective of the industrial ecosystem by carrying out a literature review on digital servitization and finding that there is still a gap on how to co-create and capture value between partners in the same ecosystem.

   In this study, the authors highlight types of BMI according to different dimensions such as solution customization, pricing, and digitalization. The maturity in which one of these dimensions can drive to five types of BMI: product-oriented service provider, industrializer, customized integrated solution provider, platform provider, and outcome provider.
3. In the third group, only two studies [2, 39] address the three concepts in an integrated way, namely BMI, servitization and digitalization. [2] build a conceptual framework that identifies the types of BMI, according to the level of servitization (smoothing, adapting, and substituting) and I4.0 (low, moderate and high levels). The combination between these levels leads to nine configurations classified into manual, digital, and industry 4.0-related services, which can concentrate on smoothing, adapting, or substituting services. Finally, [39] identify the effects on BMI arising from IoT and servitization.

Even so, it is noticeable that interactions within an ecosystem initially arise from manufactures. Surrounding SMEs benefit strongly from the interactions of the innovation ecosystem in which they find themselves, as their structures do not allow an easy access to technologies, knowledge, and complex resources independently [42, 57]. In the case of [5] and [56], the change in the supplier BM is even more evident when customers initiate changes and influence the decision to adopt new technologies.

Moreover, in all cases it is possible to see that the exchange of knowledge and resources has increased collaboration and interaction on different levels between actors of the ecosystem to boost innovation [51]. The greater the interaction with the innovation ecosystem, such as partners, suppliers, research centres and start-ups [59], the greater the innovation and creation of value. These innovative products, processes, tools, and services accelerate digitalization, change value propositions, and ultimately drive innovation in your BMs.

5 Conclusions

Industry 4.0 paradigm and its technologies have been generating significant pressure for the digitalization of companies, not only manufacturers, but also those in the value chain. This trend has created numerous possibilities for manufacturers to improve their processes, products, and services in order to respond to the market challenges.

One of such possibilities is servitization, which has been accelerated by recent technological advances. In servitization, companies associate services to their products to improve the value they add to their customers or provide the operation of their products by their customers only as a service. Since this is not a trivial change, they often have to innovate their own business models, either improving them or creating new ones to better respond to their own challenges and their customers' challenges, as well. Frequently, the BMI is conducted by companies in collaboration with other entities in their innovation ecosystems, such as universities, research organizations, suppliers, customers, among others.

In such context, we identified in the literature the need to understand the interplay between these dynamics – I4.0 digitalization and servitization – and the actors of the innovation ecosystem towards boosting BMI.

To achieve such goal, we conducted a systematic literature review using keywords related with these topics (detailed previously). In the initial steps of the methodology, a considerable number of articles were found (585), while after the following rounds of selection, 10 articles remained as contributing to answer the research question raised. However, any of these articles addressed the combination of the three topics proposed: digitalization, servitization, and the contribution of the innovation ecosystem actors, even if they showed a clear interdependence between these areas, leading to common findings impacting BMI:

1. Several I4.0 technologies are key to improve the interaction in the value chain, namely IoT and Big Data, leading to greater automation and connectivity, but increasing complexity as well, providing significant changes in companies, and this may lead to new or reskilled human resources;
2. To cope with more complex environments, companies resort to ecosystem partners such as research organizations or solution providers from which they get knowledge and technologies, and who present diverse types of BMI depending e.g. on their position in the innovation ecosystem and maturity;
3. Servitization seems to impact BMI, especially when analyzed through a perspective of particular indicators reflecting the product-service relationship and their market performance, allowing companies to innovate toward service-based business model in an integrated and sustainable manner in the long-term;

4. As a consequence of digitalization and servitization, there has to be an alignment among ecosystem actors to capture the co-created value generated by BMI;
5. The combination of BMI, servitization, and digitalization may also be structured into frameworks according to the different levels of each of those factors, thus allowing for companies to assess and position themselves in such frameworks and identify the path to follow.

In conclusion, the impacts I4.0 technologies create in companies through digitalization are evident, not only in their processes, products, and services, but also through the combination of these in servitization. To achieve this, companies have to innovate in their BMs, frequently with their ecosystem partners, where a leading company has a key role is this process and where SMEs greatly benefit from these interactions due to their limited resources or infrastructures.

Several questions can be answered in future under the topic proposed by the present work: 1) it is not evident yet the individual contribution of each of the mentioned innovation ecosystem actors to BMI developed by manufacturers and how such contribution can be explored; 2) the identification of the most relevant I4.0 technologies to BMI it is also worth investigating further, as well as the role of each of them to diverse aspects of BMI; 3) the influence of the manufacturer's digital maturity to BMI development and adoption and which kind of value does it create in their customers could also be studied; and lastly 4) how does the innovation contexts (quantified e.g. by the European Union Innovation Scoreboard, or the OECD Science, Technology and Innovation Scoreboard) in which the manufacturer operates contributes to BMI development and adoption is also worth researching.

Acknowledgments. This work is financed by National Funds through the Portuguese funding agency, FCT - Fundação para a Ciência e a Tecnologia, within project LA/P/0063/2020.

References

1. Baines, T.S., et al.: State-of-the-art in product-service systems. Proc. Inst. Mech. Eng. Part B: J. Eng. Manuf. **221**(10), 1543–1552 (2007)
2. Frank, A.G., Mendes, G.H.S., Ayala, N.F., Ghezzi, A.: Servitization and Industry 4.0 convergence in the digital transformation of product firms: a business model innovation perspective. Technol. Forecast. Soc. Change **141**, 341–351 (2019)
3. Weill, P., Woerner, S.L.: Optimizing your digital business model. MIT Sloan Manag. Rev. **54**(3), 71–78 (2013)
4. Culot, G., Orzes, G., Sartor, M., Nassimbeni, G.: The future of manufacturing: a Delphi-based scenario analysis on Industry 4.0. Technol. Forecast. Soc. Change **157**, 1–34 (2020)
5. Müller, J.M., Däschle, S.: Business model innovation of industry 4.0 solution providers towards customer process innovation. Processes **6**(12), 1–19 (2018)
6. Paschou, T., Rapaccini, M., Adrodegari, F., Saccani, N.: Digital servitization in manufacturing: a systematic literature review and research agenda. Ind. Mark. Manag. **89**, 278–292 (2020)
7. Kohtamäki, M., Parida, V., Oghazi, P., Gebauer, H., Baines, T.: Digital servitization business models in ecosystems: a theory of the firm. J. Bus. Res. **104**, 380–392 (2019)

8. Sjödin, D., Parida, V., Kohtamäki, M., Wincent, J.: An agile co-creation process for digital servitization: a micro-service innovation approach. J. Bus. Res. **112**, 478–491 (2020)
9. Reim, W., Sjödin, D.R., Parida, V.: Servitization of global service network actors – a contingency framework for matching challenges and strategies in service transition. J. Bus. Res. **104**, 461–471 (2019)
10. Forkmann, S., Henneberg, S.C., Witell, L., Kindström, D.: Driver configurations for successful service infusion. J. Serv. Res. **20**(3), 275–291 (2017)
11. Johnson, M.W.: Digital growth depends more on business models than technology. Harvard Bus. Rev. 1–14 (2018)
12. Massa, L., Tucci, C.L., Afuah, A.: A critical assessment of business model research. Acad. Manag. Ann. **11**(1), 73–104 (2017)
13. Foss, N.J., Saebi, T.: Business models and business model innovation: between wicked and paradigmatic problems. Long Range Plan. **51**(1), 9–21 (2018)
14. Al-Debei, M.M., El-Haddadeh, R., Avison, D.: Defining the business model in the new world of digital business. Paper presented at the Fourteenth Americas Conference on Information Systems, Toronto, ON, Canada (2008)
15. Birkinshaw, J., Goddard, J.: Whats is your management model? MIT Sloan Manag. Rev. **50**, 81–90 (2009)
16. Porter, M.E.: Strategy and the internet. Harv. Bus. Rev. **79**(3), 62–78 (2001)
17. Wirtz, B.W., Pistoia, A., Ullrich, S., Göttel, V.: Business models: origin, development and future research perspectives. Long Range Plan. **49**(1), 36–54 (2016)
18. Baden-Fuller, C., Haefliger, S.: Business models and technological innovation. Long Range Plan. **46**(6), 419–426 (2013)
19. Reim, W., Parida, V., Örtqvist, D.: Product-service systems (PSS) business models and tactics – a systematic literature review. J. Clean. Prod. **97**, 61–75 (2015)
20. Osterwalder, A., Pigneur, Y.: Business Model Generation: A Handbook for Visionaries, Game Changers, and Challengers. Wiley, Hoboken (2010)
21. Teece, D.J.: Business models, business strategy and innovation. Long Range Plan. **43**(2–3), 172–194 (2010)
22. Chesbrough, H.: Open Business Models: How to Thrive in the New Innovation Landscape. Harvard Business School Press, Boston (2006)
23. Beattie, V., Smith, S.J.: Value creation and business models: refocusing the intellectual capital debate. Br. Account. Rev. **45**(4), 243–254 (2013)
24. Jocevski, M., Arvidsson, N., Ghezzi, A.: Interconnected business models: present debates and future agenda. J. Bus. Ind. Mark. **35**(6), 1051–1067 (2020)
25. Salvador, R., Puglieri, F.N., Halog, A., Andrade, F.G.D., Piekarski, C.M., De Francisco, A.C.: Key aspects for designing business models for a circular bioeconomy. J. Clean. Prod. **278**, 1–14 (2021)
26. Foss, N.J., Saebi, T.: Fifteen years of research on business model innovation. J. Manag. **43**(1), 200–227 (2017)
27. Zott, C., Amit, R.: Business model innovation: how to create value in a digital world. GFK Mark. Intell. Rev. **9**(1), 18–23 (2017)
28. Geissdoerfer, M., Vladimirova, D., Evans, S.: Sustainable business model innovation: a review. J. Clean. Prod. **198**, 401–416 (2018)
29. Cavalcante, S., Kesting, P., Ulhøi, J.: Business model dynamics and innovation: (re)establishing the missing linkages. Manag. Decis. **49**(8), 1327–1342 (2011)
30. Shakeel, J., Mardani, A., Chofreh, A.G., Goni, F.A., Klemeš, J.J.: Anatomy of sustainable business model innovation. J. Clean. Prod. **261**, 1–14 (2020)
31. Vandermerwe, S., Rada, J.: Servitization of business: adding value by adding services. Eur. Manag. J. **6**(4), 314–324 (1988)

32. Baines, T., Lightfoot, H.W.: Servitization of the manufacturing firm. Int. J. Oper. Prod. Manag. **34**(1), 2–35 (2013)
33. Grubic, T.: Remote monitoring technology and servitization: exploring the relationship. Comput. Ind. **100**, 148–158 (2018)
34. Zheng, P., Wang, Z., Chen, C.-H., Pheng Khoo, L.: A survey of smart product-service systems: key aspects, challenges and future perspectives. Adv. Eng. Inform. **42**, 1–19 (2019)
35. Adrodegari, F., Saccani, N.: Business models for the service transformation of industrial firms. Serv. Ind. J. **37**(1), 57–83 (2017)
36. Kamal, M.M., Sivarajah, U., Bigdeli, A.Z., Missi, F., Koliousis, Y.: Servitization implementation in the manufacturing organisations: classification of strategies, definitions, benefits and challenges. Int. J. Inf. Manag. **55**, 1–15 (2020)
37. Tukker, A.: Eight types of product–service system: eight ways to sustainability? Experiences from SusProNet. Bus. Strateg. Environ. **13**(4), 246–260 (2004)
38. Kamalaldin, A., Linde, L., Sjödin, D., Parida, V.: Transforming provider-customer relationships in digital servitization: a relational view on digitalization. Ind. Mark. Manag. **89**, 306–325 (2020)
39. Paiola, M., Gebauer, H.: Internet of things technologies, digital servitization and business model innovation in BtoB manufacturing firms. Ind. Mark. Manag. **89**, 245–264 (2020)
40. Pirola, F., Boucher, X., Wiesner, S., Pezzotta, G.: Digital technologies in product-service systems: a literature review and a research agenda. Comput. Ind. **123**, 1–19 (2020)
41. Zangiacomi, A., Pessot, E., Fornasiero, R., Bertetti, M., Sacco, M.: Moving towards digitalization: a multiple case study in manufacturing. Prod. Plan. Control **31**(2–3) (2020)
42. Benitez, G.B., Ayala, N.F., Frank, A.G.: Industry 4.0 innovation ecosystems: an evolutionary perspective on value cocreation. Int. J. Prod. Econ. **228**, 1–13 (2020)
43. Da Silva, V.L., Kovaleski, J.L., Pagani, R.N.: Technology transfer in the supply chain oriented to industry 4.0: a literature review. Technol. Anal. Strateg. Manag. **31**(5), 546–562 (2019)
44. Dalenogare, L.S., Benitez, G.B., Ayala, N.F., Frank, A.G.: The expected contribution of Industry 4.0 technologies for industrial performance. Int. J. Prod. Econ. **204**, 383–394 (2018)
45. Frank, A.G., Dalenogare, L.S., Ayala, N.F.: Industry 4.0 technologies: implementation patterns in manufacturing companies. Int. J. Prod. Econ. **210**, 15–26 (2019)
46. Isensee, C., Teuteberg, F., Griese, K.-M., Topi, C.: The relationship between organizational culture, sustainability, and digitalization in SMEs: a systematic review. J. Clean. Prod. **275**, 1–19 (2020)
47. Coreynen, W., Matthyssens, P., Van Bockhaven, W.: Boosting servitization through digitization: pathways and dynamic resource configurations for manufacturers. Ind. Mark. Manag. **60**, 42–53 (2017)
48. Sestino, A., Prete, M.I., Piper, L., Guido, G.: Internet of things and big data as enablers for business digitalization strategies. Technovation **98**, 1–9 (2020)
49. Wei, Z., Song, X., Wang, D.: Manufacturing flexibility, business model design, and firm performance. Int. J. Prod. Econ. **193**, 87–97 (2017)
50. Webster, J., Watson, R.T.: Analyzing the past to prepare for the future: writing a literature review. MIS Q. **26** (2002)
51. Jiang, H., Gao, S., Zhao, S., Chen, H.: Competition of technology standards in Industry 4.0: an innovation ecosystem perspective. Syst. Res. Behav. Sci. **37**(4), 772–783 (2020)
52. Gomes, L.A.D.V., Facin, A.L.F., Salerno, M.S., Ikenami, R.K.: Unpacking the innovation ecosystem construct: evolution, gaps and trends. Technol. Forecast. Soc. Chang. **136**, 30–48 (2018)
53. Yaghmaie, P., Vanhaverbeke, W.: Identifying and describing constituents of innovation ecosystems. EuroMed J. Bus. **15**(3), 283–314 (2019)
54. Kastalli, I.V., Van Looy, B., Neely, A.: Steering manufacturing firms towards service business model innovation. Calif. Manag. Rev. **56**(1), 100–123 (2013)

55. Mihardjo, L.W.W., Sasmoko, S., Alamsjah, F., Elidjen, E.: Digital leadership role in developing business model innovation and customer experience orientation in industry 4.0. Manag. Sci. Lett. **9**(11), 1749–1762 (2019)
56. Müller, J.M.: Business model innovation in small- and medium-sized enterprises: strategies for industry 4.0 providers and users. J. Manuf. Technol. Manag. **30**(8), 1127–1142 (2019)
57. Müller, J.M., Buliga, O., Voigt, K.-I.: Fortune favors the prepared: how SMEs approach business model innovations in Industry 4.0. Technol. Forecast. Soc. Change **132**, 2–17 (2018)
58. Visnjic, I., Wiengarten, F., Neely, A.: Only the brave: product innovation, service business model innovation, and their impact on performance. J. Prod. Innov. Manag. **33**(1), 36–52 (2014)
59. Dressler, M., Paunovic, I.: Converging and diverging business model innovation in regional intersectoral cooperation–exploring wine industry 4.0. Eur. J. Innov. Manag. 1–28 (2020)

Development of a Procedure and Computer Application for a More Effective Management of External Logistics Services Request

Rúben D. F. S. Costa[1], Marta L. S. Barbosa[1], Francisco J. G. Silva[1,2(✉)], José C. Sá[1,2], Luís P. Ferreira[1,2], and Ana B. M. Machado[1]

[1] ISEP - School of Engineering, Polytechnic of Porto, Porto, Portugal
{fgs,cvs}@isep.ipp.pt, luispintoferreira@eu.ipp.pt

[2] INEGI - Instituto de Ciência e Inovação em Engenharia Mecânica e Engenharia Industrial, Porto, Portugal

Abstract. Manufacturers of automotive components are faced with the need to request very frequent external logistics services, sometimes with identical or very similar destinations. These services are always the subject of a quote request, which does not always present the necessary agility in terms of information flow. Thus, this work has the main objective of developing a new computer application that aims to concentrate all the information necessary for the quotation request of external logistics services, taking into account the necessary requirements. After a diagnosis of the initial situation, following an Action-Research methodology, a list of the necessary input variables for the process was developed, later the information flow diagram was established, and then the documents involved in the process were defined, allowing, based on the use of these data, to define a new computer application that was successfully developed and implemented, constituting a knowledge base for other similar cases.

Keywords: External logistics · Computer application development · Procedures' improvement

1 Introduction

Nowadays, final products are produced in an increasingly decentralized manner [1]. The owners of the main brands divide the product into components and subcontract specialized and certified companies to manufacture them, being usually responsible only for the manufacture of the core parts and the assembly of the final product [2]. A clear example of the adoption of this strategy is the automotive industry, which involves even the components' manufacturers in product development [3]. However, many other industries followed the same way, from the aeronautical [4] to the fashion industry [5]. In the automotive industry, since the personalization paradigm emerged, which put serial manufacturing aside, the diversity of components within each family has grown significantly and, in turn, orders have decreased in volume [6], creating new challenges for logistics [7]. Information systems are a perfectly indispensable platform for the management of

J. Machado et al. (Eds.): icieng 2022, LNME, pp. 128–139, 2023.
https://doi.org/10.1007/978-3-031-09360-9_11

these processes [8]. In spite of this, the specificity of each business means that each niche of companies has different needs, which makes it impossible to adopt a universal tool [9]. In fact, each company has its own internal and decision-making procedures regarding the consultation and contracting of external logistics services. Certain companies are focused only on production, subcontracting external companies that manage the finished product and direct it to customers, thus having an integration of services that benefits all parties (producer, distributor and consumer) [10]. Nevertheless, this logistics system is not suitable for small and medium-sized companies, which directly manage their deliveries to customers and have their own market consultation rules and decision-making process for assigning a given service to a set of previously selected logistics operators [11]. With the decrease in ordered batches and product diversity, this task has become truly time-consuming. Several decision support models have been investigated in order to minimize the time consumed in routine tasks of assigning external logistics services [12–14]. The development of several models, by itself, is a strong indicator that there are very well-defined and real specificities, which prevent a single model from serving the interests of a vast set of companies, even in the same sector. Usually, when companies are smaller, the process of assignment of external logistics services becomes increasingly empirical and more dependent on labour, therefore generating a greater number of nonconformities, which normally translate into economic losses, and can even degrade the image of these companies with their customers by failing to meet delivery deadlines. The solution to this problem may involve the strategic development of partnerships [15], or the creation of moderately automated systems, using computerized systems to reduce the time consumed by these tasks and minimize the probability that serious errors are made in the assignment of external logistics services [16]. Notwithstanding, as mentioned above, these solutions are usually designed according to certain needs, i.e., applications are developed based on a given set of input variables, on a set of decision support rules created for this purpose based on the assumptions that a given company or sector understands as basic for the functioning of the system. They are based as well on a set of output variables, which basically is the attribution of the service to a given company providing logistics services, within a previously defined universe of companies able to provide the service with the desired quality.

This work was developed based on the detection of a necessity by a multinational company which produces components for the automotive industry that need to establish a very specific procedure for its reality. Effectively, each of its production units need to fulfil a given delivery schedule to several assembly companies for the delivery to OEMs (Original Equipment Manufacturers), which may go out of production with different timings but should be grouped as much as possible, in order to minimize the need for transport requests and reduce costs. This work is divided into four sections: the current section aims to contextualize the reader with the problem and provide some theoretical support for the development, the second defines the methodology used, the third describes the results and the fourth summarizes the main achievements obtained.

2 Methodology

Given that this work aimed to establish a new procedure capable of solving a specific problem, but whose solution brings knowledge that can be applied in the resolution of

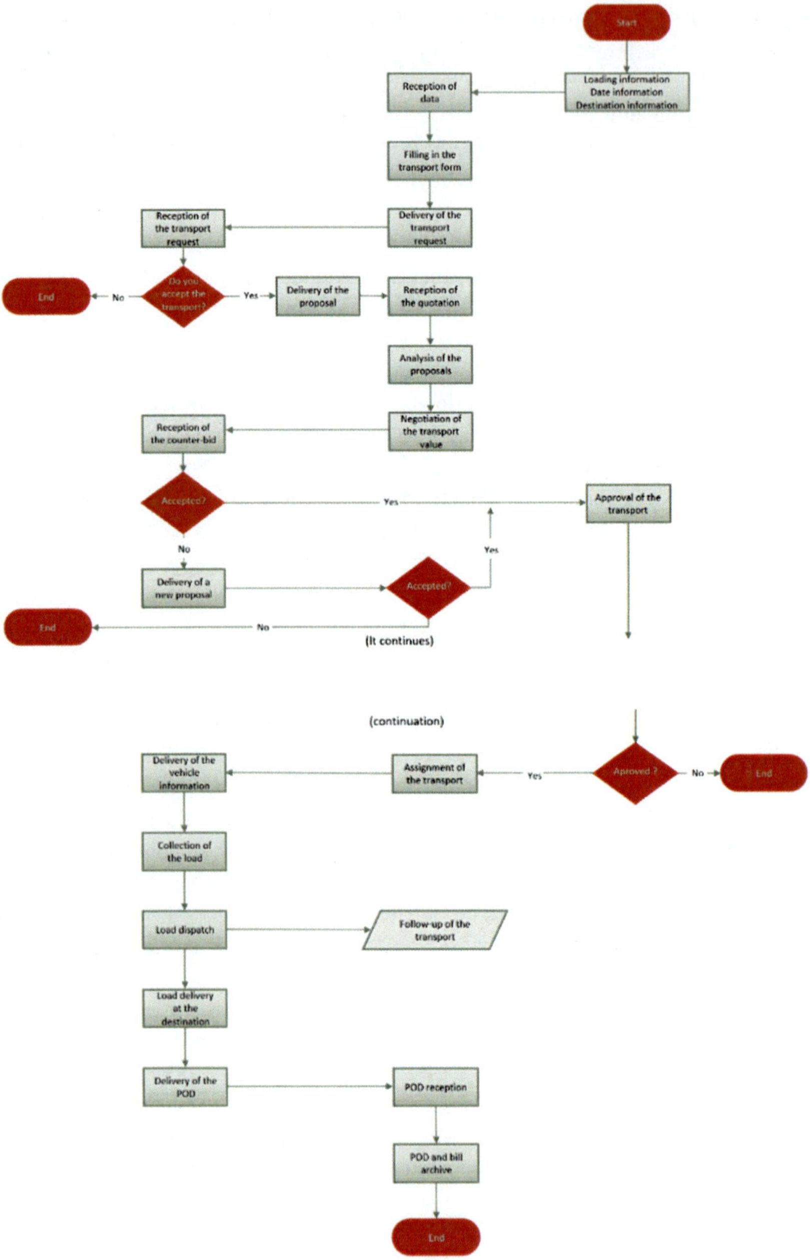

Fig. 1. Flowchart regarding the request for quotation process

similar problems and in the development of computer applications with different objectives, it was understood to apply the Action-Research method, which basically consists of the following steps [17, 18]: Initial diagnosis, Action plan, Development, Evaluation of the implemented model, and Monitoring of the model and extraction of transferable knowledge. The use of each of the phases of the methodology will be indicated as the work is presented. Regarding the Action-Research methodology, Saunders & Somekh state [19]: "Action-Research is about looking for ways to make things happen in line with one's vision, never accepting that anything is impossible, and acting strategically and politically to make new alliances and gain leverage on organization inertia". This also represents the main reason why this methodology has been used.

The diagnostic phase began with the identification of the problem of inefficiency in transport management, and the various requirements that the platform would have to have, in order to fill the shortcomings, were immediately listed: (a) Automated transport order forms; (b) Manage offers; (c) Internal transport approval; (d) Transport monitoring; (e) Storage of invoices, other documents, such as POD (proof of delivery). Currently, transport requests follow the flow diagram shown in Fig. 1.

The next action was to prepare the application prototype able to help the computer applications developers reach the level of efficiency required. Therefore, a series of steps have been done in order to establish the plan to reach the final solution in a better way, as can be seen in Fig. 2:

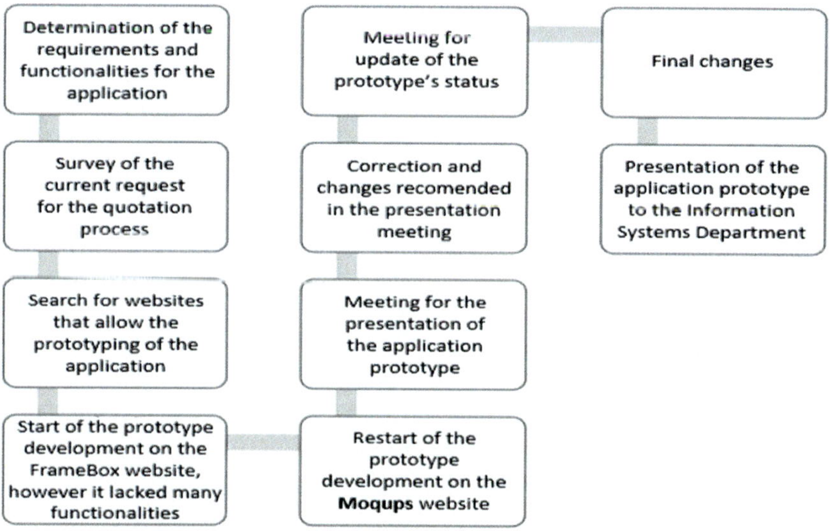

Fig. 2. Development steps to get the application prototype

The real action plan started with the definition of a prototype for the application. The development of this prototype for the external logistics request management application aimed at sharing the concept with the remaining team involved, i.e., to establish all the necessary requirements for the application. Furthermore, it allowed to save costs and time in development by showing to the IT team a prototype of a ready-to-code application and also save time in terms of application design since, at the time of programming, the aesthetics questions should already be completely defined.

Table 1. Input variables to the form for external logistics service request

Input variables	Description
Type of service	It is defined if the transport that will be carried out will be import, export or national
Transport mode	It is determined whether the transport will be by sea, road or air and whether it will be dedicated, groupage or full truckload transport
Number of unloads	If you want the vehicle to make more than one delivery or collection, the number will be placed in this field
Unloads' order	If there are several pick-ups or unloads in the field "Number of unloads", the intended order of unloading and loading will be placed
Customer's number and dispatch point	The customer's number and respective delivery point are identified
Customer's name	After placing the customer's number, the field with the customer's name is automatically filled in
Pick-up date	The day for collecting the load is defined
Delivery date	The day for delivering the load is defined
Delivery time	The time window for the delivery is identified
Unloading point	The delivery address is identified
Type of transport	The priority of the order is defined: normal or urgent
Transport responsibility	In the available fields, the responsible for the transport payment is marked: Ficosa or customer/supplier
Project code	The project of the sent references is identified
Product reference	The products are identified by their reference
Number of packages	Quantity of pallets or boxes sent
Measurements	Dimensions of the pallets sent
Weight per pallet	Weight, in kg, per pallet
Overload	Number of pallets that can be overlapped
Name of the applicant	Identification of the employee requesting the transport

(*continued*)

Table 1. (*continued*)

Input variables	Description
Department to which the applicant belongs	Identification of the employee's department who made the transport request
Transporters to send the request for quotation	From the list of transport providers, select the ones that you want to send the request to
Observations	For the transporter: in this field, add additional information that is relevant for the transporter, concerning the transport request

The work development started with the variables' definition. Input variables are those that independently stimulate the system, inducing changes and decisions. Therefore, in order to generate an external logistics service request, these input variables are introduced into the system as listed and described in Table 1.

Regarding the flow of information resulting from an order placed by the customer, the diagram represented in Fig. 3 has been drawn, which explains how information is passed between the different company departments and the corresponding stakeholders.

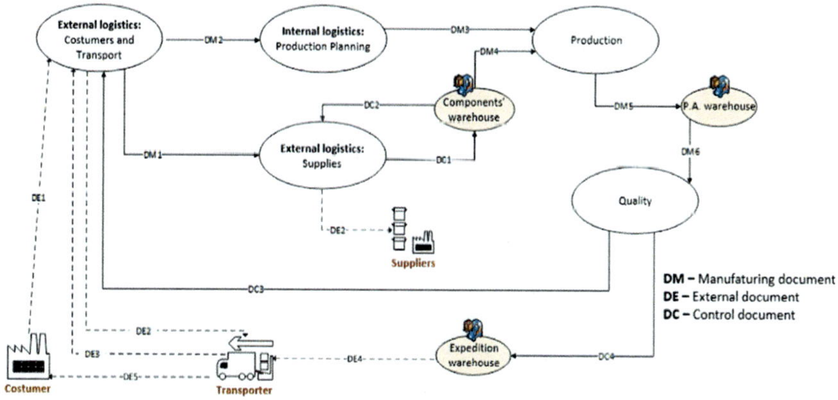

Fig. 3. Information flow diagram regarding an order request received from the customer

As can be seen from the analysis of the previous flow diagram, the beginning of the process starts with the order sent by the customer and ends with the dispatch of the order placed by the carrier. In addition, it is possible to verify the documents that flow with the information. The description of each of them is presented in Table 2.

Table 2. Description of the documents involved in the information flow

Documents	Description
DE1	It aims to transmit the order request information
DE2	It consists of sending information to request a transport quote, such as number of pallets, delivery destination and pallet dimensions and weight
DE3	It consists of sending quotations for the shipment of cargo
DE4	It aims to inform of the availability of the load pick-up, as well as transport documents (waybills)
DE5	It aims to inform about the constitution of the load (waybills)
DM1	Sending requests made by clients to external logistics
DM2	Forwarding customer orders to internal logistics
DM3	Sends the production plan
DM4	Sends information on quantities of materials
DM5	Production reports
DM6	Information on produced orders
DC1	Information about the quantities required for the production orders generated
DC2	Information about the existing quantities of materials
DC3	Validation of products
DC4	Dispatch of products approved by the quality department

3 Results and Discussion

In this step, the model development is going to be presented. Considering all the requirements and intended functionalities for the external logistics request management application, the development of the prototype was carried out. Therefore, the first menu that will appear when opening it will be the Login menu, allowing or not the access of the user to the application. After that, the next menu appears with the possible options, highlighting the external logistics service request, the status of the request, the approval of the logistics service, the consultation of the requests and, finally, the file generated. If the intention is to request a quotation for the external logistics service, the user must simply select the menu item called "Shipping Request".

Depending on whether the transport is to be carried out between the company and a customer or whether it will depart from a supplier and come to the company, the configuration and requests opened in the form will be different. Figure 4 shows the example relating to company dispatch to suppliers. However, when it comes to customers, the topic "Shipping Details" is different and more complex, appearing as seen in Fig. 5.

Once a transport request has been made, it is pending, as it awaits quotation proposals from the selected transporters. During this period, it is possible to consult the requests and their status, and it can present five different statuses:

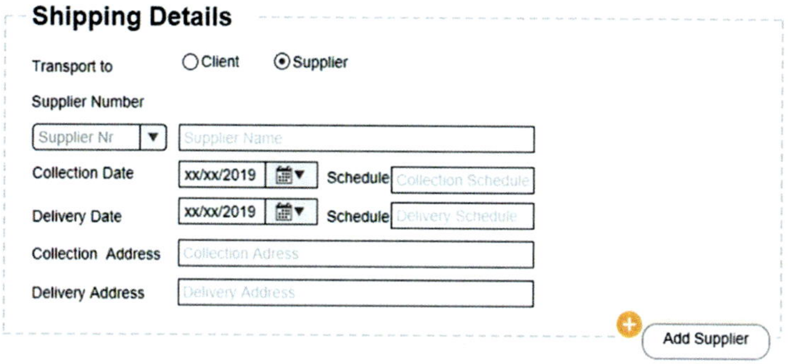

Fig. 4. Shipping details of the external logistics services request for suppliers

(a) Analyse requests - the transport applicant made the transport request and it was analysed by the logistics department;
(b) Wait for quotation - the order has been sent, but there are still no answers about quotations;
(c) Analyse proposals - answers already exist and are being analysed or negotiated;
(d) Awaiting approval - The carrier/cost for the transport has already been selected, it just needs approval from those responsible;
(e) Assigned - transport already approved and, consequently, assigned.

At the assigned point in the menu, the data entered in the transport order is placed, for it to show an overview of the orders already placed. In addition, orders that have been on the platform for more than 48 h will be deleted. Another defined requirement was the approval of transports to be made directly on the platform and, therefore, the "Approval" menu item was created, which allows access to transport data to the people involved and their immediate approval or rejection. Once the transport has been approved by the competent person, it is assigned to the carrier, followed by the collection of the cargo and, finally, dispatch.

During the transit time, it is intended for there to have updated information on transport requests, thus the "Consult Request" menu item appears. Here, information about the driver, as well as the vehicle registration and tracking number is recorded. In addition, the transport is also monitored, in order for there to exist an immediate perception of whether it will be delayed in arriving at the customer, and thus quickly solve any arising problems. The delivery state can then present three distinct options:

(a) In transit – the transport is taking place as planned, at the scheduled times;
(b) Delayed – the transport is delayed in relation to the previously agreed delivery times;
(c) Delivered – the goods are in the corresponding destination.

For further analysis, a tab called "On time?" appears, which will allow to identify the carriers that meet the agreed deadlines, and also those that usually do not meet the previously agreed deadlines.

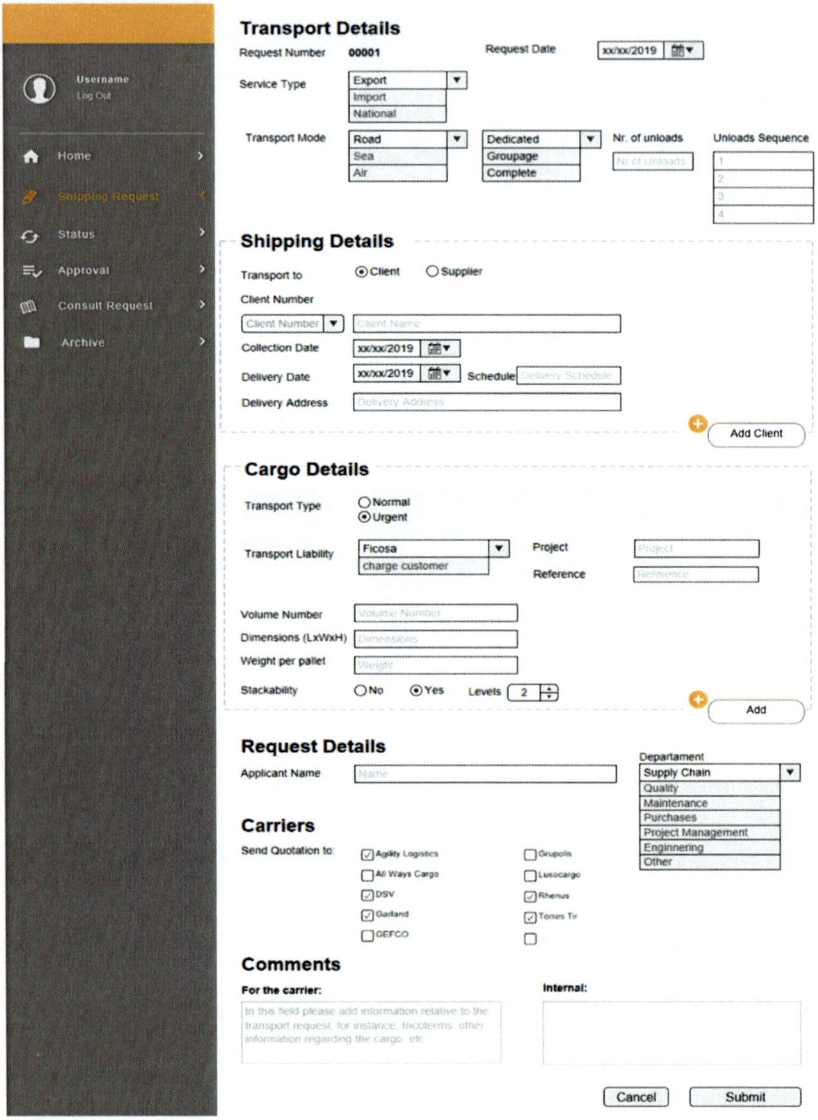

Fig. 5. External logistics services request for delivering goods for customers

Finally, the menu related to the file appears, i.e., the transport requests made are placed on this page and then the documents associated with it are attached for future reference. These documents are then the invoice and the POD. Another feature of this point is to compare costs between the assigned cost and the invoiced price, as in certain cases there are discrepancies between them. To simplify this comparison, three visual icons appear, as follows:

(a) Higher than agreed;
(b) According to the agreed;
(c) Lower than agreed.

Subsequently, a monitoring system to the application and corresponding functioning needed to be created, perceptible by the following equations.

$$PFA\ (ProcedureFailureAnalysis) = \frac{Internal\ failures\ detected\ per\ month}{Services\ requested\ per\ month} \times 100 \quad (1)$$

$$BAA\ (Budgets\ Accuracy\ Analysis) = \frac{Non - conformities\ in\ costs\ agreed}{Services\ requested\ per\ month} \times 100 \quad (2)$$

$$IDA\ (Internal\ Delay\ Analysis) = \frac{Number\ of\ delays\ in\ deliveries\ per\ month}{Processes\ conducted\ per\ month} \times 100 \quad (3)$$

The KPI's (Key Performance Indicators) previously referred are important to evaluate the application performance and atmosphere of the collaborators with it, as they identify the percentage of errors within the application use, letting the decision-makers know how to improve it. PFA should be lower than 0.5%, BAA should be lower than 1% and IDA should be lower than 2%. Attending to the fact that the main objective of this research project was the development of a computer application that aims to concentrate all the information necessary for the quotation request of external logistics services, this has been accomplished, having been successfully implemented and tested. Additionally, with this implementation, it was possible to achieve the gains shown in Fig. 6.

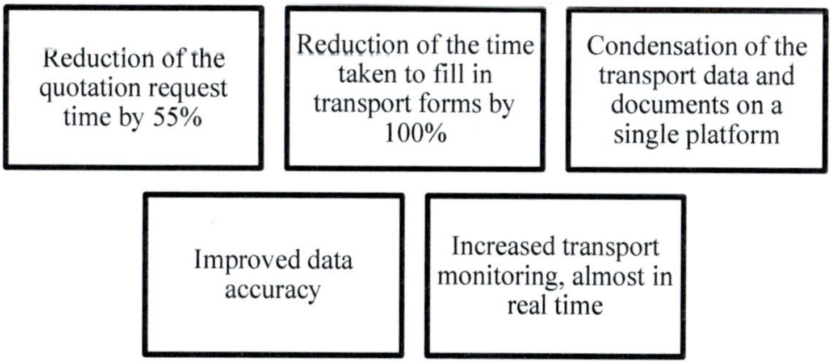

Fig. 6. Improvements with the development and implementation of the new application

Regarding the reduction of the quotation request time, the simulation carried out with 100 samples stands out a decrease of 55%, since on average the time spent without using the application is 1 min and 34 s, reducing to just 42 s. In regard to the transport request approval forms, with the implementation of the application, it is no longer necessary to fill in the approval document, since it is now carried out automatically, obtaining a 100%-time spent improvement. Otherwise, on a daily basis, the person responsible for transport requests will obtain a saving of about 45 min, so he/she can direct this time to perform other tasks. Thus, the approach here developed can be used to solve other similar cases, helping in the development of the algorithm for solving similar problems, constituting a knowledge which is transferrable to other identical situations.

This application meets the Industry 4.0 objectives, as it focuses on the automation of processes, avoiding data loss and centring all the information in one single place. This way, it allows the user to make the best decisions in every moment.

4 Conclusion

This work aimed to develop a computer application that would make it possible to speed up the request for external logistics services in a semi-automatic way, taking into account customer requests to the company, and requests for materials and components from the company to external suppliers. It should be noted that each request for external logistics services must be preceded by the respective request for quotation and corresponding analysis of the cost and compliance of dates. An initial diagnosis of the situation was made, the input and output variables of the system were identified, as well as the flow of information that involves this entire process. The time for requesting a quote was reduced by 55%, the completion of paper forms was abolished (towards the total digitization of the company), the data are now all on a single platform, there is much greater accuracy in the information, the errors are now more easily detectable, and the system allows real-time control of the situation of all external logistics. It is expected that the methodology followed here can constitute transferable knowledge for the elaboration of other computer applications that aim to solve similar problems.

References

1. Okpoti, E.S., Jeong, I.-J.: A reactive decentralized coordination algorithm for event-driven production planning and control: a cyber-physical production system prototype case study. J. Manuf. Syst. **58**, 143–158 (2021)
2. Araújo, W.F.S., Silva, F.J.G., Campilho, R.D.S.G., Matos, J.A.: Manufacturing cushions and suspension mats for vehicle seats: a novel cell concept. Int. J. Adv. Manuf. Technol. **90**(5–8), 1539–1545 (2016). https://doi.org/10.1007/s00170-016-9475-6
3. Costa, R.J.S., Silva, F.J.G., Campilho, R.D.S.G.: A novel concept of agile assembly machine for sets applied in the automotive industry. Int. J. Adv. Manuf. Technol. **91**(9–12), 4043–4054 (2017). https://doi.org/10.1007/s00170-017-0109-4
4. Freitas, F., Silva, F.J.G., Campilho, R.D.S.G., Pimentel, C., Godina, R.: Development of a suitable project management approach for projects with parallel planning and execution. Procedia Manuf. **51**, 1544–1550 (2020)
5. Fung, Y.-N., Chan, H.-L., Choi, T.-M., Liu, R.: Sustainable product development processes in fashion: Supply chains structures and classifications. Int. J. Prod. Econ. **231**, 10791 (2021)
6. Moreira, B., Gouveia, R.M., Silva, F.J.G., Campilho, R.: A novel concept of production and assembly processes integration. Procedia Manuf. **11**, 1385–1395 (2017)
7. Mourtzis, D., Doukas, M., Psarommatis, F.: Design and planning of decentralised production networks under high product variety demand. Procedia CIRP **3**, 293–298 (2012)
8. Lo, S.-C., Hall, R.H.: The design of real-time logistics information system for trucking industry. Comput. Oper. Res. **35**(11), 3439–3451 (2008)

9. Kendrick, B.A., Dhokia, V., Newman, S.T.: Strategies to realize decentralized manufacture through hybrid manufacturing platforms. Robot. Comput. Integr. Manuf. **43**, 68–78 (2017)
10. Bittencourt, G.C., et al.: A solution framework for the integrated problem of cargo assignment, fleet sizing, and delivery planning in offshore logistics. Comput. Ind. Eng. **161**, 107653 (2021)
11. Wang, Y., et al.: Collaborative multi-depot logistics network design with time window assignment. Expert Syst. Appl **140**, 112910 (2020)
12. Werner-Lewandowska, K., Kosacka-Olejnik, M.: Logistics maturity model for service company – theoretical background. Procedia Manuf **17**, 791–802 (2018)
13. Crainic, T.G., Fomeni, F.D., Rei, W.: Multi-period bin packing model and effective constructive heuristics for corridor-based logistics capacity planning. Comput Oper Res **132**, 105308 (2021)
14. Hu, X., Wang, G., Li, X., Zhang, Y., Feng, S., Yang, A.: Joint decision model of supplier selection and order allocation for the mass customization of logistics services. Transp. Res. E Logist Transp. Rev. **120**, 76–95 (2018)
15. Borgström, B., Hertz, S., Jensen, L.-M.: Strategic development of third-party logistics providers (TPLs): "Going under the floor" or "raising the roof"? Ind. Mark. Manage. **97**, 183–192 (2021)
16. Wang, J., Lim, M.-K., Zhan, Y., Wang, X.F.: An intelligent logistics service system for enhancing dispatching operations in an IoT environment. Transp. Res. E Logist. Transp. Rev. **135**, 101886 (2020)
17. Martins, R., Pereira, M.T., Ferreira, L.P., Sá, J.C., Silva, F.J.G.: Warehouse operations logistics improvement in a cork stopper factory. Procedia Manuf. **51**, 1723–1729 (2020)
18. Marinho, P., Pimentel, D., Casais, R., Silva, F.J.G., Sá, J.C., Ferreira, L.P.: Selecting the best tools and framework to evaluate equipment malfunctions and improve the OEE in the cork industry. Int. J. Ind. Eng. Manage. **12**(4), 1–13 (2021)
19. Saunders, L., Somekh, B.: Action research and educational change: teachers as innovators. In: Noffke, S.E., Somekh, B. (eds.) The SAGE Handbook of Educational Action Research, p. 193. SAGE Publications, London (2009). ISBN: 978-1-4129--4708-4

Production Process Stability: The Advantages of Going Beyond Qualitative Analysis

Anna Borucka[1] and Dariusz Mazurkiewicz[2](✉)

[1] Wojskowa Akademia Techniczna (Military University of Technology), 00-908 Warsaw, Poland
anna.borucka@wat.edu.pl

[2] Politechnika Lubelska (Lublin University of Technology), 20-618 Lublin, Poland
d.mazurkiewicz@pollub.pl

Abstract. The article proposes a method for assessing the stability of the production process using the analysis of variance ANOVA. For this purpose, factors that have a significant impact on the stability of production were analyzed and selected. The following variables were considered the best: year, month, day of the week and factors causing delays (machine failure, lack of components, overproduction, delay, others). The proposed method allows for the assessment of the current state and the implementation of solutions allowing for compensation of disturbances in the event of their identification. Ultimately, it is possible to improve the implementation of production processes, as well as to enhance the quality of final products. The method of statistical analysis of the stability of the production process used in the article has been verified on the example of a real enterprise and the processes implemented in it, which makes it utilitarian. The presented algorithm for production data analysis is a viable tool for assessing current state and implementing improvement and repair solutions. The presented study can be used in any manufacturing company as a preliminary diagnosis of the implemented production processes, which is a selection of areas requiring detailed control.

Keywords: Production process · Stability · Modelling · ANOVA

1 Introduction

Improved methods of production organization, technical and technological progress, emphasis on product quality or changes in production preparation are just a few of many manifestations conditioning the growth of systematic importance of production management area in an enterprise. The achievement of objectives in enterprises engaged in production activities is possible primarily through the proper organization of production processes, ensuring the efficient and economic production flow, and thus the stable operation of the production system [1–3].

Product and process innovations are an inherent attribute of a dynamically functioning enterprise, actively responding to market transformations and customer expectations [2, 4]. This approach is associated with changes that can disrupt the production system and cause a loss of stability in production processes. Thus, on the one hand, enterprises

J. Machado et al. (Eds.): icieng 2022, LNME, pp. 140–148, 2023.
https://doi.org/10.1007/978-3-031-09360-9_12

are faced with the need to ensure the stability of the production system and achieve the set results, and on the other hand, to quickly and effectively implement improvements to ensure growth [1, 5].

Stability, broadly defined, refers to the constancy (invariability over time) of a characteristic or sequence of states that the system goes through [6] or the system's resilience to internal or external disruptions [7]. In other words, a process is stable if the empirical time measures of backlog (delays) are bounded from above at an infinite observation horizon [1, 8, 9].

This definition of production stability means that in many cases only qualitative assessment can be used, since production stability is difficult to define in a more quantifiable way [1, 4]. Especially in the absence of IT support in the form of ERP class systems and advanced methods of production process control, analyses allowing for stability assessment using possessed empirical observations are desirable.

There are many factors that affect the stability of production, some of which can be controlled and regulated if a negative effect is found. This equilibrium means the state of the system unaffected by the transformation, i.e. none of the variables in the system change value over time. This approach to the problem allowed the authors to propose a model for assessing production stability that goes beyond qualitative analysis and shows its quantifiable assessment, which was not found in the analysed literature. The study was based on variables related to production time (year, month, day of the week) and factors conducive to delays (machine breakdown, lack of components, overproduction, etc.). It was proposed to use statistical tests to evaluate the significance of the impact of each variable on production delays. It was assumed that a stable process is one whose course is between control limits, and that the absence of significant linkages implies that production is independent of these factors and that they have no impact on the stability. The opposite situation means an undesirable state and forces the implementation of solutions restoring the balance (changes in production materials/components, share of machines, number of employees involved). The proposed method allows the identification of such a condition, so it ultimately affects the possibility of improving not only the implementation of production processes, but also the quality of the final product.

We propose a solution to the above limitation; it is a method consisting of four basic stages (Fig. 1). The first stage is a comprehensive assessment of production process empirical results in relation to the assumed production plan, along with listing all possible factors that may affect it. Their number greatly depends on the form and method of archived data. The selected factors are subjected to statistical analysis, and then their impact on the production process is assessed. If a given variable is found to be of significant effect, it is possible to establish further dependencies describing this relationship based on the knowledge about the production process.

The selection of a method of statistical analysis that allows to assess the stability of production processes is presented based on the example of an enterprise that is a global manufacturing corporation operating in the cosmetics industry. It was shown that the chosen method allows not only to assess the current state, but also to implement solutions allowing to compensate for disruptions in case of their occurrence.

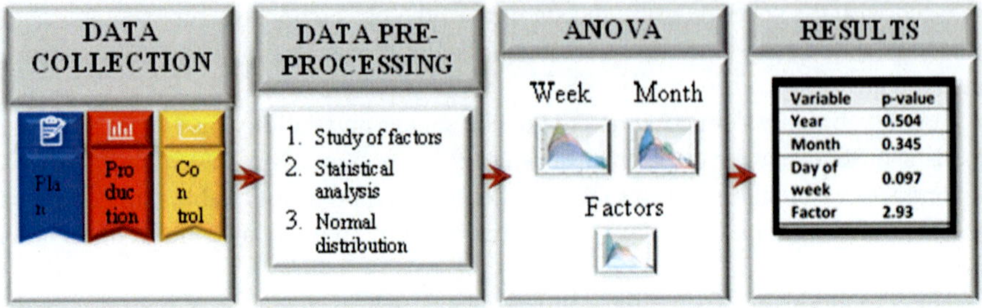

Fig. 1. Scheme of the method

The article consists of an introduction, Sect. 2 presenting the statistical methods used, Sect. 3 - in which the application of selected methods are verified on a real example and final conclusions in Sect. 4.

2 From Data to Information. A New Conversion Model

Actual production and planned production data were used to study production lines stability. This was done using statistical tests. This allowed us to verify the hypothesis of statistical significance of the differences found between the production plan and its implementation. Delays calculated as the percentage difference between the plan and the number of products actually produced were analysed. It was verified whether the impact of selected factors on the stability of production processes (delay values) is significant.

The authors analysed the effect of production year, month, and day of the week, as well as the reasons for delays, which were grouped by category:

1. **Machine breakdown –** sudden, unscheduled stoppages of a production line, lasting more than 5 min (as opposed to micro stoppages), associated with the loss of operation of more than one machine, caused by mechanical failure of parts and control failures.
2. **Overproduction –** uneven execution of the production plan manifested by production volumes higher than the scheduled completion date.
3. **Delay –** situations where there were technical problems during production other than machine failures, delays caused by suppliers.
4. **Lack of components –** situations where the value of production was less than intended due to insufficient quantities of components/raw materials to manufacture products.
5. **Other –** organizational reasons for changing production quantities: staff shortages, changing scheduling priorities, manning allocations, as well as machine start-ups and format changes between production orders.

The lack of significant differences in the groups of selected factors allows us to conclude that the enterprise studied functions properly. In other cases, the reasons causing the difference must be found. More than a dozen production lines were analysed, which allowed for a complete analysis of the enterprise studied. This paper presents results for one selected production line. The study used statistical tests to assess which factors significantly affect the production process. Research was performed according to the algorithm presented below. This procedure is appropriate when we are dealing with qualitative variables, as in the problem solution presented in [10, 11].

The testing algorithm for each production line included:

1. Calculating basic descriptive statistics for observing the values of planned and actual production in selected groups.
2. Analysing the normality and homogeneity of distributions, and selecting statistical tests.
3. Investigating the effect of the tested variables on the rate of production.

The extent of agreement between empirical and normal distributions is assessed using the Lilliefors test. For a significance level of $\alpha = 0.05$, the following statistical hypothesis are verified: H0: the distribution of the tested feature in the population is normal, H1: it differs from normal. The test statistics is of the form:

$$U_n = sup|F_n(x) - F(x)| \tag{1}$$

where $F_n(x)$ is an empirical distribution function, $F(x)$ is a distribution function of the normal distribution $N(\overline{x}, s)$. H0 is rejected and H1is accepted if the test statistics value exceeds the critical value in this test, which can be obtained from the Lilliefors Test Table.

The homogeneity of variance can be verified using Bartlett's test. The following statistical hypotheses are formulated:

$$H0 : \sigma_1^2 = \sigma_2^2 = \sigma_3^2 = \ldots = \sigma_k^2$$

$$H1 : \exists i, j : \sigma_i^2 \neq \sigma_j^2$$

The test statistics has a χ^2 distribution with $k - 1$ degrees of freedom and is of the form:

$$k = \frac{1}{c}\left[(n-k)lns^2 - \sum\nolimits_{i=1}^{k}(n_i - 1)lns_i^2\right] \tag{2}$$

where s_i^2 is the unbiased estimator of variance in all populations:

$$s_i^2 = \frac{1}{n_i - 1}\sum\nolimits_{i=1}^{n_i}(x_{ij} - \overline{x}_i)^2 \tag{3}$$

In turn s^2 is the total variance:

$$s^2 = \frac{1}{n-k}\sum\nolimits_{i=1}^{k}(n_i - 1)s_i^2 = \frac{1}{n-k}\sum\nolimits_{i=1}^{k}\sum\nolimits_{i=1}^{n_i}(x_{ij} - \overline{x}_i)^2 \tag{4}$$

while c denotes the correction:

$$c = 1 + \frac{1}{3(k-1)}(\sum\nolimits_{i=1}^{k}\frac{1}{n_i - 1} - \frac{1}{n-k}) \tag{5}$$

If at the significance level α the empirical value of the statistics K exceeds the critical value K_{α}, then the null hypothesis is rejected.

In light of the above, it is possible to compare groups distinguished by selected factors using the ANOVA analysis of variance, where [10, 12]:

$$H_0 : \sigma_1 = \sigma_2 = \cdots = \sigma_j = \cdots = \sigma_k$$

$$H_1 : \exists j_1 j_2 : \sigma_{j1} \neq \sigma_{j2}$$

This analysis uses the F Snedecor test. Let σ_1^2 and σ_2^2 be independent mean squares from normally distributed populations. If the hypothesis H0: $\sigma_1^2 = \sigma_2^2$ holds true, then the statistics:

$$F = \frac{S_1^2}{S_1^2} \tag{6}$$

where $S_1^2 > S_2^2$ has the F Fisher-Snedecor. If $F_{obl} > F_{\alpha,\mathrm{u,v}}$, then H0: $\sigma_1^2 = \sigma_2^2$ is rejected; the variances in the tested populations are significantly different. If the above condition is not satisfied, then there are no grounds for rejecting the null hypothesis.

3 Verification of the Model

In the analysed enterprise, all production lines have been fully automated, which means that all activities related to the production process are performed by machines, and the human factor is limited solely to the control of the production and supply of appropriate components for the correct course of the process. The analysed line is designed for the production of hygiene and body care products (gels). A scheme of the manufacturing process is shown in Fig. 2.

The production process starts with the bottle feeder (1), then bottles are fed to the pouring and capping machine (2). At this point, the bottle is filled with the appropriate liquid and then the proper cap is attached to the bottle. The product is then scanned and given a corresponding batch label (3). When one of the components of the manufactured product is a tube, the manufactured product is fed to the tube sealer (4) [13].

The product is inspected during each phase – if it does not meet certain quality standards (e.g., the label is damaged, there is too little liquid in the bottle, etc.), it is automatically rejected as it cannot be distributed and sold. If the product meets all requirements, it is formed in the individual case packer (5), then foiled (6) and packed (7). At the very end, the packed products are stacked by the machine on a palletizer (8). Once the correct number of assortment pieces have been stacked, the pallet is picked up by a production worker.

As mentioned, the study examined delays calculated as the percentage difference between the plan and the number of products actually produced. Planned values result from the adopted production assumptions, taking into account market demand, placed customer orders as well as production possibilities. Production is the values reported from the completed production process (actual/empirical values).

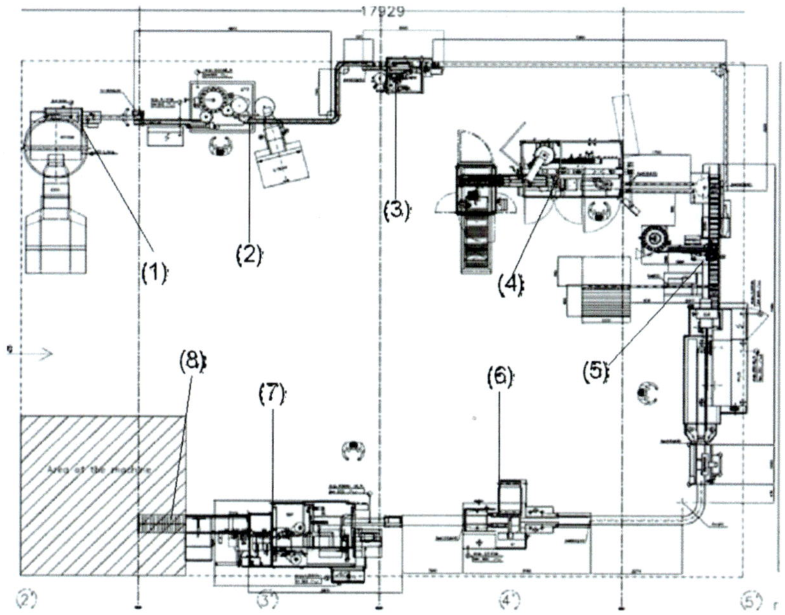

Fig. 2. Scheme of the enterprise's manufacturing process Source: Company's internal material

Table 1. Basic descriptive statistics of the production line under study

	Plan	Production
MIN	0	0
Q1	55.430	27.216
MEDIAN	69.614	72.228
AVERAGE	61.423	71.534
Q3	85.326	110.820
MAX	100.309	204.708

The empirical data were for the years 2018–2020. The impact of selected factors on these differences was analysed. First, the basic descriptive statistics of the plan and production variables for the line under study are presented in Table 1.

The results obtained show differences in the studied period which may result from disruptions during production and affect the stability of the process. The study of factors contributing to delays began by assessing the goodness of fit of the distributions (Table 2). The year, month, day of the week variables, and factors causing delays (machine failure, component shortage, overproduction, delay, other) were analysed. These variables were selected due to the availability of such information only. The studied variables are qualitative data, therefore, to analyse their impact, it is necessary to apply appropriate statistical tests, as presented in detail in chapter one.

Table 2. Results of testing normality of distributions by Lilliefors test

	Variable	Test statistic	p-value		Variable	Test statistic	p-value
Year	2018	0.0403	0.2574	Month	January	0.0831	0.2266
	2019	0.0331	0.6347		February	0.0669	0.566
	2020	0.0793	0.0013		March	0.0909	0.0988
Day of the week	Monday	0.0478	0.7439		April	0.0627	0.6391
	Tuesday	0.0704	0.1618		May	0.0679	0.5543
	Wednesday	0.0724	0.0995		June	0.0974	0.061
	Thursday	0.0742	0.1116		July	0.1016	0.0424
	Friday	0.0422	0.8528		August	0.0815	0.2875
	Saturday	0.0625	0.2764		September	0.0559	0.843
	Sunday	0.0628	0.4788		October	0.0894	0.4722
Factor	Machine failure	0.1502	$1.2 * 10^{-1}$		November	0.0722	0.7292
	Lack	0.1177	$9.4 * 10^{-2}$		December	0.1066	0.2256
	Other	0.2154	$4.1 * 10^{-8}$				
	Overproduction	0.1186	$10 * 10^{-3}$				
	Delay	0.0809	$6.2 * 10^{-2}$				

At the significance level $\alpha = 0, 05$, all distributions follow a normal distribution. Next, Barlett's test was performed to verify the homoscedasticity of variance. The results confirmed the fit of the variance in the year (K = 3.335, df = 2, p-value = 0.189), month (K = 8.729, df = 11, p-value = 0.647), week (K = 5.248, df = 6, p-value = 0.512), and factors (K = 3.939, df = 4, p-value = 0.414) groups. Confirmation of the normality of the distributions and homogeneity of the variances allowed the use of ANOVA analysis of variance, the results of which are presented in Table 3.

Table 3. ANOVA results obtained for the analysed production line

Variable	Number of de gress of freedom	F	p-value
Year	2	0.687	0.504
Month	11	1.12	0.345
Day	6	1.808	0.097
Factors	4	8.13	$2.93\ 10^{-6}$

The result obtained is statistically significant for the adopted level of significance $\alpha = 0.05$ only for the factors variable. The other results are not statistically significant;

the variables do not contribute to delays in production. This signifies the stability of the production line and demonstrates the lack of influence of calendar factors on the differences in planned and realized values. This is particularly important because calendar factors may contribute to the seasonality of production resulting not only from changing demand but also from, for example, holiday and vacation periods when staffing availability is very limited. The last months of each year, when stock-taking and settlement for the entire year of operations is carried out, are also conducive to disruptions in the production process. To the extent studied, the enterprise functioned very well under changing conditions by adjusting needs to its capabilities. The problem arises in situations where there are random factors which have a significant impact on delays. The biggest delays occurred in the area of purchasing, supporting processes (related, among others, to changeover and start-up of machines) and production itself. Quality management tools can improve the quality of the production system and enhance the production profile. They allow for supervision and stabilization of production processes, as well as their better understanding through, among others, introduction of a system archiving data from production. Process control is enhanced by tools (e.g., control charts, process diagrams, analysis sheets, Ishikawa diagrams, Pareto-Lorenz diagrams, etc.) that emphasize continuous process improvement, which translates into improved production stability, as well as cost reduction or the reduction or elimination of the need for control in the supply chain. In addition, delays were caused by the uneven operation and efficiency of the line, which involved, in addition to machine failures, their changeover and start-ups. This can be reduced by applying standardization of operations (SMED), or so-called chronoanalysis. The proposed methods and tools can be successfully implemented in the studied enterprise, and their application would allow the avoidance of many additional costs.

4 Conclusions

In this article, based on the example of a manufacturing company, a method of statistical analysis has been selected that allows for the assessment of the stability of production processes. It was assumed that the chosen method would allow not only the assessment of the current state, but also the implementation of solutions allowing the compensation of disturbances in the event of their identification. This hypothesis was confirmed in the course of the study.

This paper presented a novel model for production process stability assessment that makes it possible not only to estimate deviations between planned and actual production, but also to control factors that may cause disturbances. This, in effect, allows for quick and effective assessment of production process stability by eliminating the necessity of detailed analyses of non-destabilizing variables. The achieved advantages of going beyond qualitative analysis can be assumed to provide a measurable assessment of production process stability. The presented study is a preliminary diagnosis of the enterprise and its production processes. The deficiencies found at this stage in further analyses should be subject to detailed control.

References

1. Jasiulewicz-Kaczmarek, M., Antosz, K., Gola, A.: From lean to sustainable manufacturing-an overview. Eur. Res. Stud. **24**, 291–300 (2021)
2. Borucka, A., Grzelak, M.: Application of logistic regression for production machinery efficiency evaluation. Appl. Sci. **9**(22), 4770 (2019)
3. Zaborowski, M.: Data processing in self-controlling enterprise processes. Bull. Pol. Acad. Sci. Tech. Sci. **67**(1), 3–20 (2019). https://doi.org/10.24425/bpas.2019.127333
4. Grzelak, M., Rykała, M.: Modeling the price of electric vehicles as an element of promotion of environmental safety and climate neutrality: evidence from Poland. Energies **14**(24), 8534 (2021)
5. Stączek, P., Pizoń, J., Danilczuk, W., Gola, A.: A digital twin approach for the improvement of an autonomous mobile robots (AMR's) operating environment—a case study. Sensors **21**(23), 7830 (2021)
6. Bubnicki, Z.: Analysis and decision making in uncertain systems. Springer Science & Business Media (2013)
7. Łampika, Ł, Burduk, A., Chlebus, T.: A model of production process stability measurement and control with use of shewhart control charts. In: Hamrol, A., Grabowska, M., Maletic, D., Woll, R. (eds.) MANUFACTURING 2019. LNME, pp. 218–230. Springer, Cham (2019). https://doi.org/10.1007/978-3-030-17269-5_16
8. Ehrlinger, L., Lettner, Ch., Himmelbauer, J.: Tackling semantic shift in industrial streaming data over time. In: DBKDA, Lisbon, Portugal (2020)
9. Lu, P., Jiang, P.: Sensitivity analysis-based process stability evaluation for one-of-a-kind production. J. Mech. Eng. Sci. **233**(1), 63–77 (2019)
10. Borucka, A.: Forecasting of fire risk with regard to readiness of rescue and fire-fighting vehicles. In: Interdisciplinary Management Research XIV, Croatia, pp. 397–395 (2018)
11. Kozłowski, E., Mazurkiewicz, D., Sęp, J., Żabiński, T.: The use of principal component analysis and logistic regression for cutter state identification. In: Machado, J., Soares, F., Trojanowska, J., Ivanov, V. (eds.) Innovations in Industrial Engineering. Lecture Notes in Mechanical Engineering, pp. 396–405. Springer, Cham (2022). https://doi.org/10.1007/978-3-030-78170-5_34
12. Gołasiewicz, M.: Stability of manufacturing processes as an element of control production system, engineering thesis (2021, unpublished)
13. Kozłowski, E., Borucka, A., Świderski, A., Skoczyński, P.: Classification trees in the assessment of the road-railway accidents mortality. Energies **14**(12), 3462 (2021)

Using Simulation to Cell Design

Bárbara Santos[1], Daniela Almeida[1](✉), Sara Machado[1], Tatiana Guimarães[1], and José Dinis-Carvalho[2]

[1] Integrated Master's in Industrial Management and Engineering, School of Engineering, University of Minho, Guimarães, Portugal
daniela_almeida97@hotmail.com

[2] Production and Systems Department, School of Engineering, University of Minho, Guimarães, Portugal
dinis@dps.uminho.pt

Abstract. With the popularization of Lean Thinking and Continuous Improvement, there is an increasing motivation for companies to find more efficient ways of producing. However, it is common for companies to test improvements in their processes in a real context, which implies stopping production or allocating extra resources. These experiments can be easily tested using simulation software, which allows the user to compare a large number of variables acting on the assembly cells and how they will influence production results. This way, less time and money are spent in searching for the best solution. Therefore, this article intends to emphasize the importance of this type of software with an example of its application in a project developed in an electrical devices production industry. The main goal of the project was to through simulation models find ways to increase the production capacity of an assembly cell, taking into consideration the variation in demand as well as in the number of workers available.

Keywords: Simulation · Lean production · Cell design · Standard work

1 Introduction

This paper was written in the context of the Integrated Project in Industrial Management and Engineering II, one of the subjects of Integrated Master's in Industrial Management and Engineering at University of Minho, in the north of Portugal. This article presents a case study on a multinational company that produces devices for low voltage electrical installations, more specifically circuit breakers for electrical panels.

The project took place on a specific manufacturing cell in which the main target was to increase its production rate. Furthermore, it was also aimed to increase workers flexibility and versatility and to reduce ergonomic problems associated with the cell's current situation. Throughout this document, it will be demonstrated how company's goals were achieved, what the methods used were and the problems and difficulties found during this study.

Moreover, this article enforces the importance of using simulation software to test different scenarios virtually, using less company's resources. It also shows its usefulness

J. Machado et al. (Eds.): icieng 2022, LNME, pp. 149–161, 2023.
https://doi.org/10.1007/978-3-031-09360-9_13

when dealing with problems which have many variables and parameters that are not easy to face. As so, it is proved that the use of simulation software is a money and time saver method that helps solving this kind of problems.

The structure of the paper has been divided into six sections. The first section presents an introduction on the subject studied. The second section is devoted to a literature review on lean manufacturing, key performance indicators and simulation modeling. The third section exposes the current condition of the cell under analysis. The fourth section presents a proposal development to the problem described. The fifth section reports the software simulation results. Finally, the sixth section exposes some conclusions on the importance of this study and highlights the advantages of the presented proposal.

2 Literature Review

2.1 Lean Manufacturing

Nowadays, with the growth of technology, competitiveness between companies is increasing and arising to a high level. When it comes to manufacturing, it is essential that costs are minimized [1], so that products can be sold by the lowest price. This is only possible if the companies' efficiency is maximized. Lean Production is a way of thinking that has been helping many companies achieving their goals.

Lean concept was mainly born from the Toyota Production System (TPS) [2], which supports that costs can be reduced by eliminating waste [3]. The real capacity of a production line is an equation where waste is added to the actual work [1]. Since waste is an activity that consumes resources but does not create any value [4] and the customer does not pay for it, it must be eliminated. Lean Thinking [4] is a way to fight *muda*[1], by seeking value and doing more with less – less time, less space, among others.

Each day, Lean Production becomes more familiar in manufacturing enterprises. Product diversity is a difficult brain teaser to work with, when it comes to its production. There are several common concepts attached to Lean that help to deal with this problem, such as cellular manufacturing and teamwork. Cellular Manufacturing [5, 6] is one of the best ways to physically arrange the equipment and workstations in the context of manual assembly of small products. This layout has many advantages, such as waste elimination, minimization of operating cycle time and the increase of workflow. These benefits can only be achieved with teamwork, the key that turns any implementation into a success. Standard Work is also an important concept in order to establish normalization in a cell and each work instruction must be brief and clear [1]. All these perceptions are connected and can be found especially in successful multinational corporations.

2.2 Key Performance Indicators and Simulation Modeling

In order to monitor the success of an implementation, it is crucial to use Key Performance Indicators (KPIs). These indicators are critical to measure the performance of an organization and, consequently, its current and future success [7]. To guarantee its efficiency, KPIs must be measured frequently and focused on a specific activity [7]. Also,

[1] Japanese word meaning waste.

it is essential that all the staff involved understands them. Thus, manufacturing KPIs, such as Rework, Inventory KPIs and Capacity Utilization, will help maintain control of the processes and increase the production quality.

Continuous Improvement is a Lean concept that all companies should never forget, since it is the main key to its development. When rethinking a better way to produce a certain product, theoretical concepts and equations are not enough. With the purpose of checking the new idea veracity, it must be put into practice, so that external factors are considered, such as transportation and waiting times. However, companies are not always available to physically rearrange a production cell just to test a new idea, as it consumes time and money.

Simulation Modeling is a good and cheaper way to test solutions and to determine the defined KPIs, especially when they involve many parameters and conditions. The first step is to create a model of the existing reality such as layout, production processes, sequence of operations, number of people, processing times. Then the model is tested with the existing data to check if the results are equivalent to the existing reality and validate the model. Different scenarios can be tested with the validated simulation model to see the effect of different changes in the variables. This is very practical because it allows to test ideas without stopping and changing the shop floor.

SIMIO is the simulation modeling software used in this study and is based on intelligent objects [8]. It allows users to create 2D and 3D models, without programming [8]. As it is a very visual tool, it is possible to see the reactions to change in the system created. For companies to deal with process variability, this software is a powerful, interactive and money saver tool. By using it, it will no longer be necessary to stop the production or allocate extra resources.

3 Problem Description

The following project is integrated in a differential blocks production cell which, for reasons of confidentiality, will be called as Electrical Device 1 (ED1) and Electrical Device 2 (ED2). ED1 represents around 96% of the cell's daily production and ED2 only 4%. Each electrical device could be of a different type.

In the cell there were workstations dedicated to the assembly of ED1, other workstations where prepared to assembly ED1 and ED2 and, finally, some of the workstations were devoted to assembly ED2. Thus, the company proposed that all workstations of both types of devices could be integrated into a new layout.

There was not precise information on the timing of operations related to this second type of device. So, the only requirement concerned to ED2 was to integrate its workstations completely in the new cell's proposal layout, as one of its workstations was outside the current cell's layout. When it comes to ED1, the goal was to not only redesign cells layout but also to adjust cells operating mode in order to achieve the goals mentioned above.

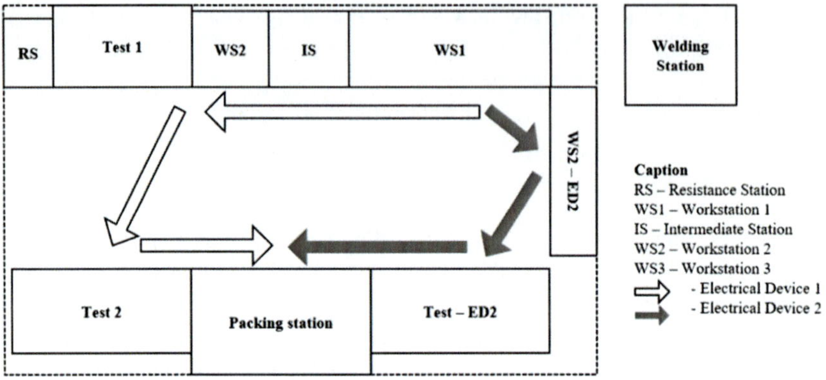

Fig. 1. Current layout

Figure 1 represents the current layout of the mentioned cell as well as the workstation "Welding Station", responsible for producing ED2, that was not part of the cell. In the figure above it is possible to understand the current work sequence of ED1 and ED2.

Regarding the initial situation, some performance indicators were calculated to allow a future comparison with the solution proposed to the company. These indicators are presented in Sect. 4.

Therefore, the cell had a Production Rate of approximately 16 devices per hour, which with two workers means a Productivity of approximately eight devices per Man-Hour as it is expressed in Table 1. As for the Occupation Rate, it has a value of 67%, which reflects that about 33% of the time is wasted.

Table 1. Performance measurements for the current situation (two workers on the cell).

Cycle time	215.92 s
Production Rate	16.67 products/hour
Productivity	8.34 products/hour.men
Occupation rate	67.45%

In other words, although wasted time is not visible in direct observation of the stations, it exists in high values mostly because operators spend a large part of their time in correcting defects instead of producing a larger number of devices that could match the demand.

Given the goals associated with this project, the group focused on the production process of the devices and devoted many hours of study to it, since these were very detailed parts and activities, which required a high degree of attention and precision.

It was during this direct analysis of the workstations that inconsistencies were found with the data presented by the company, such as the number of workers per day present in the cell and the work sequences of each station. The tasks did not have a standardized order of execution. For example, sometimes they followed the production of batches in a part of the device, other times they followed the philosophy of Lot-For-Lot. In addition, the time duration of some operations was out of date. Besides that, there were testing machines in the cell that did not require workers to remain on the machine while they were working. However, their processing time was variable according to the type of electrical device tested. For these reasons, workstations were poorly balanced.

Due to the few certainties associated with the process, there was then a need for a new measurement of times and study of the feasibility of disaggregating some operations. It was in the course of this study that one of the biggest problems found in this cell detected, the high rework with about 40% of the devices produced per day needing repair, which ended up occupying about 9% of the working time, in a shift of 473 min.

As previously mentioned, rework affected the efficiency of the cell and it was a problem not yet diagnosed by the company. Therefore, there was a need of studying the main causes associated to it and find ways to reduce it and possibly eliminate it in the new cell's proposal.

Thus, in order to achieve the main objective defined by the company, to increase daily capacity, there was a need to provide a solution that would be able to deal not only with the problem of rework, but also with the remaining cells' issues, such of ergonomic problems associated with continuous work in sitting positions.

4 Proposal Development

As it was mentioned before, in order to achieve a better work balance in the cell, it was made an exhaustive study of all the operations necessary to assembly the product. With this step it was possible to break down existing operations into operations with shorter duration. For that, a study of the times of the operations was carried out so that they were updated.

Subsequently, work areas were balanced (Fig. 2) in order to satisfy the demand of the cell which was 150 products. Three work areas came up by balancing workstations. The term Work Area (WA) was used to represent a group of tasks that were possible to execute until *Takt Time*[2] was reached. In this balance, it was ensured that the worker that was accountable for the automatic testing machines had other tasks to perform while these automatic tests were occurring.

Having into consideration the demand mentioned above and available time in a working day, the following performance measurements were determined (Table 2).

One of the company's requirements for this cell was the possibility to adjust the cell's production accordingly to demand. For that reason, the next step was to find out what the maximum productive capacity of the cell was, theoretically. By increasing cells demand to its maximum and having into consideration this line's balancing and the available time in a working day, it was determined that, at most, the cell could respond to a demand of 200 products per day (Fig. 3).

The goal of this study was to provide the company the possibility to produce exactly what was necessary when it was needed by implementing a different operating mode.

[2] Takt Time sets pace on production since it establishes the speed at which products should be produced. As so, takt time is given by the time available divided by demand in a time period [9].

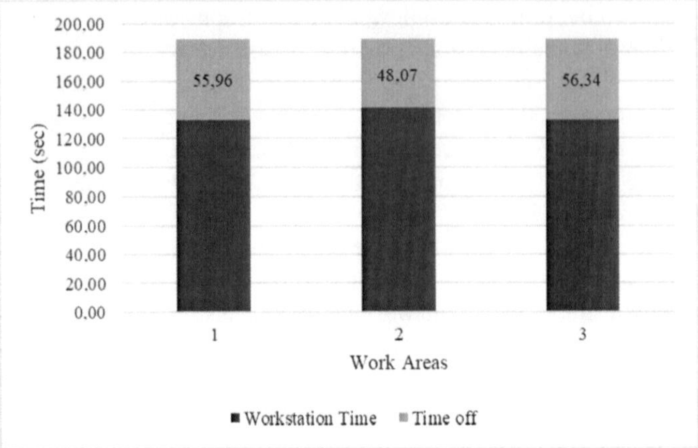

Fig. 2. Work areas balancing for a demand of 150 products.

Table 2. Performance measurements for a demand of 150 products.

Cycle time	141.13 s
Production rate	25.51 products/hour
Productivity	8.50 products/man.hour
Occupation rate	71.75%
Efficiency	96.18%

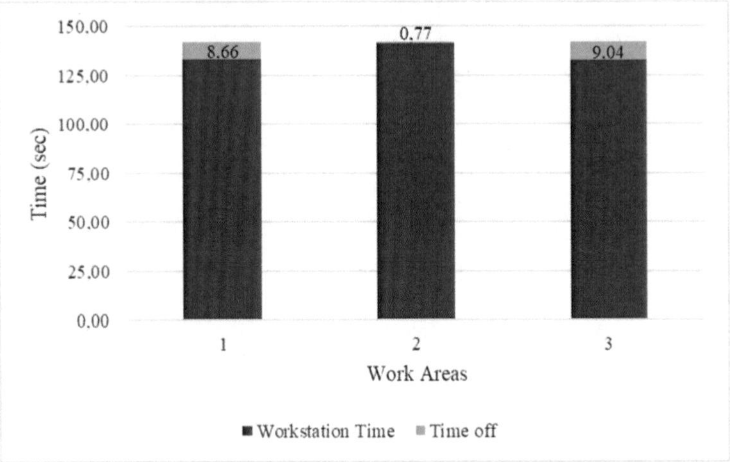

Fig. 3. Work areas balancing for a demand of 200 products.

As so, the company would be able to produce more than 150 products in one day and less than 150 products the following day instead of always producing an average of 150 products.

As shown in Table 3, occupation rate increased about 24%. However, the performance measurements determined are theoretical since they do not consider transportation and waiting times and the workers' movements, which affect the actual cell's production. To face these requirements, it was presented an alternative that is explained below.

Table 3. Performance measurements for a demand of 200 products.

Cycle time	141.13 s
Production rate	25.51 products/hour
Productivity	8.50 products/man.hour
Occupation rate	95.66%
Efficiency	96.18%

4.1 Operating Mode

Afterwards, and to make it possible to produce a variable quantity of products each day, it was necessary to understand how workers would operate in this cell. Since two of the problems identified were the existence of defects and over processing, it is proposed that workers would follow the product throughout the production process. Subsequently, workers are responsible for the products they have assembled, which would reduce the impact of the problems stated above in daily production, namely the rework. As workers follow their products, this operating mode allows workers to be introduced or removed from the cell as required by the company. In this way, the company can adapt the number of workers in the cell to according to demand. In addition, with this operating mode, workers develop a high adaptability, which reduces the impact of absenteeism and increases their independence in the execution of their tasks.

Since company has a Kanban system of 12 products, the worker assembles batches of 12 devices in each work area. However, in WA3 there are two automatic test machines where workers only need to load and unload the products once test machines test them automatically. Since Test Machine One has a limit of two devices to be tested and Test Machine Two has a limit of four devices, it was defined that within WA3 workers can operate a maximum of two products at a time.

4.2 Layout

Since the proposed operations' balancing brought changes in the workstations that previously existed, it was necessary to reconfigure the layout of this production cell. Thus, it was developed a layout that would minimize transportation and the workers' movements between workstations, namely in the WA3 because it is in this area that workers perform more movements. At this stage, the integration of the workstations of ED2 produced in this cell was also taken into consideration, as required by the company. That said, the following layout was proposed.

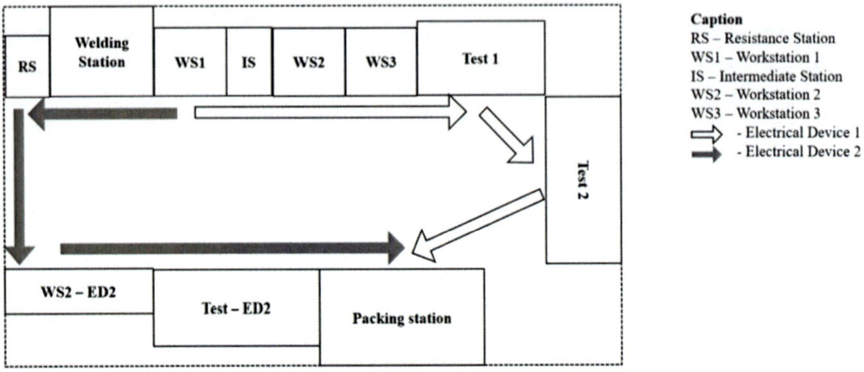

Fig. 4. Proposed layout

As it is possible to see in Fig. 4, the cell's layout includes all workstations of both device types.

In order to quantify the influence that movements have on the production process and to obtain more viable performance measurements, simulation software was used.

The software used was SIMIO 11.0, in which processes were modeled according to the specifications of this project regarding the workflow and its constraints, processing times, cell's capacity, among other properties (Fig. 5).

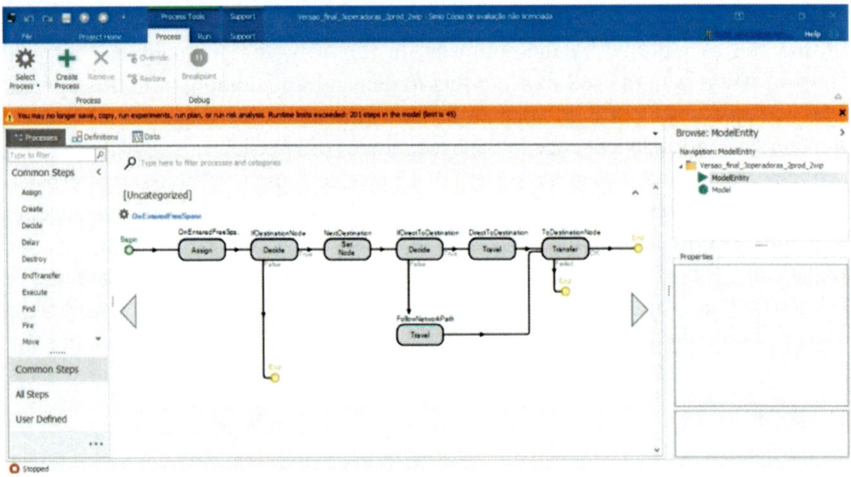

Fig. 5. Process modeling on SIMIO 11.0

Thus, the dimensions of the workstations, the dimensions of the cell, an average displacement speed of workers of 1m/s and the execution time of each operation were considered.

In addition, using this software allowed to test different scenarios. Therefore, it was possible to test the quantity of products that the worker would operate at a time in WA3. That is, the possibility of producing one or two products at a time between workstations in WA3 was evaluated.

By using simulation, it was possible to analyze how the number of workers in the cell would affect daily production (Fig. 6).

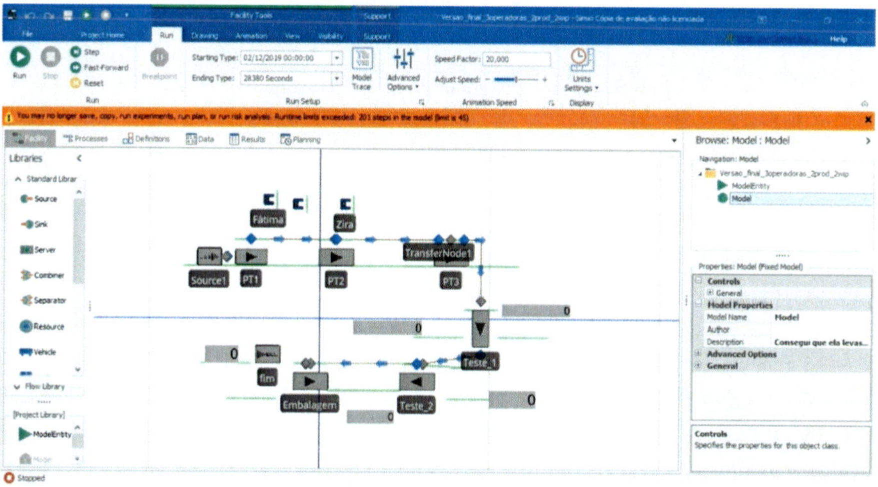

Fig. 6. Facility view on SIMIO 11.0 for a 3-worker simulation

Moreover, with the software it was also possible to understand how the cell's production was affected when more workers were added to it. This is because, although the three work areas are balanced, there is an inevitable slight discrepancy between them. That said, six different combinations were tested: the possibility of operating one product at a time for one, two or three workers and the possibility of operating a batch of two products at a time for one, two or three workers.

5 Results

As was mentioned in the previous section, by using SIMIO 11.0 it was possible to get the present results for different scenarios, under different conditions, as shown in Table 4 and Table 5.

Table 4. Scenario 1: one-unit transportation in WA3

Number of workers	Number of devices produced	Travelled distance
1	62 devices	515 m
2	123 devices	538 m
3	141 devices	433 m

Table 5. Scenario 2: two-unit transportation in WA3

Number of workers	Number of devices produced	Travelled distance
1	64 devices	289 m
2	126 devices	272 m
3	178 devices	423 m

As it can be seen in both tables, when the worker uses a Two-Unit batch, the production level is higher in the three different conditions: for either one, two or three workers, a greater number of devices were produced. On the other hand, also for the Two-Unit batch, the worker always travels a shorter distance.

Therefore, it was decided that the most beneficial scenario would be the worker transporting a batch of two products, among WA3. Thus, in this work area the worker performs standardized tasks that were established having into consideration the two automatic machines' test duration. This way, waiting periods are minimized and the rate of labor utilization is maximized. This standardized work sequence is described in Fig. 7.

The use of simulation was also crucial to realize that when there is more than one worker on the cell, the worker must leave four products in the automatic Test Machine Two that will be completed by the coworker who succeeds him. Thus, in this operation mode, the workers will all start a batch of 12 devices but four of their devices will be packed (last workstation in the WA3) by the coworker who succeeds him. This happens in order to minimize waiting periods caused by automatic machines.

In conclusion, when comparing these results with the initial cell information, where there was a daily production of 100 devices with two workers, it is possible to affirm that this transformation will increase production by 26%.

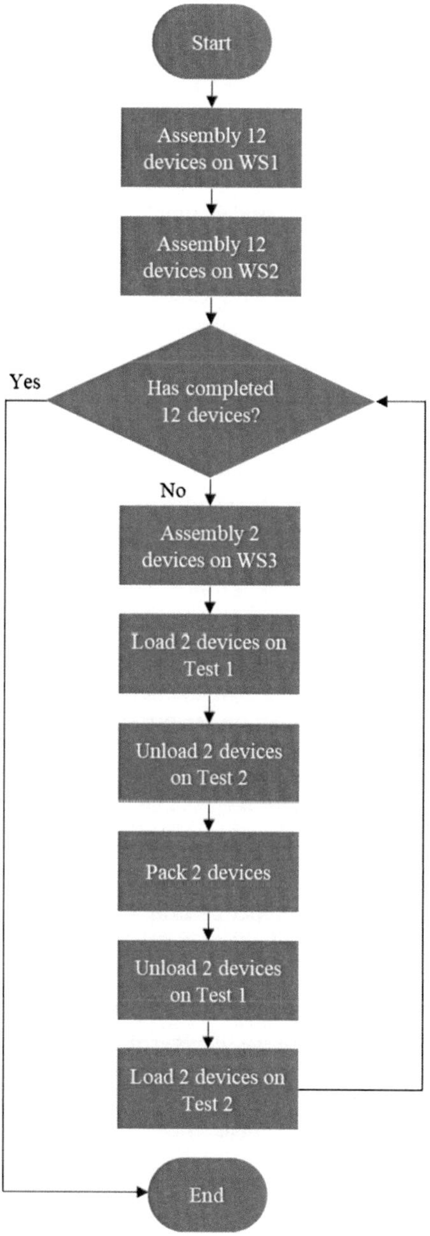

Fig. 7. Process flowchart for work sequence standardization.

6 Conclusion

Analyzing the complete proposal, it is possible to highlight some advantages that its implementation can bring to the mentioned company. Depending on the demand, the production manager can easily introduce or remove workers to reach the target values for each day. Besides, the adoption of a Standard Work will reduce the clutter between workers and decrease the WIP previously accumulated between all workstations. Also, with the introduction of the new operating mode, where workers carry out all the operations of their product sequentially, not only will greater versatility and stimulation of the workers' capacities be achieved, but also a reduction in rework and an increase in product quality. Finally, with the replacement of sitting positions for standing and sitting positions, it is possible to prevent some ergonomic problems, one of the company's goals already mentioned in Sect. 3.

However, it is necessary to take into account the challenges that this implementation can bring. With the introduction of a different operating mode, it will be necessary to invest in new training, since not all workers are trained to carry out every task of the process. Also, it is necessary to increase the travelled distance, since in the initial situation workers remained in the same workstation during the shift and in the new layout, they will have to go through the entire cell. All these changes can cause some difficulties for workers to adapt, which can translate into resistance to change, a matter that the company will have to be prepared to deal with and manage.

The goal of increasing the capacity in a cell with several characteristics in the production process, such as the time spent on rework or the need to introduce or remove workers according to demand levels, made it difficult to choose a single solution. The choice of the most suitable solution for the intended objective was possible to select only by the utilization of software.

Thus, it is understood that to reach the most efficient solution, it is important to consider the various combinations of variables, a task that proves to be quite complicated when it is not possible to test the scenarios in a real context. And that is where the importance of simulation tools comes into the whole process of designing production cells. Therefore, it is essential to invest more in the development of software that allows this application to be easier and more intuitive, so that companies do not have to spend resources testing the various possibilities in a real situation.

Acknowledgments. This work has been supported by FCT – Fundação para a Ciência e Tecnologia within the R&D Units Project Scope: UIDB/00319/2020.

References

1. Ohno, T.: Toyota Production System. Productivity Press (1988)
2. Womack, J.P., Jones, D.T., Roos, D.: The machine that changed the world. Bus. Horiz. (1992). https://doi.org/10.1016/0007-6813(92)90074-J
3. Sugimori, Y., Kusunoki, K., Cho, F., Uchikawa, S.: Toyota production system and Kanban system materialization of just-in-time and respect-for-human system. Int. J. Prod. Res. (1977). https://doi.org/10.1080/00207547708943149

4. Womack, J.P., Jones, D.T.: Lean thinking—banish waste and create wealth in your corporation. J. Oper. Res. Soc. **48**, 1148 (1997). https://doi.org/10.1038/sj.jors.2600967
5. Reorganizing the factory: competing through cellular manufacturing. J. Manuf. Syst. (2002). https://doi.org/10.1016/s0278-6125(02)80175-3
6. Dinis-Carvalho, J., Alves, A.C., Sousa, R.M.: Moving from job-shop to production cells without losing flexibility: a case study from the wooden frames industry. S. Afr. J. Ind. Eng. (2014). https://doi.org/10.7166/25-3-864
7. Parmenter, D.: Key Performance Indicators (KPI) (2010)
8. Simio: Simulation, Production Planning and Scheduling Software. Simio
9. Linck, J., Cochran, D.S.: The importance of takt time in manufacturing system design. SAE Tech. Pap. (1999). https://doi.org/10.4271/1999-01-1635

Sectorization of a Parcel Delivery Service

Mafalda Mostardinha[1], Pablo Escobar[2], Cristina Lopes[3,5], and Ana Maria Rodrigues[3,4](✉)

[1] FCUP, U.PORTO, Porto, Portugal
up201606213@fc.up.pt
[2] Facultad de Ciencias Matemáticas, U. VALENCIA, Valencia, Spain
paesher@alumni.uv.es
[3] CEOS.PP, ISCAP, Polytechnic of Porto, Porto, Portugal
cristinalopes@iscap.ipp.pt
[4] INESC TEC, Porto, Portugal
ana.m.rodrigues@inesctec.pt
[5] LEMA - Laboratório de Engenharia Matemática, Porto, Portugal

Abstract. This paper explores the problem of sectorization of a parcel delivery service that wants to assign an action region to each of its teams, regarding the number of deliveries scheduled for each zone, so that there is a balanced service amongst sectors, covering contiguous zones, and considering limited capacities for the teams. Besides being relatively easy to model, the available optimization tools and software provide poor results when dimension increases in these types of problems, with computational capacity exceeding. In this paper an integer programming model, combined with an heuristic to return a faster solution, was implemented to solve a sectorization problem in two different situations. The main advantage of the strategy proposed, compared to previous ones, is its simplicity and easy implementation while still returning an optimal solution.

Keywords: Sectorization · Heuristics · Integer programming

1 Introduction

The problem of sectorization, also known as the territory design problem, arises when a geographic space that was previously divided in basic units (BUs), needs to be arranged into several regions, or sectors, in a way that the grouping of the BUs is as efficient as possible according to relevant planning criteria. Initially, the sectorization problem arised to deal with the legislative districting problem [1,2], which consisted in subdividing a state into a specific number of compact and contiguous districts of nearly equal population. Nowadays sectorization has a wide range of applications that vary among sales territory design, power districting and public services such as school districting [3,4], water distribution networks [5], solid waste collection [6], healthcare districting optimization [7] or electrical power distribution [8].

J. Machado et al. (Eds.): icieng 2022, LNME, pp. 162–174, 2023.
https://doi.org/10.1007/978-3-031-09360-9_14

This paper was developed as a follow-up to the VII Iberian Modeling Week and focus on the sectorization of a parcel delivery service. Let's consider the problem in which a parcel delivery service wants to distribute an action region among its teams. For that purpose, the action region is split into several basic units, according to their postal code. Territories have to be designed in order to minimize the total travelled distance while balancing workload of teams. On the other hand, a team should not be required to cross another team's territory in the course of its route, which means that some kind of contiguity must be ensured between the constituent units of the same sector.

The BUs considered initially are rectangle-shaped, and the action region is bidimensional, which makes the problem much simpler but far less realistic. Following this initial example the heuristic approach was tested with real geographical data. In order to establish which teams should cover each sector these are the factors that need to be taken into account:

- All the teams should have at least one basic unit assigned.
- All the teams should have a similar amount of total deliveries assigned, also known as workload (**balance criterion**).
- All the sectors assigned to a particular team should be compact zones in order to optimize delivery time (**compactness criterion**).
- There is a limited capacity in the workload that each team can ensure.
- Each basic unit has an amount of deliveries to be fulfilled.

The paper is organized as follows. In Sect. 2 the mathematical problem is formally described, with a full description of all parameters, decision variables and constraints. Then, Sect. 3 proposes an algorithm to reduce the number of possible solutions tested, dealing with the overall computational problem. Next, in Sect. 4 two practical examples are tested, for both rectangle-shape data and the geographical data, and the solutions are evaluated. Finally, Sect. 5 gathers some final remarks and future work ideas.

2 Integer Programming Model

There is a total of n Basic Units (BUs) to be distributed among k teams. Let $i = 1, ..., n$ be the index for BUs and $j = 1, ..., k$ the index for teams. The number of deliveries scheduled for each unit i, w_i, is given *a priori*, as well as each team's maximum working capacity, m_j, and the location of each BU in the bidimensional geographic coordinate system - latitude and longitude.

2.1 Measuring Distance

It is important to define a measure to determine how distant each BU is from all the others. There are several ways to do this, but for this paper each BU will be represented by a single point, that corresponds to its centroid, and euclidean distance is used to calculate distance between each pair of BUs.

Thus, it is possible to compute the distance matrix $(d_{il}) \in M_{n\times n}(\mathbb{R})$, where d_{il} represents the distance from BU i to BU l, for $i, l \in \{1, ..., n\}$. The symmetry property allows to restrict the computation only to values on the lower triangular half of the matrix, and because the diagonal only has zeros, there is a total of $\frac{n(n-1)}{2}$ entries to calculate, instead of n^2.

The euclidean distances between centroids are measured with GPS coordinates. For that reason the distance values themselves don't make any real sense, because latitude and longitude are not coordinates in a Cartesian coordinate system - however, they are used for comparison purposes.

2.2 Decision Variables

The decision variables for this problem are binary and provide information regarding the attribution of BUs to each team.

$$x_{ij} = \begin{cases} 1 & \text{if BU i is assigned to team j} \\ 0 & \text{otherwise} \end{cases} \qquad i = 1, ..., n; j = 1, ..., k \tag{1}$$

2.3 Constraints

After defining all the model's parameters and decision variables, it is time to write the constraints. All the constraints for this model are linear.

- **Exclusivity**
  Each unit must be assigned to one (and only one) team.

$$\sum_j x_{ij} = 1, \forall i \in \{1, ..., n\} \tag{2}$$

- **Capacity**
  The total number of deliveries must not exceed the capacity of each team.

$$\sum_i x_{ij} w_i \leq m_j, \forall j \in \{1, ..., k\} \tag{3}$$

- **No empty sectors**
  Each team is assigned to at least one BU.

$$\sum_i x_{ij} \geq 1, \forall j \in \{1, ..., k\} \tag{4}$$

- **Balance**
  To assure that the workload is balanced among teams because it makes no sense that one of the teams is almost at its maximum capacity of daily deliveries while another one has very few. The total workload is $W = \sum_i w_i$, which gives an average of W/k deliveries/sector. To ensure that for each team there is a leveled number of deliveries, this number is restricted to an interval. If α

is set to 0.1, for example, it is guaranteed that the total number of deliveries for each team does not deviate more than 10% from the average workload.

$$\frac{W}{k}(1-\alpha) \leq \sum_{i} x_{ij} w_i \leq \frac{W}{k}(1+\alpha) \quad ,\forall j \in \{1,...,k\} \tag{5}$$

Note that when a team's capacity m_j, $j \in \{1,...,k\}$ is greater than $\frac{W}{k}(1+\alpha)$, this constraint makes the capacity constraint redundant.

2.4 Objective Function - Compacity

Balance in the number of deliveries is not the only important concern to look out for when designing the model. Compacity is an equally important characteristic, since there should not be any BUs isolated from the other BUs of the same sector, and also because BUs in the same sector should not be too far apart. Figure 1 shows two ways of partitioning the same region into 6 sectors. In Fig. 1(a) the BUs in each sector are contiguous, but in Fig. 1(b) if the team wants to leave BU43, for example, it necessarily needs to cross another team's territory.

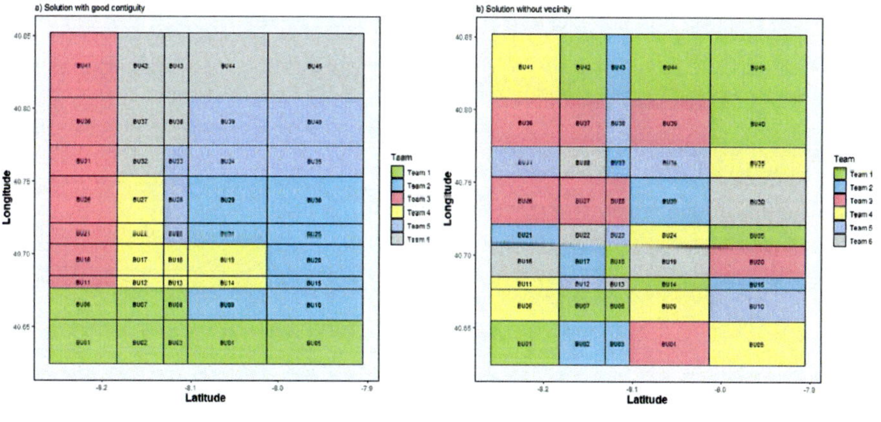

Fig. 1. A solution with good contiguity (a) and a solution without vicinity (b).

One way to try to achieve compacity is to minimize the total distances between blocks in the same sector, and following this reasoning the non linear objective function is given by:

$$\min \sum_{j} \sum_{i=1}^{n-1} \sum_{l=i+1}^{n} x_{ij} x_{lj} d_{il} \tag{6}$$

2.5 Nonlinear Optimization Model

The nonlinear integer programming model is the following:

Minimize:

$$\sum_{j}\sum_{i=1}^{n-1}\sum_{l=i+1}^{n} x_{ij}x_{lj}d_{il} \tag{7}$$

Subject to:

$$\sum_{j} x_{ij} = 1, \forall i \in \{1, ..., n\}$$

$$\sum_{i} x_{ij} \geq 1, \forall j \in \{1, ..., k\}$$

$$\sum_{i} x_{ij}w_i \leq m_j, \forall j \in \{1, ..., k\}$$

$$\sum_{i} x_{ij}w_i \leq \frac{W}{k}(1+\alpha), \forall j \in \{1, ..., k\}$$

$$\sum_{i} x_{ij}w_i \geq \frac{W}{k}(1-\alpha), \forall j \in \{1, ..., k\}$$

$$x_{ij} \in \{0, 1\};\ i = 1, ..., n;\ j = 1, .., k$$

With the current model, there are nk decision variables, $n + 4k$ restrictions and $\frac{n(n-1)}{2} + n + k$ nonzero parameters.

2.6 Linear Model

It is possible to transform the current model into a linear model by adding extra variables y_{ilj} $y_{ilj} =$

$$\begin{cases} 1 & \text{if BUs i and l are in team j} \\ 0 & \text{otherwise} \end{cases} = \begin{cases} 1 \text{ if} & x_{ij} = x_{lj} = 1 \\ 0 & \text{otherwise} \end{cases}, i, l = 1, ..., n; j = 1, ..., k \tag{8}$$

In this case the objective function would be:

$$\min \sum_{j}\sum_{i=1}^{n-1}\sum_{l=i+1}^{n} y_{ilj}d_{il} \tag{9}$$

Additional restrictions can now be added in order to establish the relationship between variables x and y:

$$x_{ij} + x_{lj} - 1 \leq y_{ilj} \tag{10}$$

With the linear model approach the problem's computational complexity increases even further. In fact, this model has n^2k more variables than the nonlinear one. Due to the additional decision variables and constraints the computer couldn't run the model in a reasonable amount of time, therefore it is not viable to choose this model over the nonlinear one.

3 Heuristic

Nonlinear methods often return an optimal solution. Unfortunately, due to the combinatorial complexity of the nonlinear model, there is a high computational cost, and so the practical use of these methods is limited.

Because of the problem's dimension, commercial softwares take a long time to run, as will be seen in Sect. 4. In fact, for each possible sector design, the computer tests $k!$ solutions, corresponding to all the possible assignments of sectors to teams. This is insignificant for the problem at hand - the order of the teams in the sectors doesn't matter.

In order to reduce the number of possible solutions tested by the software we propose applying a heuristic strategy before running the nonlinear model. The heuristic consists in assigning previously different BUs to some teams so that the number of possible solutions tested by the software is considerably lower and it doesn't evaluate redundant solutions.

3.1 Distance

To spare some computational time, one could previously assign the two BUs that are farthest apart to two different teams, because it does not seem reasonable that both of them belong in the same sector. The farthest points between a group of points are $I, J \in \{1, ..., n\}$ such that:

$$d_{IJ} = d_{\max} = \max_{i,j=1,\ldots,n} d_{ij} \tag{11}$$

After finding I and J we would add the following restrictions to the optimization problem: $x_{I1} = 1$ and $x_{J2} = 1$.

In small instance problems this may be enough to find a satisfactory solution in a reasonable amount of time. However, as n and k increase, some extra manipulation is required. The problem of finding k points from a point cloud, so that they are as far possible apart from each other can be solved with a technique called *Farthest Point Sampling* (FPS) [9] and the algorithm is very simple:

1. Choose one random starting point and add it to the set of solutions S.
2. Find the farthest point to all points in S from the remaining set of points.
3. Do this until k points are selected.

3.2 Concentration

To ensure that the sectors are balanced, the algorithm should not be based in distance alone. Let A_i be the area of BU i. Then, the corresponding work concentration is $c_i = \frac{w_i}{A_i}$. The k points previously fixed before running the algorithm should not only be far apart from each other, but also have a high work concentration to ensure work balance.

The procedure proposed in this paper is to create a subset with the points that have the higher concentration values, say the top 40%, and apply the FPS only to that set of points. Furthermore, the starting point for the algorithm will be the point with the higher concentration value, instead of being chosen at random. The FPS modified algorithm is:

1. Let T_{40} be the set of the top 40% points in 1,...,n in respect to c_i and $j := 1$.
2. Choose as starting point the $i \in T_{40}$ with maximum c_i and add it to the set of solutions S, let $x_{ij} = 1$. Remove point i from T_{40}.
3. Let $j := j + 1$. Select the point $i \in T_{40}$ that is farthest from all elements in S, and add the constraint $x_{ij} = 1$ to the optimization model. Remove point i from T_{40}.
4. Repeat step 3 until $\#S = k$.

4 Application Examples

To test the proposed strategy, two different data sets were used: first, a set of 45 rectangle-shaped data with simulated workload values and 6 teams responsible for the delivery service of the area. Then we introduced real geographical data, again with 45 BUs with different simulated workload values and 6 teams.

The implementations that follow were performed in a 12 GB RAM computer, with a Intel(R) Core(TM) 7-9750H CPU 2.60 GHz processor, using RStudio software [10] for the heuristic implementation, and AMPL software [11] to solve the optimization problem.

4.1 Example with Rectangle-Shaped BUs

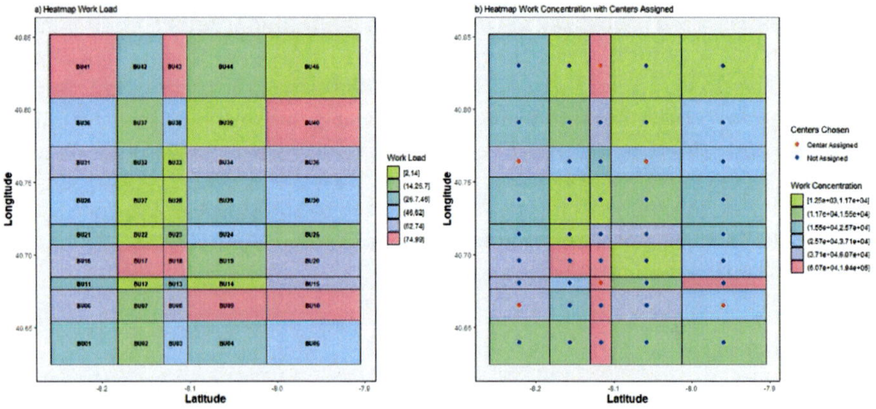

Fig. 2. (a) Workload (on the left) versus (b) concentration (on the right) for each BU with centroids chosen in red

Let's consider an example where $n = 45$ BUs need to be assigned to $k = 6$ teams. Along with geographical coordinates, the capacity $m_j = 500$ of each team is given, as well as the demand/workload, w_i, for each BU. It is thus possible to compute the concentrations c_i from the given data. Figure 2 shows the shape and position of all BUs, colored from green to purple according to it's workload and concentration (left and right heatmaps, resp.)

The average that comes from W/k is 337.5 deliveries/sector. Considering $\alpha = 0.1$ (10%) then $m_j > \frac{W}{k}(1+\alpha), \forall j \in \{1, ..., k\}$, which means that the capacity constraint is unnecessary.

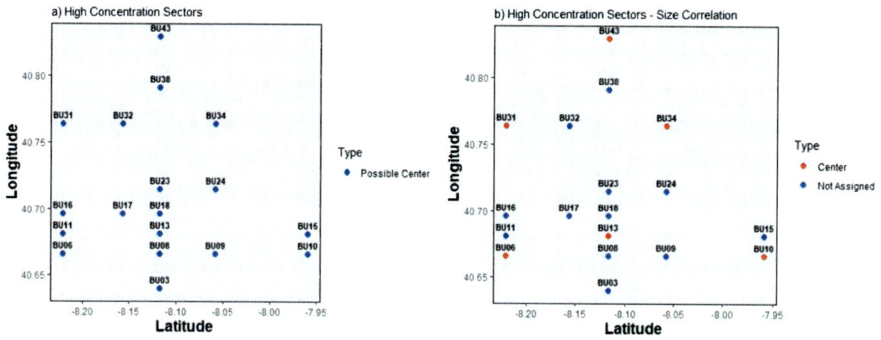

Fig. 3. (a) On the left, points in T_{40} - top 40 % work concentrations. (b) On the right, the size of each point $i \in T_{40}$ is proportional to c_i.

Let's apply the heuristic. After reducing each BU to its centroid, we get a set of 45 points and the distance matrix $(d_{il}) \in M_{45\times 45}(\mathbb{R})$. The subset T_{40} has the 18 points with higher concentrations, and the starting point for the algorithm is BU13 (see Fig. 3) which has the maximum c_i for $i \in T_{40}$. The FPS technique is now applied to this set of points, with $k = 6$. The farthest points that resulted from the implementation of this heuristic are, in descending order, BUs 13, 10, 43, 31, 06 and 35, as can be seen in Fig. 2(b) in red. Finally, we solve the optimization model as described in the previous section, with $n = 45$, $k = 6$, $\alpha = 0.1$ and $S = \{6, 10, 13, 31, 34, 43\}$. All results were obtained using AMPL software and are presented in Fig. 4. Four different variations were tested for the same problem:

- Nonlinear integer programming model
  As expected, the nonlinear model provides the optimal solution (Fig. 4(a)) - no assumptions are made, which turns out to be the same solution as the one obtained by fixing BUs 1 and 45 (Fig. 4(d)). Moreover, the nonlinear algorithm took more than 8 h to run, which is a long time. It is clear that with bigger instance problems the time due to find an optimal solution is not viable, and therefore there's a need for an heuristic.
- Model with FPS implementation
  After determining which BUs are in T_{40} two different approaches were tested. First (Fig. 4(c)) FPS was performed in T_{40} with BU13 as starting point, and then (Fig. 4(b)) FPS was performed in T_{40} with a random starting point.

The distribution of BUs amongst sectors is very similar, and the only differences are in the attributions of BUs 7 and 38. However, standard deviations for workloads are approximately 22.8 and 39.5 for each method (respectively) which means that the sectors are more balanced in the first case, even if the objective function, compacity, is slightly worse.

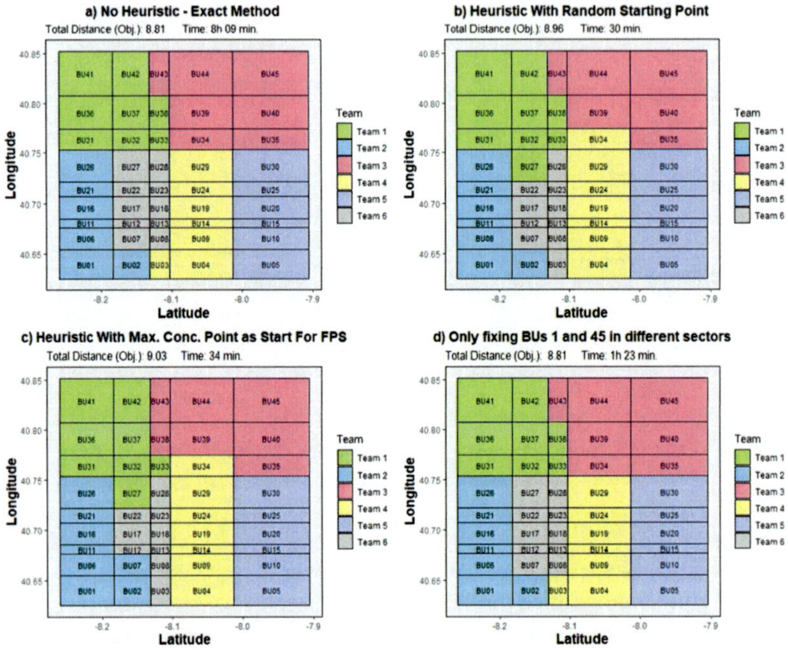

Fig. 4. Results for the problem using the nonlinear integer programming model without any heuristic (a), with usual FPS (b), with the FPS crossed with concentration info (c), and fixing only the two farthest BUs 1 and 45 (d).

- Model with only two additional restrictions
  It seems reasonable to assume that at least BUs 1 and 45 are assigned to different sectors (Fig. 4(d)). In fact, for each possible sector design, the computer tests $6! = 720$ solutions, corresponding to all the possible assignments of sectors to teams. This is insignificant for this problem - the order of the teams in the sectors does not matter. By fixing these two BUs, some unnecessary calculations are spared. Because it has less assumptions, in this case this heuristic's results are in fact the optimal solution, but it still takes a long time to run, which would increase as the problem's dimension increases.

For this application in particular, the use of an heuristic has decreased computational time by far as stated in Fig. 4, and the results are quite similar to the optimal solution. To get faster results, one could try to decrease the value of α, thus restricting the condition for the total labor equilibrium of each sector.

4.2 Example with Real Geographical Data: The Murcia Implementation

In order to test the algorithm's performance in a more realistic set, let's consider the south-eastern spanish region of Murcia. This region, much like the previous example, integrates 45 municipalities, or BUs, but this time they have very different shapes and areas. The same number of 6 sectors is considered, and demand for each BU was randomly generated with values from a uniform distribution in [1, 99]. The values for both workload and concentration of workload are represented in Fig. 5(a) and (b) respectively.

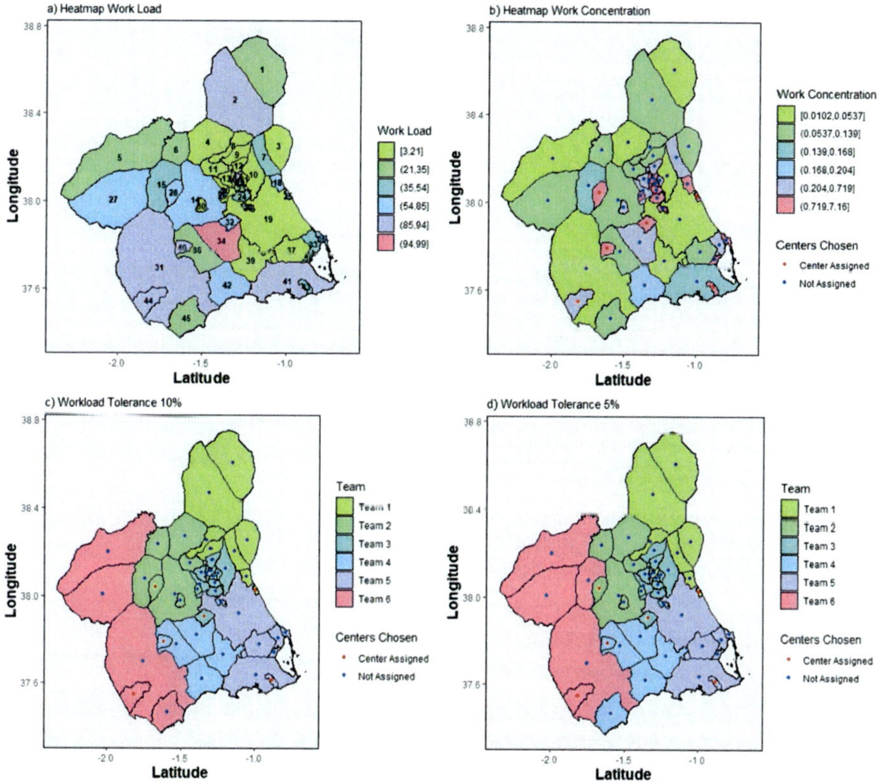

Fig. 5. Workload is represented in (a) while work concentration is represented in (b). Meanwhile, (c) and (d) represent the BUs assigned to each sector for both $\alpha = 0.1$ and $\alpha = 0.05$. In (b), (c) and (d) the centroids chosen for each sector are painted in red.

This time, two values for α were tested - 0.1 and 0.05 - and results are presented in Fig. 5(c) and (d). The first immediate remark to make is that running time has significantly decreased by the use of $\alpha = 0.05$ - 2 min 30 s versus 16 min 41 s. Besides that, it is clear that the model on Fig. 5c) has more balanced sectors, work wise, and the objective function result was approximately the same for both.

However, both models show non-contiguous sectors, i.e., there are some BUs that are not in the same agglomerate as the rest of the BUs in that sector. This is explained by the shape and size of the BUs for this example: some centroids can be closer to the centroid of another BU than to its own actual limits. As areas are reduced to a single point, some important geographic information is ignored. Overall, the algorithm performs well and in a perfectly good time frame (Fig. 6).

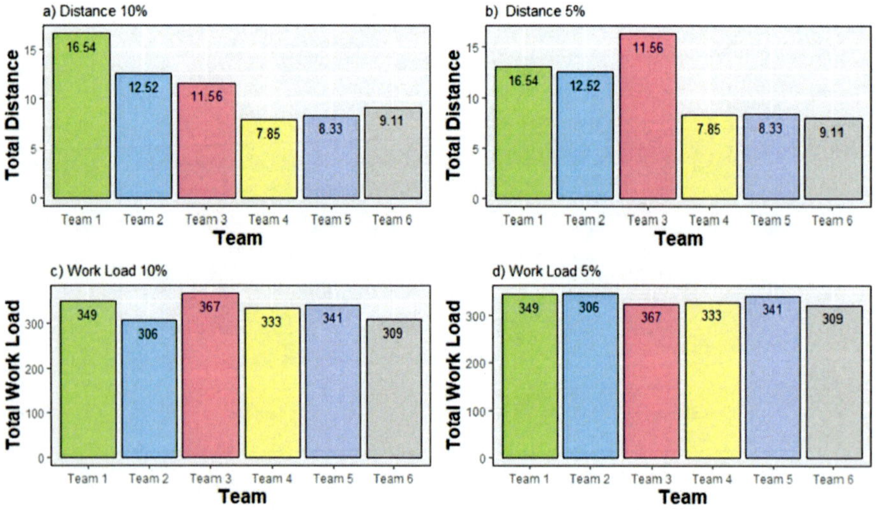

Fig. 6. Differences in total distance [(a) for $\alpha = 0.1$, (b) for $\alpha = 0.05$] and total workload assigned to each team [(c) for $\alpha = 0.1$, (d) for $\alpha = 0.05$].

5 Conclusions and Future Work

Heuristics always make assumptions based in empirical reasoning, and for that reason the optimal solution always assured by nonlinear integer programming models is usually lost. However, in real life situations the time due to solve territory design problems with exact methods is not viable. In this paper we proposed using an FPS algorithm to reduce the number of possible solutions analysed by nonlinear model, therefore reducing the computational cost. In previous papers some better solutions, in terms of balance and total distance for each team, have been proposed, but they are much more complex. In contrast, the strategy proposed in this paper is easy to implement and easy to understand, while still producing great solutions for the example applications.

In applications of very high dimensions, let's say 10 000 BUs to be distributed amongst 50 teams, for example, this heuristic would still not be enough to solve the problem in a reasonable amount of time. As future work, the solution could be to add an extra step after the FPS implementation - to allocate to each of the selected points in S its contiguous neighbours, thereby creating 50 point clusters that would be assigned to the 50 different teams as a starting point to the exact method. We could even follow the same reasoning over and over again with really big dimension problems, and ultimately the optimization model would only serve as criteria for BUs in the agglomerations boundaries.

Finally, the problem that arises in the Murcia implementation regarding non-contiguous sectors could be corrected incorporating an adjacency matrix with a constraint that would force all of the BUs in one sector to be contiguous. Nonetheless, this solution would create a much more restrictive model and the results could be worse in terms of total distance and workload distribution.

Acknowledgments. This work is financed by the ERDF - European Regional Development Fund through the Operational Programme for Competitiveness and Internationalisation - COMPETE 2020 Programme and by National Funds through the Portuguese funding agency, FCT - Fundação para a Ciência e a Tecnologia within project 'POCI-01-0145-FEDER-031671'.

References

1. Hess, S.W., Samuels, S.A.: Experiences with a sales districting model: criteria and implementation. Manag. Sci. **18**(18), 41–54 (1971). https://doi.org/10.1287/mnsc.18.4.P41
2. Ricca, F., Scozzari, A., Simeone, B.: Political districting: from classical models to recent approaches. Ann. Oper. Res. **204**(1), 271–299 (2013). https://doi.org/10.1007/s10479-012-1267-2
3. Caro, F., Shirabe, T., Guignard, M., Weintraub, A.: School redistricting: embedding GIS tools with integer programming. J. Oper. Res. Soc. **55**(8), 836–849 (2004). https://doi.org/10.1057/palgrave.jors.2601729
4. Bouzarth, E.L., Forrester, R., Hutson, K.R., Reddoch, L.: Assigning students to schools to minimize both transportation costs and socioeconomic variation between schools. Socioecon. Plann. Sci. **64**, 1–8 (2018). https://doi.org/10.1016/j.seps.2017.09.001
5. Hajebi, S., Roshani, E., Cardozo, N., Barrett, S., Clarke, A., Clarke, S.: Water distribution network sectorisation using graph theory and many-objective optimisation. J. Hydroinf. **18**(1), 77–95 (2015)
6. Rodrigues, A.M., Ferreira, J.S.: Sectors and routes in solid waste collection. In: Almeida, J.P., Oliveira, J.F., Pinto, A.A. (eds.) Operational Research. CSMS, vol. 4, pp. 353–375. Springer, Cham (2015). https://doi.org/10.1007/978-3-319-20328-7_20
7. Emiliano, W., Telhada, J., do Sameiro Carvalho, M.: Home health care logistics planning: a review and framework. Procedia Manuf. **13**, 948–955 (2017). https://doi.org/10.1016/j.promfg.2017.09.165
8. Bergey, P.K., Ragsdale, C.T., Hoskote, M.: A decision support system for the electrical power districting problem. Decis. Support Syst. **36**(1), 1–17 (2003). https://doi.org/10.1016/S1344-6223(02)00033-0

9. Kamousi, P., Lazard, S., Maheshwari, A., Wuhrer, S.: Analysis of farthest point sampling for approximating geodesics in a graph. Comput. Geom. **57**, 1–7 (2016). https://doi.org/10.1016/j.comgeo.2016.05.005
10. RStudio Team: Rstudio: Integrated development environment for R (2020). http://www.rstudio.com/
11. Fourer, R., Gay, D., Kernighan, B.: AMPL: a modeling language for mathematical programming (2003). https://ampl.com/resources/the-ampl-book/chapter-downloads/

Decision Making Models for Sustainable Supply Chain in Industry 4.0: Opportunities and Future Research Agenda

Divanshu Sharma[1], Anbesh Jamwal[1], Rajeev Agrawal[1](✉), Jinesh Kumar Jain[1], and Jose Machado[2]

[1] Department of Mechanical Engineering, Malaviya National Institute of Technology, J.L.N. Marg, Jaipur, Rajasthan, India
ragrawal.mech@mnit.ac.in

[2] MEtRICs Research Centre, University of Minho, Braga, Portugal

Abstract. Managing supply chain activities have become a critical issue for manufacturing organization due to globalization and challenges faced by sustainability. Performance of supply chain can be evaluated by indicators and other factors. The evaluation of these factors becomes difficult due to lack of benchmarks and evaluation method. The supply chain has wide range of opportunities in the Industry 4.0 which is driven by many key enabling technologies such as: artificial intelligence, machine learning, deep learning and blockchain technology. These technologies have a positive impact on the sustainability performance of the supply chain activities. The decision-making models plays an important role in the evaluation of the supply chain strategies. The present study aims to identify and report the different models used in the sustainable supply chain management. These models are multi-criteria decision-making models, multi-objective-based models, artificial intelligence-based models. Bibliometric analysis is done with the VOSviewer and R studio using the bibliographic data of digital scientific databases. The content analysis is done after the bibliometric analysis based on the cluster analysis done in R studio. The main research areas and future research directions have been reported in the end of study. It is expected that this study can be helpful for the practitioners, policymakers and academia working in the area of sustainable supply chain management.

Keywords: Supply chain management · Sustainability · Decision making models · Industry 4.0 · Meta heuristics

1 Introduction

With globalization, supply chains are becoming more complex and difficult to manage in Industry 4.0. With increasing complexity, the factors affecting the sustainability of a supply chain are also increasing, making it difficult to analyse and quantify their effect on the supply chain [1]. Without, a thorough analysis of these factors the supply chains become susceptible to disruption. Traditional methods are becoming ineffective or very

J. Machado et al. (Eds.): icieng 2022, LNME, pp. 175–185, 2023.
https://doi.org/10.1007/978-3-031-09360-9_15

time inefficient [2]. Moreover, with increasing importance of sustainability aspect in every domain of Industrial management, integration of sustainability with Supply Chain Management has become very difficult for the industry professionals.

A great deal of computational power and non-traditional decision-making approaches are required to efficiently analyse the relative importance of these factors and their effect in the achieving sustainability in supply chain management. Some new methods have surfaced with advancement in Decision making Models and Artificial Intelligence techniques [2]. Multi Criteria Decision Making (MCDM) models and Meta-heuristic Search Algorithms can prove to be tools of great importance for mathematical optimization of the above factors. Although, these are emerging technologies in the field of supply chain management and very less research has been done in application of these technological tools for achieving sustainability in supply chain but they have the potential to transform Sustainable Supply Chain Management (SSCM).

This paper aims to understand the research growth and feasibility of these tools in supply chain management and enhance supply chain sustainability using past published research articles in this field. Since this research area is fairly novel, all research papers from three popular databases i.e. Scopus published in this field are considered. Bibliometric analysis is very helpful in evaluating published articles, books, research papers and chapters published in journals, magazines and conferences. It is very effective in measuring the influence of a scientific article in the scientific community. Given this Sustainable supply chain management literature, it is very important to assess, evaluate and organize the research related to different decision-making techniques to understand their relationship with SSCM. It will also help us in identifying the research gaps and direction for future research.

2 Methodology

2.1 Data Collection

This study has considered Scopus database for extraction of previous literatures because of the comprehensive coverage in scientific fields. The essential key-words used for data extraction is "Sustainable Supply Chain Management" and "SSCM". Since two types of decision-making models are considered in this analysis to different strings of auxiliary keywords are combined to cover the widen the scope as much as possible. After extracting the data using above keywords, relevant research regarding the field of "Application of Decision-making models in Sustainable Supply Chain Management" was sorted out.

2.2 Statistical Analysis

Figure 1(a) shows the trend of research papers published in this field from 1997 to 2022. It is evident that since 2009, the amount of research articles published in field of Sustainable Supply Chain Management in relation to decision-making models has increased continuously. The amount of literature published before 2010 as less than 10, which increased to 111 alone in year 2021. This increasing trend in the number of articles, reveals that increasing number of researchers are paying attention to this particular field resulting in abundance of relevant results for further use.

3 Bibliographic Coupling Analysis

Bibliographic coupling is the method to find the clusters for future research specialization. It was introduced by Kessler in 1963. It occurs when two articles cite a common third article (Document A and Document B cite Document C). As shown in Fig. 1(b), bibliographic coupling of this database has revealed 5 major clusters on the basis of common citation. This study follows the approach, according to which name of each cluster is given based on the subjective judgement of the author, which is similar to approach used in factor analysis. In this paper, first clusters are identified then aiming of clusters is done.

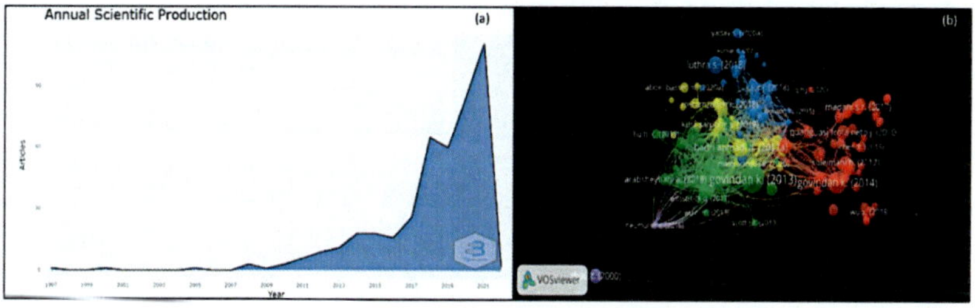

Fig. 1. **(a)** The trends of research paper published **(b)** Bibliographic coupling

4 Analysis of Clusters

4.1 Artificial Intelligence Based Optimization of SSCs

[3] has optimized a joint location-inventory problem for reduction in carbon emission using Genetic algorithm (GA), further they have tested the applied GA through sensitivity analysis and validated the accuracy of the algorithm using GAMS. [4] addressed the issue of optimizing a hierarchized sustainable supply chain network design (SSCND), using bi-level programming (mixed-integer linear programming) to create the model, and solved using an optimization-based algorithm based on simulated annealing.

[5] optimized a multi-objective and multi-period sustainable blood supply chain model by first converting the multi-objective problem to single-objective problem using ϵ-constraint method, further simulated annealing (SA) is used to solve the model, moreover compared its performance with harmony search (HS) algorithm. [6] has optimized the of impact of perishable food supply chin on environment and society using a novel method for optimization of multi-objective models through a hybrid of two multi-objective algorithms i.e. multi-objective particle swarm optimization (MOPSO) and adapted multi-objective variable neighbourhood search (AMOVNS).

[7] optimized and compared the price and the extant of environmental friendliness of two substitute products, one green and other non-green, under centralized and decentralized scenarios using Game theory. [8] used Game Theoretic approach to compared solutions for the power structure of three supply chains to study the impact of power-based relationship on supply chain results and their performance in field of sustainability.

These optimization techniques have played a very significant role in driving the research forward for a long time. Artificial intelligence-based techniques like game theory, hybrid metaheuristic algorithms, evolutionary strategies, etc. have been used for optimization of supply chains, but very limited use of these strategies has been seen for optimization of supply chain models with respect to sustainability. Moreover, very limited research has been done on carbon emission optimization in a supply chain. This has provided a way forward for future research in this field using more artificial intelligence-based techniques.

4.2 Supplier Selection for SSCM

[9] suggested a technique called multi-criteria group decision making (MCGDM), which is based on analytical network process (ANP) and VIKOR (ViseKriterijumskaOptimizacijaKompromisnoResenje) methods. He used triangular neutroscopic numbers to address the problem of supplier selection and validated the model using genetic algorithm. [10] addressed the prominent problem of supplier selection for sustainable supply chain by using a fuzzy multi-objective optimization model based on the ratio analysis (fuzzy MOORA) and used failure mode and effects analysis (FMEA) for supplier risk analysis. Similar problem was also addressed by [11] to select sustainable suppliers in telecom industry of Iran, for which he used the combination of analytical hierarchy process (AHP) and improved grey relational analysis (IGRA) techniques. [12] has also proposed the model for sustainable supplier selection but with respect to social aspect of sustainability. For that, he introduced a decision support tool using two grey-theory based MCDM methods in combination i.e. best-worst method' (BWM) and TODIM (TOmada de DecisãoInterativa e Multicritério – in Portuguese 'Interactive and Multicriteria Decision Making'). [13] used fuzzy best worst method (F-BWM) and integration of Bonferroni mean function with combined compromise solution (CoCoSo'B) for sustainable supplier selection to enhance flexibility in decision making, eliminate impact of inconsistent data and easy analysis of robustness of results. [14] integrated triple bottom line approach of sustainability for supplier selection and used ANP to evaluate the proposed model. [15] integrated triple bottom line (TBL) approach and fuzzy TOPSIS (Technique for Order Preference by Similarity to Ideal Solution) for prioritizing suppliers on basis of their qualitative performance.

From above cluster it can be clearly inferred that a lot of work is done towards the optimization of supplier selection process for a sustainable supply chain management. It is also clearly evident that MCDM methods have been the go-to methods for this type of optimization. For future research application of artificial intelligence-based models can prove to be a fruitful step towards discovering new results. Moreover, Since, MCDM methods have been very successful in supplier selection and evaluation, there is a high possibility that it can be helpful in other types of SSCM optimization problems as well.

4.3 Analysis of Factors Impacting the Implementation of Sustainability in a Supply Chain Environment

[16] analyzed the performance of Circular economy and flexible SSC using objective-DEMATEL method. [17] analysed the enablers in maintaining efficient supply chain in

agricultural sector in India, using Interpretive structural modelling (ISM), DEMATEL (to establish relationship among them) and ANP (to select best supply chain). After analysis of various criteria influencing the sustainability of a supply chain using ANP and ISM [18] found that IT enabled support systems, awareness among supply chain partners, social issues and limited natural resources were most significant among them. [19] provided a framework to identify significant barriers to integrate circular economy with supply chain management in food sector, using DEMATEL (to establish causal-effect relationship between them). [20] is one of the few authors who studied oil and gas supply chain in context of sustainability. He used ISM (to establish mutual relationship among the factors) and structural equation modelling (SEM) to analyse the data related to factors of SSCM and their impact on Operational and Business Performance (OPR). [21] explored compliance of central firm's corporate sustainability standards (CSS) by its sub-suppliers as the new dimension of SSCM, by analysing past sub-supplier management studies and their critical success factors (CSFs) using DEMATEL. [22] used DEMATEL to investigate barriers related to implementation of green supply chain management (GSCM) in electronics industry in Canada. [23] proposed a structural framework to investigate the key enablers in implementation of SSCM in Indian industries using ISM and MICMAC (to separate enablers on the basis of driving and dependence power).

From this cluster it was inferred that lot of research has been done on the analysis of various factors involved in adoption of sustainable supply chain practices. Very limited research is present related to analysis of factors influencing the adoption of Industry 4.0 techniques for creating a sustainable supply chain management environment. It as also realized that industry specific researches are dominant which shows that its time to focus macro-level research will be mor appreciated in future. It is also clear that MCDM techniques have dominated in the tools and techniques, used in analysis of influencing factors, research can also turn towards AI based optimization tools for this purpose.

4.4 Analysis of Risks Associated with Sustainable Supply Chain Implementation

[24] used a combination of plithogenic MCDM technique based on TOPSIS and Criteria Importance Through Inter-criteria Correlation (CRITIC) techniques. [25] analysed the financial risk associated to uncertainty in combining sustainability and supply chain financing, using TODIM and TOPSIS techniques based on BWM. [26] reviewed literatures related to supplier risk management related to sustainability, and found that MCDM are most dominant tools for analysis, but an increasing trend is also observed in the usage of fuzzy heuristics. He also discovered that risks related ethical conduct for businesses and CSR can is an emerging field. [27] proposed a model covering assessment of sustainability and recent significant fields like life cycle impact assessment (LCIA) and social life cycle assessment (SLCA), using AHP. [28] investigated forty-four risk factors associated with leather industry divided among all five dimensions of sustainability, using BWM for purpose of making decisions. [29] proposed a model to quantitatively evaluate the social risk factors associated with global supply chains of German automotive industry, by combining Leontief's Input Output model with risks associated with countries and intensity of labour specific to sector, by using fuzzy-AHP.

After analysing this cluster, it was discovered that most of the studied have used MCDM approaches for analysing the risk factors. So, for future metaheuristic-based optimization techniques can be used to develop sustainable supply chain networks by overcoming the risk factors found in these studies. Most of the studies are related to particular industry, so there is a need for studies which provide a generalized and holistic approach towards a macro-level scenario for the analysis of risks associated to sustainable supply chain management issues.

4.5 Industry 4.0 and Sustainable Supply Chain Management (SSCM)

Author systematically reviewed the literature related to industry 4.0 and sustainable supply chain management to identify the Industry 4.0 enablers for SSCM implementation provide a framework for future research. He concluded that social aspect of sustainability is ignored in previous research which can be a possible research direction. He identified 13 key enablers for Industry 4.0 implementation for achieving sustainability in supply chain [30]. [31] has summarized the literature related to application of Blockchain and Industry 4.0 in enhancing sustainability in supply chain management especially in environmental and social aspects. He identified three main capabilities of Industry 4.0 and 4 main capabilities of Blockchain for achieving sustainability in SCM efficiently. Developed a holonic framework for Adaptive and Integrated Sustainable Supply Chain Management (AISSCM) at multilevel and multiscale systems, to support social metabolism smartly, using adaptation and integration which can support co-evaluation of sustainable supply chain management for the environment. [32] evaluated the impact and feasibility of Industry 4.0 barriers and drivers, with respect to application in multi-tire supply chain to increase sustainability in manufacturing sector. [33] proposed a framework to overcome barriers in integrating Industry 4.0 technologies with SSCM. For this he used a hybrid MCDM approach by combining BWM and ELimination and Choice Expressing REality (ELECTRE) techniques and validated using expert opinions.

Through the analysis of these papers, it has been established with conformity that Integration of Industry 4.0 technologies with all domains of industrial management especially in supply chain management is the future of Industries. Also, there is very limited research in analysing the feasibility of application of Industry 4.0 technologies in MSME sector of Industries. Less amount of research papers in this clusters with increasing trends, suggest there is a wide scope of research for integration of Industry 4.0 technologies with supply chain management to enhance the supply chain sustainability and overall sustainability.

5 Conceptual Framework for SSCM Implementation

According to the literature review is being proposed SSCM implementation. It is inferred that a shift from traditional supply chain practices towards Industry 4.0 integrated SSCM practices will transform the sustainability aspect of not only the supply chain and its partners but also of the industry as a whole. Framework shown in Fig. 2 will target to improve sustainability by focusing on all three aspects of sustainability according to triple bottom line approach i.e. environmental, financial, and social using the disruptive technologies of Industry 4.0. Integration of Industry 4.0 with SSCM practices will integrate whole supply chain through IoT and cyber physical systems (CPS), which in turn will make it more flexible and resilient. But first step towards integration of Industry 4.0 technologies with SSCM is assessing the present situation of supply chain and the feasibility and viability of upgradation.

First level of this framework represents all the dimensions of sustainability according to triple bottom line approach. Here we can see that sustainability of a supply chain can be measured in three dimensions i.e. social, economic and environmental. Here it is also clear that every approach leading towards supply chain management is impacted by sustainability in all dimensions.

Second layer discusses the technical tools and techniques which help in optimizing the sustainability criteria. Different models and techniques have different benefits in different scenarios. From the literature five most common and effective optimization techniques are identifies which can help in driving future research.

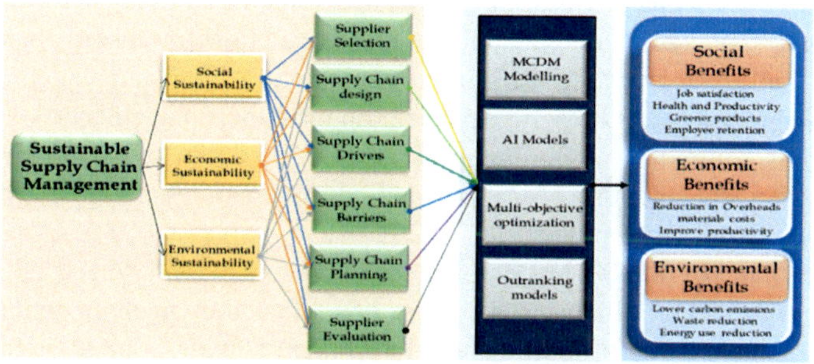

Fig. 2. Conceptual framework for SSCM implementation

Third layer shows the benefits which are possible to achieve if we work according to a planned approach guided through this framework. The benefits are also loosely classified in three dimensions of sustainability. Some of the benefits which have been observed by the past researchers and practitioners have been shown and further more varied benefits are possible to attain.

This framework is not the only approach to attain sustainability in supply chain, rather a guideline for future researchers and practitioners aiming for achieving an overall sustainable supply chain management.

6 Research Implications

6.1 For Future Researchers

According to the literature review some broad research areas have been identified which could be addressed in future.

- **Artificial Intelligence based Optimization for SSCM:** With increasing complexity of supply chains, it has become very difficult to optimize supply chain problems using conventional methods and equations. Due to the increase in number of variables and factors effecting a supply chain, the supply chain optimization problems have come in the category of np-hard problems. It is very difficult and time consuming to solve these problems using traditional optimization techniques. Artificial intelligence-based optimization methods have proved to be very effective in solving np-hard problems.

- **Application of Industry 4.0 technologies for SSCM**: It is evident from past literature that emergence of Industry 4.0 has provided new research opportunities for upcoming researchers and experts in the field of reconfigurable and flexible manufacturing, efficient and sustainable supply chains. Industry 4.0 has introduced many new and disruptive technologies to the industrial sector. Every new technology comes with its limitations and capabilities. Evaluating these attributes of every technology w.r.t. industrial management perspective has become one of the most impactful research opportunities for future researchers.
- **Assessment of factors affecting SSCM:** Assessment of factors affecting SSCM implementation has been going on from a long time but most of these assessments were done using MCDM techniques. [33] explained some limitations of various MCDM techniques. AHP is sometimes unable to solve linear equations, based on possibilistic measures and is subjective to the response providers. ANP is unable to handle large number of factors and is heavily dependent on the expert judgement. BWM method is limited by a 9-point scale for comparison. TOPSIS doesn't account the difference between the ideal solutions of positive and negative nature and the attribute values vary monotonically.
- **Supplier selection for SSCM:** Supplier selection for SSCM has been and still is one of the most difficult problems to handle as it is affected by many in comprehensible factors. The estimation of these factors in problem solving is pretty subjective and highly inaccurate. Using fuzzy set theory has somewhat tried to comprehend these factors but the responses are still subjective to the expert's knowledge and experience. In past literature studies related to artificial intelligence, machine learning and big data are rather theoretical so, integration of Artificial intelligence, Machine learning and Big data for application purpose can open doors towards new findings in this field.
- **Risk factor assessment for SSCM:** Adoption of any new idea like sustainability always possesses some associated risks, so assessment of these risks is very important to decide whether to adopt the idea or not. Risk assessment for SSCM is studied and applied in past but it is largely associated with risks possessed for a certain industry specific supply chain. Macroscopic research is still not present for risk assessment. Most of the literatures have used MCDM techniques for assessment purpose but sensitivity analysis has been used rarely. Sensitivity analysis is useful for analysing the reliability of results.

6.2 For Industry Professionals and Decision-Makers

- Emerging technologies like blockchain, Big data, Machine learning, IoT, have found their way in the industrial sector, but people working in industrial sectors are not yet equipped to embrace these technologies according to their requirements. A huge gap has been identified between present knowledge base of the industry professionals and the skill and knowledge required for adoption of these technologies for according to the requirement of the industry. If an enterprise needs to compete with future global market competitions it needs to take the training and education of its employees, especially in emerging technologies, very seriously.

- Every new technology possesses its own risks and benefits. Management needs to be a bit risk taking in order to improve the efficiency of its present SSCM systems and compete in global market. Using Industry 4.0 tools professionals will be able to exercise a greater level of control over the supply chain, especially the sustainability aspect.

7 Conclusion

Governments all over the globe are now understanding the importance of sustainability for overall growth of the nation's economy. Industrial sector is the backbone of any nation's economy, so a high impact of this is being seen in norms and regulations made by government for the industrial sector. Industries need to focus on the environmental as well as social aspects of sustainability along with the economic aspect to achieve balanced sustainable development. It has been already established that impact of un-regulated supply chain has immense adverse effects on environment, people and economics of the stakeholder organizations in long run. This study has discussed a framework to provide guidelines for achieving and enhancing sustainability in supply chain management. The bibliometric analysis and the review from Scopus database, performed in this study will provide clarity to readers about the present trends in research on this field. In this bibliometric analysis past literature is analysed on the on the basis of time-frame, authorship, first author's affiliation, country, citations, and keywords. Clustering of data to classify the present research and understand future research directions is done for emerging researchers and professionals in this field. Five major clusters are identified according to bibliographic coupling of the articles. This study is limited to only one database i.e. Scopus, so for future research other databases like Web of science may be used for bibliometric analysis using more software like Gephi and BibExcel. The framework proposed in this study doesn't provide empirical insights, which can also be addressed in future. Industry 4.0 still remains a limitedly explored territory for researchers till now, but is gaining importance at a rapid pace.

References

1. Sharma, M., Kamble, S., Mani, V., Sehrawat, R., Belhadi, A., Sharma, V.: Industry 4.0 adoption for sustainability in multi-tier manufacturing supply chain in emerging economies. J. Clean. Prod. **281** (2021). https://doi.org/10.1016/j.jclepro.2020.125013
2. Arrais-Castro, A., Varela, M.L.R., Putnik, G.D., Ribeiro, R.A., Machado, J., Ferreira, L.: Collaborative framework for virtual organisation synthesis based on a dynamic multi-criteria decision model. Int. J. Comput. Integr. Manuf. **31**, 857–868 (2018)
3. Diabat, A., Al-Salem, M.: An integrated supply chain problem with environmental considerations. Int. J. Prod. Econ. **164**, 330–338 (2015). https://doi.org/10.1016/j.ijpe.2014.12.004
4. Kaboli Chalmardi, M., Camacho-Vallejo, J.-F.: A bi-level programming model for sustainable supply chain network design that considers incentives for using cleaner technologies. J. Clean. Prod. **213**, 1035–1050 (2019). https://doi.org/10.1016/j.jclepro.2018.12.197

5. Eskandari-Khanghahi, M., Tavakkoli-Moghaddam, R., Taleizadeh, A.A., Amin, S.H.: Designing and optimizing a sustainable supply chain network for a blood platelet bank under uncertainty. Eng. Appl. Artif. Intell. **71**, 236–250 (2018). https://doi.org/10.1016/j.engappai.2018.03.004
6. Govindan, K., Jafarian, A., Khodaverdi, R., Devika, K.: Two-echelon multiple-vehicle location-routing problem with time windows for optimization of sustainable supply chain network of perishable food. Int. J. Prod. Econ. **152**, 9–28 (2014). https://doi.org/10.1016/j.ijpe.2013.12.028
7. Jamali, M.-B., Rasti-Barzoki, M.: A game theoretic approach to investigate the effects of third-party logistics in a sustainable supply chain by reducing delivery time and carbon emissions. J. Clean. Prod. **235**, 636–652 (2019). https://doi.org/10.1016/j.jclepro.2019.06.348
8. Chen, X., Wang, X., Chan, H.K.: Manufacturer and retailer coordination for environmental and economic competitiveness: a power perspective. Transp. Res. Part E Logist. Transp. Rev. **97**, 268–281 (2017)
9. Abdel-Baset, M., Chang, V., Gamal, A., Smarandache, F.: An integrated neutrosophic ANP and VIKOR method for achieving sustainable supplier selection: a case study in importing field. Comput. Ind. **106**, 94–110 (2019). https://doi.org/10.1016/j.compind.2018.12.017
10. Arabsheybani, A., Paydar, M.M., Safaei, A.S.: An integrated fuzzy MOORA method and FMEA technique for sustainable supplier selection considering quantity discounts and supplier's risk. J. Clean. Prod. **190**, 577–591 (2018). https://doi.org/10.1016/j.jclepro.2018.04.167
11. Badri Ahmadi, H., Kusi-Sarpong, S., Rezaei, J.: Assessing the social sustainability of supply chains using Best Worst Method. Resour. Conserv. Recycl. **126**, 99–106 (2017). https://doi.org/10.1016/j.resconrec.2017.07.020
12. Bai, C., Kusi-Sarpong, S., Badri Ahmadi, H., Sarkis, J.: Social sustainable supplier evaluation and selection: a group decision-support approach. Int. J. Prod. Res. **57**, 7046–7067 (2019). https://doi.org/10.1080/00207543.2019.1574042
13. Ecer, F., Pamucar, D.: Sustainable supplier selection: a novel integrated fuzzy best worst method (F-BWM) and fuzzy CoCoSo with Bonferroni (CoCoSo'B) multi-criteria model. J. Clean. Prod. **266** (2020). https://doi.org/10.1016/j.jclepro.2020.121981
14. Faisal, M.N., Al-Esmael, B., Sharif, K.J.: Supplier selection for a sustainable supply chain: triple bottom line (3BL) and analytic network process approach. Benchmarking **24**, 1956–1976 (2017). https://doi.org/10.1108/BIJ-03-2016-0042
15. Govindan, K., Khodaverdi, R., Jafarian, A.: A fuzzy multi criteria approach for measuring sustainability performance of a supplier based on triple bottom line approach. J. Clean. Prod. **47**, 345–354 (2013). https://doi.org/10.1016/j.jclepro.2012.04.014
16. Bai, C., Sarkis, J., Yin, F., Dou, Y.: Sustainable supply chain flexibility and its relationship to circular economy-target performance. Int. J. Prod. Res. **58**, 5893–5910 (2020). https://doi.org/10.1080/00207543.2019.1661532
17. Chauhan, A., Kaur, H., Yadav, S., Jakhar, S.K.: A hybrid model for investigating and selecting a sustainable supply chain for agri-produce in India. Ann. Oper. Res. **290**(1–2), 621–642 (2019). https://doi.org/10.1007/s10479-019-03190-6
18. Digalwar, A., Raut, R.D., Yadav, V.S., Narkhede, B., Gardas, B.B., Gotmare, A.: Evaluation of critical constructs for measurement of sustainable supply chain practices in lean-agile firms of Indian origin: a hybrid ISM-ANP approach. Bus. Strategy Environ. **29**, 1575–1596 (2020). https://doi.org/10.1002/bse.2455
19. Farooque, M., Zhang, A., Liu, Y.: Barriers to circular food supply chains in China. Supply Chain Manag. **24**, 677–696 (2019). https://doi.org/10.1108/SCM-10-2018-0345
20. Gardas, B.B., Raut, R.D., Narkhede, B.: Determinants of sustainable supply chain management: a case study from the oil and gas supply chain. Sustain. Prod. Consum. **17**, 241–253 (2019). https://doi.org/10.1016/j.spc.2018.11.005

21. Grimm, J.H., Hofstetter, J.S., Sarkis, J.: Interrelationships amongst factors for sub-supplier corporate sustainability standards compliance: an exploratory field study. J. Clean. Prod. **203**, 240–259 (2018). https://doi.org/10.1016/j.jclepro.2018.08.074
22. Kaur, J., Sidhu, R., Awasthi, A., Chauhan, S., Goyal, S.: A DEMATEL based approach for investigating barriers in green supply chain management in Canadian manufacturing firms. Int. J. Prod. Res. **56**, 312–332 (2018). https://doi.org/10.1080/00207543.2017.1395522
23. Kausar, K., Garg, D., Luthra, S.: Key enablers to implement sustainable supply chain management practices: an Indian insight. Uncertain Supply Chain Manag. **5**, 89–104 (2017). https://doi.org/10.5267/j.uscm.2016.10.005
24. Abdel-Basset, M., Mohamed, R.: A novel plithogenic TOPSIS- CRITIC model for sustainable supply chain risk management. J. Clean. Prod. **247** (2020). https://doi.org/10.1016/j.jclepro.2019.119586
25. Abdel-Basset, M., Mohamed, R., Sallam, K., Elhoseny, M.: A novel decision-making model for sustainable supply chain finance under uncertainty environment. J. Clean. Prod. **269** (2020). https://doi.org/10.1016/j.jclepro.2020.122324
26. da Silva, E.M., Ramos, M.O., Alexander, A., Jabbour, C.J.C.: A systematic review of empirical and normative decision analysis of sustainability-related supplier risk management. J. Clean. Prod. **244** (2020). https://doi.org/10.1016/j.jclepro.2019.118808
27. Kolotzek, C., Helbig, C., Thorenz, A., Reller, A., Tuma, A.: A company-oriented model for the assessment of raw material supply risks, environmental impact and social implications. J. Clean. Prod. **176**, 566–580 (2018). https://doi.org/10.1016/j.jclepro.2017.12.162
28. Moktadir, M.A., et al.: Analysis of risk factors in sustainable supply chain management in an emerging economy of leather industry. J. Clean. Prod. **283** (2021). https://doi.org/10.1016/j.jclepro.2020.124641
29. Zimmer, K., Fröhling, M., Breun, P., Schultmann, F.: Assessing social risks of global supply chains: a quantitative analytical approach and its application to supplier selection in the German automotive industry. J. Clean. Prod. **149**, 96–109 (2017). https://doi.org/10.1016/j.jclepro.2017.02.041
30. Sousa, R.A., Varela, M.L.R., Alves, C., Machado, J.: Job shop schedules analysis in the context of industry 4.0. In: 2017 International Conference on Engineering, Technology and Innovation (ICE/ITMC), pp. 711–717. IEEE (2017)
31. Esmaeilian, B., Sarkis, J., Lewis, K., Behdad, S.: Blockchain for the future of sustainable supply chain management in Industry 4.0. Resour. Conserv. Recycl. **163** (2020). https://doi.org/10.1016/j.resconrec.2020.105064
32. Yadav, S., Singh, S.P.: An integrated fuzzy-ANP and fuzzy-ISM approach using blockchain for sustainable supply chain. J. Enterp. Inf. Manag. **34**, 54–78 (2020). https://doi.org/10.1108/JEIM-09-2019-0301
33. Jamwal, A., Agrawal, R., Sharma, M., Kumar, V.: Review on multi-criteria decision analysis in sustainable manufacturing decision making. Int. J. Sustain. Eng. **14**(3), 202–225 (2021)

Value Measurement in Health Care Delivery Process for a Paediatric Hospital in Guinea-Bissau

J. Antas Aguiar[1(✉)], J. Telhada[2(✉)], M. Sameiro Carvalho[2(✉)], and J. C. Sá[3,4(✉)]

[1] Laboratório de Análises Clínicas, Manuel Pimenta, Lda, 4990-013 Ponte de Lima, Portugal
antasdebarros@engenheiros.pt
[2] ALGORITMI Research Centre, Department of Production and Systems, University of Minho, 4710-057 Braga, Portugal
{telhada,sameiro}@dps.uminho.pt
[3] ISEP – School of Engineering, Polytechnic of Porto, 4200-072 Porto, Portugal
cvs@isep.ipp.pt
[4] INEGI - Institute of Science and Innovation in Mechanical and Industrial Engineering, Rua Dr. Roberto Frias, 400, 4200-465 Porto, Portugal

Abstract. This project is the result of a volunteer program that took place in the Republic Guinea-Bissau (RGB) at the Paediatric Hospital. The Paediatric Hospital was working with the same technologic skills as Portugal in the 70's years. There was no wired network between departments, no informatic system and all work was handmade. The physicians received incomplete clinical reports and, sometimes, with gross errors. There was no triage neither outpatient short list during the medical urgency period. The outpatients needed to wait for long time to get the analytical report and most time without meal or single bread. On the Clinical Analysis Laboratory, the biologic sample identification and the manual data report were lower changes to be credible and there was no final checklist. This paper describes the improvements made by the authors in the hospital logistical processes and in the development and implementation of an information system that was started in March 2018. These improvements were developed based on lean manufacturing principles, on methods and best practices for some referenced case studies from America, Saudi Arabia, Italy and England. As a result, the hospital reduced outpatient waiting time, improved Hospital, Laboratory and outpatient outcomes, increased confidence work for physicians and laboratory technics.

Keywords: Electronic medical records · Lean tools · Lean healthcare · Sustainability · Volunteer program

1 Introduction

This paper describes the improvements made by the authors in the hospital logistical processes, in the development and implementation of an information system that was

J. Machado et al. (Eds.): icieng 2022, LNME, pp. 186–196, 2023.
https://doi.org/10.1007/978-3-031-09360-9_16

started in March 2018. These improvements were developed based on lean manufacturing principles, on methods and best practices for some referenced case studies.

As a result, the hospital reduced outpatient waiting time, improved outpatient outcomes and increased confidence work for physicians and laboratory technics. Additionally, increased efficiency has been achieved for all staff, and, in general, the hospital has now an improved management system.

This paper has a special focus on improving the IT infrastructure and on identifying the main and urgent logistics problems that causes long waiting times and health problems, in particular to children and pregnant women.

This paper will contribute to the United Nations goals for sustainable development[1], in particular:

- Goal 3, Good health and well-being, to reduce the maternal mortality rate and end preventable deaths of new-borns and childrens;
- Goal 8, Decent work and economic growth, to achieve higher levels of productivity of economies through diversification, technological modernization and innovation.

As a secondary objective, it is intended to implement a new logistics system that can improve health care system. Such a system must be previously validated, easy to understand and must demonstrate its practical utility.

To achieve the above-mentioned objectives, it is proposed to: Study the opportunities to improve processes adopting lean logistics principles and tools; Create algorithms and communication flows with laboratory equipment's and stakeholders; Train native technicians to have skills to operate and participate, in order to drive the logistic implementations to another upgrade steps; Shape procedures and work habits that foster a prosperity balance of patient needs and simultaneously guarantee quality of care, timely interventions and ensure equity of access and well-being to all citizens; Develop an Automated Hospital Management System (AHMS) that will explore the best practices for reorganizing and coordinating health care, improving process efficiencies.

Specific goals of this paper include: Standard working methods; Unique outpatient identification for all cycle care; Eliminate the manual transcription errors; Match clinical resources to clinical processes; Avoid medical errors due to error-prone analytical reports; Reduce complete cycle care and improve resource utilization; Reduce time to prepare and delivery analysis clinic report, Report haemoglobin statistics; Send alerts if clinical analyses report is not all completed; Send reports to e-mail; Medical statistics and administrative statistics – no paper required; Provide the Hospital and Laboratory to have access to free software without setup or maintenance cost.

The structure of the paper is as follow: Sect. 2 presents a literature review of the main topics of the study; Sect. 3 describes the materials and methods applied; Sect. 4 presents the main results obtained, that are then discussed in Sect. 5; finally, Sect. 6 presents the main conclusions.

---

[1] https://www.un.org/sustainabledevelopment/sustainable-development-goals/.

2 Literature Review

This section resumes the advantages of shifting from a manual to an AHMS, addresses the limitations and good practices, how to move from this forward, and tools that drives to more medical efficiency using lean concepts.

Cases from Kingdom of Saudi Arabia, Italy and USA have similar issues to the issues of the system under study in this paper. The purpose is to understand the difficulties that affect their organizations and how they improved and they developed the strategies to grow up and how they quantified their outcomes.

A sustainability model and a volunteer perspective are issues that are also here considered to ascertain how to engage the community to support the success of the initiatives.

2.1 The Importance of Moving from Manual to Automatized Medical Records

Manual records can be destroyed, disappear or deteriorate relatively easily [1].

Manual processing of outpatient information may lead unnecessary duplication of the data. Inconsistency of data may occur since records are kept in more than one location [2].

Throughout time paper-based systems have proven to become more and more inefficient and are continuously failing to meet the care provider's needs [3].

Simultaneous access from multiple location, legibility, variety of views on data and sharing care support [4].

Information and Communication Technology (ICT) has become a major tool in delivery of health services and has had a revolutionary impact on how we live and perceive the world. ICT enables people to interact and communicates no matter the distance, also makes it easy for us to obtain goods and services in convenient ways [5].

One overall objective of EMR is to improve patient education and to enhance the patient experience through the use of Information Technology (IT) so as to facilitate the sharing of information between providers and their patients [6].

It is important that Electronic Medical Record (EMR) should be easy to develop and modify, works for clinical personnel and saves waiting time. The EMR should be intuitive, requiring little or no training [7].

Eiseman and Fossum propose that any global health resource tracking system would contain valid, detailed data (who, what, where, how much) [8].

2.2 Holistic View for Different Approach

In recent years, there has been an increase in the adoption of lean manufacturing principles, as well as tools, in multiple industrial sectors. This management philosophy, which is already well-established, has been applied successfully in other various contexts than just production [9].

Lean Thinking is a management and leadership model with a view to continuous improvement, encouraging people to think and solve problems, creating value [10].

The Lean tools application allowed to increase the efficiency of the production system, reducing on average 33% of cycle times, reducing energy consumption by 38%

and reducing scrap by 66% which demonstrates that Lean tools application leads companies to environmental and economic gains and contributes to better environmental management [11].

This work will use nine LM tools: Kaizen philosophy, the 5 S's, Standard Work, the Heijunka or Levelled Programming, the Value Stream Mapping, the Pull System, the Three MU's, the Process Flow and the Visual Management. It was found that Lean Thinking and its tools are an added value for improving work conditions [11].

A series of methods designed to minimize the waste of material and labour, makes for an organization that's better equipped, improved efficiency, frees up employees and resources for innovation and quality control that would have previously been wasted, while maintaining or increasing levels of production.

Time-driven activity-based costing (TDABC) methodology calculates a cost for each activity. This study is one transparent cost analysis of patient care and an inexpensive way to learn how all resources are currently being used, determine excess capacity and measure the cost of unused capacity, resource by resource.

Mayo Clinic [12] is a real example that used TDABC costing methodology.

Providers are achieving saving 25% or more by tapping opportunities such as better capacity utilization, more-standardized processes, better matching personnel skills to tasks, locating care in the most cost-effective type of facility and many others [13].

TDABC captures all the resource costs incurred over a patient's complete care cycle for a specific medical condition [14].

Organizations must find ways to recruit and engage volunteers that are committed to sustainable and lasting practices and that will truly engage the community in which they serve [15].

The National Health System (NHS) of England is based on a sustainability model that is used from the outset of an improvement initiative as it can provide a valuable understanding on which area to focus on to maximizing sustainability [16]. Here we identify three dimensions: Process, Staff and Organization. Inside each dimension there are general issue categories.

Both the individual making a decision to volunteer and the organization making a decision to use volunteer labour face benefits and costs of their actions yet these costs and benefits almost always remain unarticulated, perhaps because the common perception of the do-good volunteer who contributes his or her labour for free discourages rational benefit cost calculus [17].

The most important resource that volunteer groups have is the volunteers themselves and the effort they forth [18].

Many hospitals around the world have proposed solutions to overcome the long waiting time problem in outpatient clinics. The following paragraphs will explain their solutions and the progress of their improvements and the inspiration for this thesis.

The case study "KSA Hospital" [19] pretend to identify and solve the reasons behind the long waiting time in the outpatient clinic which affects the level of patient satisfaction, the quality of service, resource utilization and overall clinic management.

Mayo Clinic [12] describes that documentation by physicians is expensive and limits their ability to see additional patients.

They learn that scribes decrease documentation time by 33% or an hour on average and $488 per shift by decreasing the physician time needed for these tasks and eliminating transcription needs.

In their study [14] assist physicians and staff in the appropriate allocation of resources at their own institution. TDABC captures all the resource costs incurred over a patient's complete care cycle for a specific medical condition. The total costs drop down 24.8% less for nonemergent eye complaints.

In Italy the unmet health care needs statistics refers that 21.4% of population reported to have renounced to treatment due long waiting list [20].

The modified process resulted in a 16% reduction in process time, a 12% decrease in costs for technical staff and a 67% reduction in costs for professional staff [14].

With new logistic approach with TDABC method, BCH handle the same volume and mix of patients while spending $126,810 less on office visits.

A belief that EMR could save time and money, has potential for increasing safety and productivity and EMR systems allow connections between services [7].

One of the main drawbacks in outpatients clinics is long waiting time for patients – which affects the level of patient satisfaction and the quality of services [19].

3 Materials and Methods

A swot analysis was implemented to support the diagnosis stage concerning the activity at the Hospital and the Clinic Analyse Laboratory.

Then a visual outpatient logistic process inside the Hospital and Laboratory to join all logistic process that pretend to analyse and a Value Stream Map for each stage (Fig. 1).

For each stage we collected and wrote the important issues. For the Laboratory, was drew a logistic process and use red colour to identify pain process. We also drew a Value Stream Map to the complete outpatient cycle care (Fig. 2).

The Time-Driven Activity Based Cost (TDABC) methodology calculates a cost for each activity by determining the time spent and the actual cost of each resource used for that activity and allows a much more accurate determination of the true cost of providing care.

A TDABC study allows to identify what procedures truly cost and what drives those costs. This incorporates a broad of evidence base as possible to make evidence-informed decisions and now this method has now been adopted in the management of operations.

TDABC helps providers find numerous ways to substantially reduce costs without negatively affecting outcomes (and sometimes improving them).

We used 5W2H development tool (What, Why, Where, When, Who, How, How Much) with a set of questions used to compose action plans quickly and efficiently. This tool defines effective tasks and follow-up in a visual, agile and simple manner, facilitate the definition, investigation and resolution of a problem.

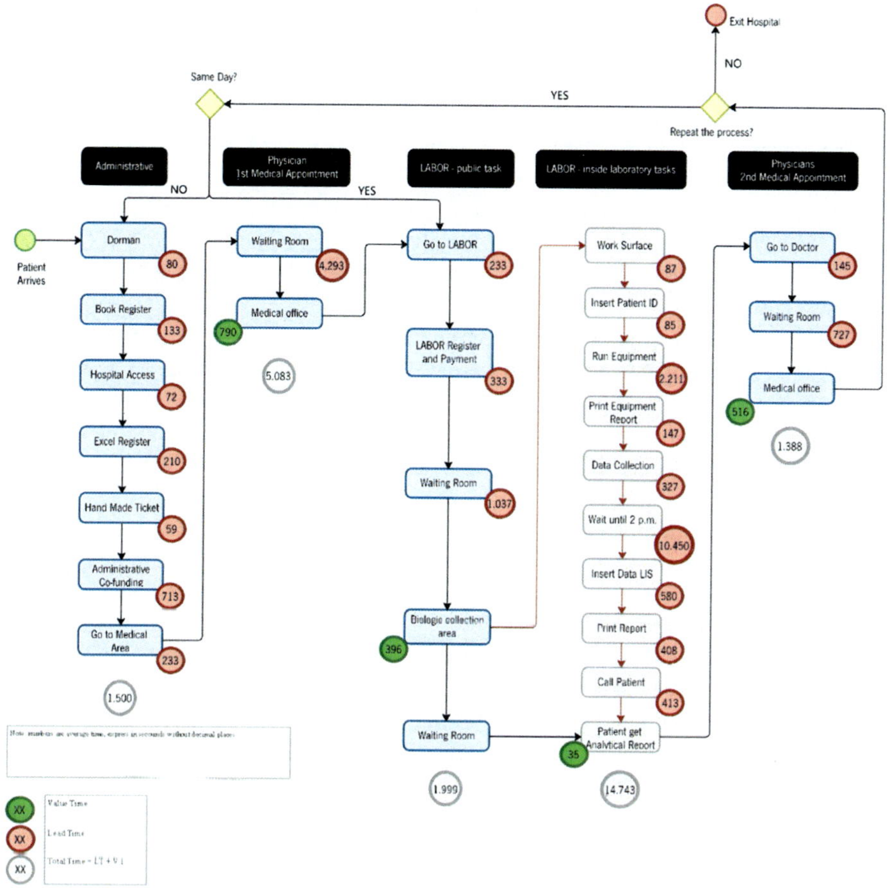

Fig. 1. Value Stream Map for complete outpatient cycle care

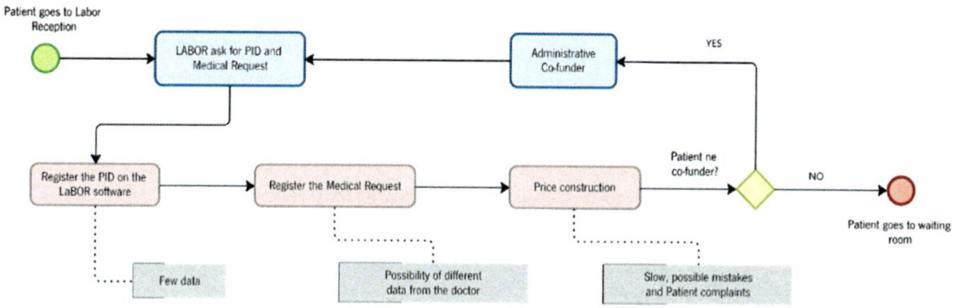

Fig. 2. Logistic process for outpatient laboratory admission

4 Results

With Manchester triage the medical staff saw online how many outpatients were waiting for care and how serious their health problems were. This standard procedure saved time and manpower and decreased the time that physicians and care providers taken to communicate. Physicians can send directly to Laboratory the medical prescription.

The laboratory receives automatic electronic Patient Identification (PID) from reception and electronic medical requisition directly from the physician, reduce 100% time to introduce PID, reduce 100% PID errors and reduce 100% medical gaps or analysis change.

Now the Laboratory print the invoice with all medical prescription detail. The outpatient came with the feeling that they were well treated and no confusion with analysis prices.

New label with unique barcode prints the PID and the collect sample time and this label allowed to reduce the errors of biological sample exchanges by 100%, to ask and confirm PID on the biologic collect moment.

The Laboratory Equipment (LE) scanners were used to read barcode from biologic sample. This has reduced by 100% sample typing errors on the LIS (Laboratory Information System). Physician's had PC access automatically with no transcribe errors. This upgrade reduced transcribe errors by 100%.

Send data from equipment directly to LIS and visual management applications were implemented.

Optimize workflows and efficiencies, improving outpatient outcomes and clinician satisfaction, enhance outpatient engagement with complete reports and correct results, analyse and quickly mobilize clinical data for reporting, real-time assessment and ongoing quality improvement, collect data from most important laboratory equipment and manual techniques and share them instantly to the physicians as needed. Then, analysis and equipment's statuses can be followed live by all the Laboratory techniques. Productivity can also be analysed instantly from Portugal, provide the solutions on many different devices to improve flexibility and effectiveness: computers, tablets and smartphones. All of them are connected to a central network infrastructure by wired or wi-fi.

5 Discussion of Results

One of the most totted benefits of voluntourism is the potential for increased or faster economic development [21].

This project is an ongoing, continuous project and will grow for further skills like pharmacy logistics and a more integrated supply chain (including to external parties).

For the Hospital, it already increased levels and quality of services, as well as reduced staff workload.

There are authors, like [22], that describes the main driver for volunteers is the enhancement of their Curriculum Vitae (CV), gaining new skills and field experience, in order to find employment within their desired sectors or needed the experience as a credit in his university degree or doctoral research.

Other benefits, not tangible, are the training, great work environment, respect of your attitude, feeling appreciated, large and genuine friendships are made, great days always with huge levels of happiness, pull of international adventure and a break from corporate burnout, the human interaction and a deeper understanding of personal privilege on the side of the voluntarist. We come home with a sense of fulfilment that we added goodness to the world [23].

For general outcomes, the TDABC model proved to be a catalysing agent for cost-conscious care redesign. The costs have decreased 31.87% and the time has been shortened 66.25%. This was a huge upgrade for outpatient health care.

For administrative outcomes, the value reduction is less 28.31% and administrative time cycle before was 1,500 s and now is 740 s, less 760 s (less 50.67%).

The physician first appointment outcomes was evident the best outcomes for patient due to a more then 70% of time reduction.

The Laboratory was a great potential cost-savings from using Manchester Triage, EMR and LIS rather than analogic data and unsorted outpatient. After technician end biologic analysis process, their average time was 14,743 s and moved, with the same resources and a better LIS, to 3,021 s.

The annual unused capacity XOF ratio was between 40.55% and 47.31% and moved to 0,00%. The problem was solved with new logistic approach.

The gross transcribes errors cost and time moved to zero.

The most common errors were misspelled. The LABOR misunderstands the request or the analytical report was incomplete. Now no errors found, so all the costs and time costs were completed eliminated.

That represent the annual physician's capacity to see more 164 to 205 outpatients without spend human resources needs or facilities.

Considering outpatient outcomes, this model reduced waiting times significantly, as shown in Fig. 3. The best reduction they have got was 66.25% with the same people, the same service, the same resource, and three times more service capacity, included reduced transcriptions errors, administration times and cost savings related to decreased stock, workforce costs and waste.

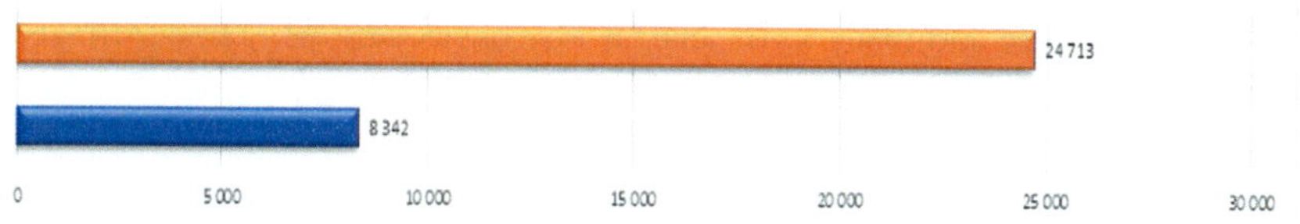

Fig. 3. Outpatient waiting times (seconds) before (orange) and after (blue) (Color figure online)

5.1 Cost Benefits

The software was free to develop and was free to install, had no fee and has no mensal fee to pay, no number of workstation limitation, free backup's automatic diary and free cloud space.

5.2 Software Architecture Advantages

Assent logistical and procedures that were the best to HPSJB, assent medical requirements, speed (less than one second to show reports or statistics), joined the laboratory and hospital needs without duplicate task or information input, possibility of connecting new hospital or laboratory equipment that use HL7 standard protocol or other, no user's limitation to work at same time, not a complex tool that was necessary robust computers to work or a super server. Just software functions that were necessary to use, draw architecture database by the book and just with useful information and space.

6 Conclusions

Note that this study is based on a volunteer project, and there is no link, no participation and no sponsorship and no kind of agreement with any external company, organization, politician, institution or lobbying.

This project started to fill a gap in health care. The Hospital and Laboratory has had no standard work habits, no informatic network, no LIS, no organized outpatients' queues and a long list of logistical problems. The only way forward was the possibility of developing a volunteer project that started on March 2018 by studying and understanding the reality and design a road map of how to get the goals which includes organization, standard work, data communication, organized outpatient queue by health care condition and shorten.

In fact, during this project the Hospital and the Laboratory moved from a confused situation to an extraordinarily complex situation and, as a result of this project, they moved from that complex situation to a better organized Hospital and Laboratory.

Basically, all objectives and goals defined to this project were successfully achieved, as planned. This was perfect for inspiration and upskilling.

The study was carried out using a holistic approach which applied a variety of analysing and development tools such as swot analysis, value stream mapping, 5W2H and TADBC methodology. Hospital spaces were adequately organized and visual signed, outpatients flow and medical processes were organized, digitalized and automatized as much as possible, outpatients medical tracking and historical data are now stored in the hospital database and ready to be analysed by physicians. Manchester triage was implemented to improve medical assistance to outpatients.

The main outcomes includes: elimination of the manual transcription errors, match of clinical resources to clinical processes, avoidance of medical errors due to error-prone analytical reports, alerts if clinical analyses report is not all completed, reports sent by e-mail, and further gains like reduction of cycle care time, correct PID, reduction cost treatment and resources, better outcomes for the patient, much more productivity, promote the institutional trust and patient health status, and all of this without increasing human resources, nor applying huge hardware nor software investments.

Additionally, other statistical tools are now available to administration, like haemoglobin, medical and administrative reports.

References

1. Williams, F., Boren, S.: The role of the electronic medical record (EMR) in care delivery development in developing countries: a systematic review. J. Innov. Health Inform. **16**(2), 139–145 (2008)
2. John-Otumu, M.A., Eriata, F.U., Obahiagbon, K.: Framework for developing and implementing an automated hospital management system on an intranet. Nigerian Ann. Nat. Sci. **13**(2), 039–049 (2018)
3. Kalogriopoulos, N.A., Baran, J., Nimunkar, A.J., Webster, J.G.: Electronic medical record systems for developing countries: review. In: Annual International Conference of the IEEE Engineering in Medicine and Biology Society, 53706, pp. 1730–1733 (2009)
4. Musen, M.A., Van Bemmel, J.H.: Handbook of Medical Informatics. Springer, Houten/Diegem (1997)
5. Idowu, P., Cornford, D., Bastin, L.: Health informatics deployment in Nigeria. J. Health Inform. Dev. Countries **2**(1), 15–23 (2008)
6. Leonard, K.J.: The role of patients in designing health information systems: the case of applying simulation techniques to design an electronic patient record (EPR) interface. Health Care Manag. Sci. **7**(4), 275–284 (2004). https://doi.org/10.1007/s10729-004-7536-0
7. Øvretveit, J., Scott, T., Rundall, T.G., Shortell, S.M., Brommels, M.: Implementation of electronic medical records in hospitals: two case studies. Health Policy **84**(2–3), 181–190 (2007)
8. Clifford, G.D., Blaya, J.A., Hall-Clifford, R., Fraser, H.S.F.: Medical information systems: a foundation for healthcare technologies in developing countries. BioMed. Eng. OnLine **7**(1), 18 (2008). https://doi.org/10.1186/1475-925X-7-18
9. Teixeira, P., Sá, J.C., Silva, F.J., Santos, G., Fontoura, P., Coelho, A.: Lean contribution to the companies' sustainability. In: Powell, D.J., Alfnes, E., Holmemo, M.D.Q., Reke, E. (eds.) Learning in the Digital Era. ELEC 2021. IFIP Advances in Information and Communication Technology, vol. 610, pp. 400–408. Springer, Cham (2021). https://doi.org/10.1007/978-3-030-92934-3_42
10. Cordeiro, P., Sá, J.C., Pata, A., Gonçalves, M., Santos, G., Silva, F.J.G.: The impact of lean tools on safety—case study. In: Arezes, P.M., Santos Baptista, J., Barroso, M.P., Carneiro, P., Cordeiro, P., Costa, N., Melo, R.B., Sérgio Miguel, A., Perestrelo, G. (eds.) Occupational and Environmental Safety and Health II. SSDC, vol. 277, pp. 151–159. Springer, Cham (2020). https://doi.org/10.1007/978-3-030-41486-3_17
11. Silva, S., Sá, J.C., Silva, F.J.G., Ferreira, L.P., Santos, G.: Lean green—the importance of integrating environment into lean philosophy—a case study. In: Rossi, M., Rossini, M., Terzi, S. (eds.) ELEC 2019. LNNS, vol. 122, pp. 211–219. Springer, Cham (2020). https://doi.org/10.1007/978-3-030-41429-0_21
12. Heaton, H.A., Nestler, D.M., Barry, W.J., Helmers, R.A., Sir, M.Y., Goyal, D.G.: A time-driven activity-based costing analysis of emergency department scribes. Mayo Clin. Proc. Innov. Qual. Outcomes **3**(1), 30–34 (2019)
13. Porter, M., Lee, T.: The strategy that will fix health care. Harv. Bus. Rev. **91**, 1–18 (2013)
14. Chou, J., Shah, M., Watts, A., Gardiner, M., Kaplan, R., Miller, J.: TDABC cost analysis of ocular disorders in an ophthalmology emergency department versus urgent care: clinical experience at massachusetts eye and ear. J. Acad. Ophthalmol. **10**(01), e55–e60 (2018)
15. Smith, N.L., Cohen, A., Pickett, A.C.: Exploring the motivations and outcomes of long-term international sport-for-development volunteering for American Millennials. J. Sport Tourism **19**(3–4), 299–316 (2016)
16. NHS England and NHS Improvement: Sustainability Model (2010)

17. Handy, F., Mook, L.: Volunteering and volunteers: benefit-cost analyses. Res. Soc. Work. Pract. **21**(4), 412–420 (2011)
18. Chinman, M.J., Wandersman, A.: The benefits and costs of volunteering in community organizations: review and practical implications. Nonprofit Volunt. Sect. Q. **28**(1), 46–64 (1999)
19. Almomani, I., AlSarheed, A.: Enhancing outpatient clinics management software by reducing patients' waiting time. J. Infect. Public Health **9**(6), 734–473 (2016)
20. Cavalieri, M.: Geographical variation of unmet medical needs in Italy: a multivariate logistic regression analysis. Int. J. Health Geogr. **12**(27), 1–11 (2013)
21. Stein, Y.R.: Volunteering to colonize: a cost-benefit analysis of the impacts of voluntourism. Portland State University (2017)
22. Daldeniz, B., Hampton, M.P.: Charity-based voluntourism versus "lifestyle" voluntourism: evidence from Nicaragua and Malaysia (2010)
23. Q and A: This is how stakeholder capitalism can help heal the planet. https://www.weforum.org/agenda/2020/01/stakeholder-capitalism-environment-planet/. Accessed 14 Dec 2021

Meteorological Time Series: An Exploratory Statistical and Critical Analysis

A. Manuela Gonçalves[1], F. Catarina Pereira[1], Marco Costa[2], and Celina P. Leão[3(✉)]

[1] Department of Mathematics and Center of Mathematics, University of Minho, Braga, Portugal
mneves@math.uminho.pt, id9976@alunos.uminho.pt

[2] Águeda School of Technology and Management and Center for Research and Development in Mathematics and Applications, University of Aveiro, Aveiro, Portugal
marco@ua.pt

[3] ALGORITMI Centre, University of Minho, Guimarães, Portugal
cpl@dps.uminho.pt

Abstract. Increasingly, reduction of water availability has been a reality, and population growth, pollution, and climate change have contributed to exacerbating this problem. Dry periods, which occur when precipitation is lower than expected in a given territory, have become more frequent and prolonged, and therefore it is crucial to efficiently manage water use in response to environmental concerns. The main challenge in this work is to present the irrigation problem as an optimal control problem along with the presentation of preliminary results based on an exploratory statistical and critical analysis of daily meteorological variables. The variables considered are: maximum air temperature, minimum air temperature, and total precipitation recorded during the last ten years (2010–2019). The methodology followed, based on state-space models, shows flexibility to allow the integration of new data, updating in real time the model, and the incorporation of covariates that are important to explain the process in analysis.

Keywords: Exploratory data analysis · Meteorological variables · Imputation · Irrigation · Time series

1 Introduction

The multiple effects of climate change on the planet are a growing concern because dry periods have become more frequent, and the increasing threats associated with meteorological phenomena have deeply affected water availability and the water cycle, which has a fundamental role in regulating the ecosystem.

The rise in temperatures that have occurred in recent years and the drought episodes that have been faced constitute, in a global context, economic, social, and environmental unbalances.

J. Machado et al. (Eds.): icieng 2022, LNME, pp. 197–208, 2023.
https://doi.org/10.1007/978-3-031-09360-9_17

Agriculture is one of the sectors undergoing great impact from the disruption of weather systems because this economic sector is directly dependent on weather conditions: approximately 69% of the world's freshwater is used in agriculture [1]. There is much that can be done to save water, and this collective effort is of the utmost importance to our planet.

Some solutions have been proposed to increase water availability, such as desalination. However, this solution has a major disadvantage because, in most cases, this technique uses a lot of energy and thus contributes to a increasing greenhouse gas emissions if the energy source is not renewable [2,3].

The excess of water in the soil is responsible for significant water waste, and therefore we have to create solutions to mitigate the dwindling of water resources. Understanding the behavior of humidity in the soil allows, among other advantages, efficient planning of water use, particularly in irrigation systems.

This study is carried out in the context of the project "TO CHAIR - The Optimal Challenges in Irrigation" (`systec.fe.up.pt/projects/FCT-TOCHAIR/`), which aims to understand and analyze the behavior of humidity in the soil in order to find optimal solutions to improve the efficiency of daily water use in irrigation systems [4].

The main goal of this work is to study the behavior and perform an exploratory statistical and critical analysis of three daily weather variables, namely maximum air temperature, minimum air temperature, and total precipitation, to particularly determine evapotranspiration [5] (related to the irrigation planning problem). The ultimate goal is to predict these environmental variables at a given location (in this case, a farm in the Portuguese district of Bragança), within a time horizon of 7 days.

In environmental variables, it is common to encounter missing values [6,7]. In this work we will use two approaches to impute these missing values, namely the linear interpolation [8], and a linear regression analysis [9] where the independent variable has a strong association with the dependent variable.

2 Exploratory Data Analysis

Our data source are the records of the three weather variables (maximum air temperature, minimum air temperature, and total precipitation) observed at a farm located in the municipality of Bragança in northern Portugal, recorded at the farm by a portable station in the period from January 1st, 2010 to December 31st, 2019 on a daily basis.

In Fig. 1 are represented the time series of the meteorological variables under study. To begin with, the presence of a strong seasonal pattern in the series of maximum and minimum air temperatures is noticeable. In addition, the series of total precipitation shows great variability with several extreme values. Furthermore, all three meteorological variables present missing values that have to be addressed.

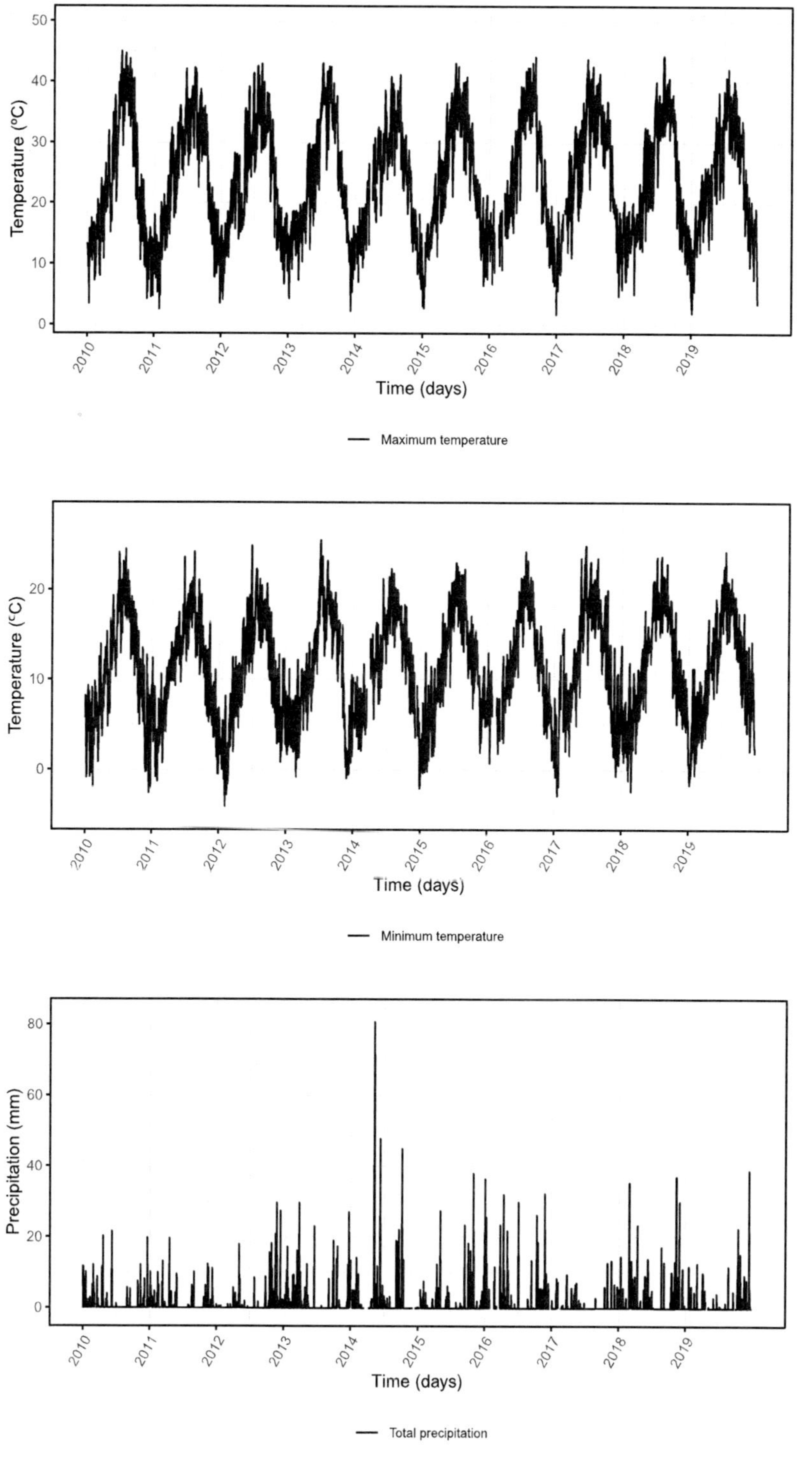

Fig. 1. Graphical representation of daily meteorological time series: maximum temperature, minimum temperature, and total precipitation.

The boxplots of each of the three variables under study are shown in Fig. 2, where both variables of maximum and minimum air temperature seem to have a symmetrical distribution, besides not presenting outliers, unlike the variable of total precipitation, which presents many outliers, highlighting the highest value on the order of 80 mm, and seems to have a positively skewed (right-skewed) distribution, which can be seen in the boxplot of total precipitation without outliers (Fig. 2, in the lower right corner). In the study, these outliers of total precipitation were not removed, they were just not displayed in Fig. 2 to be able to visualize and analyze the boxplot.

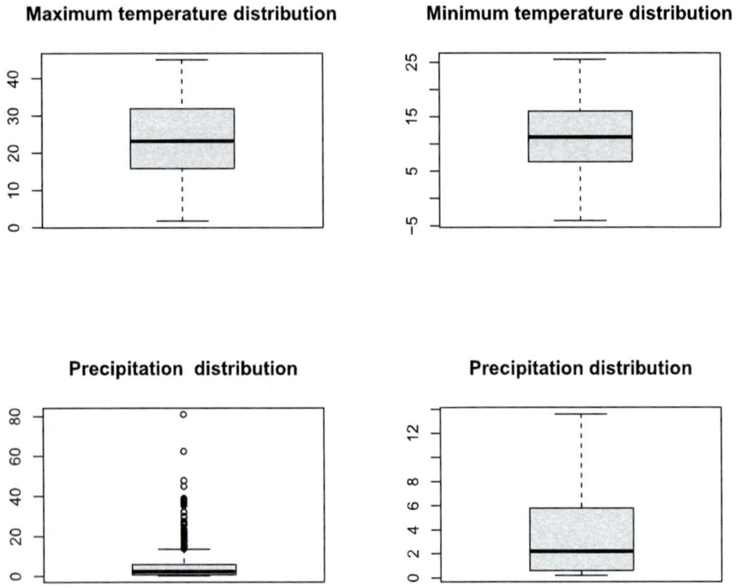

Fig. 2. Boxplot of meteorological time series: maximum temperature (°C), minimum temperature (°C), and total precipitation (mm) with and without outliers.

2.1 Exploratory Data Analysis by Year

Table 1 shows the descriptive statistics of maximum air temperature by year. In general, the maximum temperature averages between 2010 and 2019 do not present great differences, with 2017 being the hottest year in this period recording an average temperature of 25.34 °C. The highest temperature was recorded in 2010, a year which also presented a higher standard deviation (10.79 °C). 2018 was the only year where there are no gaps, unlike 2016, which presents 43 missing values (NA's). In total, this variable presents 122 NA's.

Table 1. Descriptive statistics of maximum air temperature by year.

Year	Min.	Max.	Range	1st Quart.	Median	3rd Quart.	Mean	Std deviation	NA's (122)
2010	3.40	45.00	41.60	14.15	20.65	33.48	23.32	10.79	5
2011	2.50	42.30	39.80	16.00	25.60	32.50	24.19	9.51	4
2012	4.10	43.00	38.90	16.30	22.80	31.20	23.54	9.12	1
2013	2.20	43.10	40.90	14.88	21.90	32.15	23.20	9.86	9
2014	5.90	41.10	35.20	16.75	25.30	30.35	24.09	8.33	38
2015	2.70	43.10	40.40	15.75	23.00	31.75	23.52	9.56	2
2016	3.20	44.10	41.40	16.65	23.10	32.75	24.52	9.70	43
2017	1.70	43.80	42.10	18.15	26.20	33.10	25.34	9.36	10
2018	5.00	44.30	39.30	15.20	21.10	32.30	23.37	9.33	0
2019	1.90	42.10	40.20	16.45	22.70	30.70	23.43	9.00	10

As regards the minimum temperature, it appears that 2012 presented the lowest minimum temperature recorded (−4.10 °C). Also in 2012, about 25% of the observations were below 5.70 °C. The pattern about missing values repeats itself as in the maximum temperature (see Table 2). The exploratory analysis of total precipitation (Table 3) only considered the days on which precipitation occurred. There was a total 882 days (25.75% of the data). From Table 3, the minimum value recorded was 0.20 mm for all observed years. 2014 stands out as presenting the maximum value of precipitation (80.80 mm), which is clearly visible in Fig. 1; the annual standard deviation indicates a higher variability in the year 2014 as well. In 2018, approximately 50% of the days had total precipitation greater than 2.90 mm.

Table 2. Descriptive statistics of minimum air temperature by year.

Year	Min.	Max.	Range	1st Quart.	Median	3rd Quart.	Mean	Std deviation	NA's (122)
2010	−2.60	24.50	27.10	6.68	10.65	16.10	10.99	6.08	5
2011	−1.30	24.20	25.50	6.90	11.40	15.00	11.10	5.36	4
2012	−4.10	24.90	29.00	5.70	10.50	15.70	10.30	6.18	1
2013	−1.00	25.50	26.50	5.80	10.10	15.80	10.56	6.28	9
2014	−2.10	22.30	24.40	7.95	12.10	16.10	12.01	5.06	38
2015	−1.50	23.00	24.50	7.15	11.20	15.90	11.29	5.84	2
2016	0.70	24.20	23.50	7.90	11.80	16.45	11.96	5.29	43
2017	−2.90	24.90	27.80	6.45	11.40	16.90	11.40	6.52	10
2018	−2.40	23.70	26.10	6.80	10.80	16.80	11.27	5.78	0
2019	−1.70	24.20	25.90	5.90	11.40	15.75	10.99	5.99	10

Table 3. Descriptive statistics of precipitation by year.

Year	Min.	Max.	Range	1st Quart.	Median	3rd Quart.	Mean	Std deviation	NA's (122)
2010	0.20	21.60	21.40	0.80	2.10	4.90	3.71	4.57	5
2011	0.20	19.60	19.40	0.40	1.60	4.70	3.18	3.79	4
2012	0.20	29.60	29.40	0.40	1.20	3.50	3.65	6.02	1
2013	0.20	29.60	29.40	0.60	2.60	7.60	5.10	6.36	9
2014	0.20	80.80	80.60	0.80	2.60	6.75	7.17	13.77	38
2015	0.20	38.00	37.80	0.40	1.90	4.50	4.05	6.43	2
2016	0.20	36.40	36.20	0.80	2.80	8.20	6.41	8.24	43
2017	0.20	13.40	13.20	0.80	2.20	5.15	3.39	3.28	10
2018	0.20	37.20	37.00	1.00	2.90	7.90	5.31	6.72	0
2019	0.20	38.80	38.60	0.60	2.00	4.70	3.94	2.50	10

2.2 Exploratory Data Analysis by Month

Another way to analyze time series data is to consider monthly subseries. Table 4 shows the descriptive statistics of maximum air temperature in degrees, by month. The average maximum temperature is higher in spring/summer. On the other hand, January presents the lowest average maximum temperature (12.07 °C). The standard deviation indicates higher variability in May, June, and October.

Table 4. Descriptive statistics of maximum air temperature by month.

Month	Min.	Max.	Range	1st Quart.	Median	3rd Quart.	Mean	Std deviation	NA's (122)
Jan.	1.70	20.50	18.80	9.80	12.70	14.70	12.07	3.80	15
Feb.	2.50	24.90	22.40	12.62	15.25	17.70	15.20	3.60	32
Mar.	7.90	30.30	22.40	15.70	18.40	22.22	18.85	4.30	30
Apr.	10.80	32.80	22.00	18.12	22.00	25.98	22.18	4.89	14
May	15.10	38.00	22.90	22.50	27.30	30.50	26.67	5.13	1
June	19.00	43.80	24.80	27.60	31.40	35.50	31.47	5.37	0
July	22.80	45.00	22.20	32.95	35.85	38.60	35.60	4.22	0
Aug.	24.60	44.30	19.70	32.82	36.15	38.70	35.78	3.94	0
Sept.	20.20	44.10	23.90	29.10	32.30	34.90	31.91	4.53	2
Oct.	10.60	36.80	26.20	20.60	24.30	28.00	24.52	5.29	10
Nov.	4.20	23.80	19.60	13.90	16.65	18.80	16.34	3.56	0
Dec.	2.20	21.40	19.20	9.70	12.85	14.93	12.33	3.70	18

Regarding amplitudes, October stands out with an amplitude of 26.20 °C. Only the months of June, July, August, and November have no missing values. Regarding the minimum air temperature (Table 5), January and February present the lowest average minimum temperature (4.41 and 4.38 °C, respectively). In July, about

25% of the days registered minimum temperatures above 20.30 °C. February presented the lowest minimum temperature recorded (−4.10°C). Again, the pattern of missing values repeats itself as in the maximum temperature.

Table 5. Descriptive statistics of minimum air temperature by month.

Month	Min	Max	Range	1st Quart.	Median	3rd Quart.	Mean	Std deviation	NA's (122)
Jan.	−2.90	13.80	16.70	1.70	4.40	6.90	4.41	3.47	15
Feb.	−4.10	15.70	19.80	2.10	4.35	6.57	4.38	3.46	32
Mar.	0.70	13.90	13.20	4.70	6.55	8.43	6.59	2.67	30
Apr.	3.00	16.80	13.80	7.90	9.50	11.38	9.53	2.36	14
May	4.80	23.50	18.70	10.50	12.40	14.40	12.34	2.93	1
June	8.70	24.90	16.20	13.38	15.45	17.80	15.67	3.08	0
July	10.70	25.50	14.80	16.92	18.60	20.30	18.60	2.65	0
Aug.	12.40	24.50	12.10	17.20	18.50	20.00	18.50	2.20	0
Sept.	10.20	22.90	12.70	14.43	16.30	17.88	16.00	2.39	2
Oct.	5.00	20.00	15.00	10.38	12.40	14.50	12.34	2.90	10
Nov.	−1.10	17.40	18.50	5.77	8.20	10.50	8.02	3.45	0
Dec.	−2.60	14.20	16.80	2.90	5.60	7.60	5.38	3.30	18

Table 6 shows the descriptive statistics of total precipitation by month. Similar to the analysis of this variable by year, only the days on which precipitation occurred were considered. Variability is higher in February, with a standard deviation of 13.46 mm. Regarding the amplitudes, May stands out with an amplitude of 80.60 mm. The highest average values of total precipitation were recorded in May and September (6.39 and 6.57 mm, respectively). The behavior of the missing values is repeated, with the months of February and March standing out as having the highest number of NA's.

Table 6. Descriptive statistics of total precipitation by month.

Month	Min.	Max.	Range	1st Quart.	Median	3rd Quart.	Mean	Std deviation	NA's (122)
Jan.	0.20	36.40	36.20	0.40	2.00	5.25	3.81	3.87	15
Feb.	0.20	35.40	35.20	0.80	2.20	6.30	3.97	13.46	32
Mar.	0.20	29.60	29.40	0.80	1.80	5.80	4.15	8.27	30
Apr.	0.20	32.00	31.80	0.75	2.00	4.60	4.01	6.11	14
May	0.20	80.80	80.60	0.80	2.40	6.80	6.39	4.44	1
June	0.20	47.80	47.60	0.60	1.60	6.20	5.04	6.44	0
July	0.20	29.80	29.60	0.60	1.40	3.55	3.90	6.78	0
Aug.	0.20	17.20	17.00	0.60	1.40	3.10	3.28	6.95	0
Sept.	0.20	23.40	23.20	1.05	4.90	9.85	6.57	7.51	2
Oct.	0.20	45.00	44.80	0.55	2.90	7.60	5.94	6.98	10
Nov.	0.20	38.00	37.80	0.70	2.20	5.80	5.20	6.28	0
Dec.	0.20	38.80	38.60	0.60	2.00	4.50	4.12	5.85	18

Figure 3 shows the boxplots of maximum air temperature, minimum air temperature, and total precipitation by month. It can be seen that precipitation shows several discrepant values, unlike the maximum/minimum temperature variables. The monthly subseries of maximum/minimum air temperatures show a clear seasonal pattern, reaching higher temperatures in July and August. The median of the maximum/minimum air temperatures gradually decreases from summer to winter. The median of total precipitation is higher in September.

The analysis revealed the existence of random missing values. This situation is quite common in real data and it may derive from multiple factors - e.g., sensors malfunction at the weather station, data loss, etc. Dealing with time series with missing data can be quite problematic because most modeling approaches are designed for complete time series. To solve this problem, a possible solution is to do imputation, which consists in filling these missing values based on some properties of the data.

2.3 Predictive Model

In this work, two types of imputation were chosen. One of them was conducted through a database with daily records of maximum air temperature and minimum air temperature between January 1 to December 31, 2019 in Vila Real, located about 50 km from the site under study (i.e., a farm in the Portuguese district of Bragança). The Vila Real database was selected because these are a good predictor of the location of the farm. The interpolation was done through a linear interpolation model fitted to a **linear regression model** given by $Y_t = \alpha + \beta X_t + e_t$, where Y_t and X_t are the maximum/minimum air temperature recorded at the farm and in Vila Real, respectively, and t is the time, in days. This method was not considered for imputation of missing data for total precipitation since there is no strong correlation for this variable between the two locations: the farm and Vila Real.

Therefore, the model was adjusted for the maximum and minimum air temperatures variables and the coefficients of determination obtained were 0.9340 and 0.9089, respectively. It should be noted that the imputation of data through this method was done only for the year 2019 given the availability of data. For all other missing values (between 2010 and 2018 for maximum/minimum air temperature, and for all total precipitation data), it was decided to impute through **linear interpolation** [8,10].

Figure 4 shows the series of maximum air temperature, minimum air temperature, and total precipitation with the imputed values (in red). Therefore, for the series of the maximum/minimum air temperatures from 2010 up to and including 2018, the imputation was done through linear interpolation, and in 2019 through the linear regression model between the farm data and the Vila Real data. For the total precipitation variable, imputation of all the missing data was done through linear interpolation.

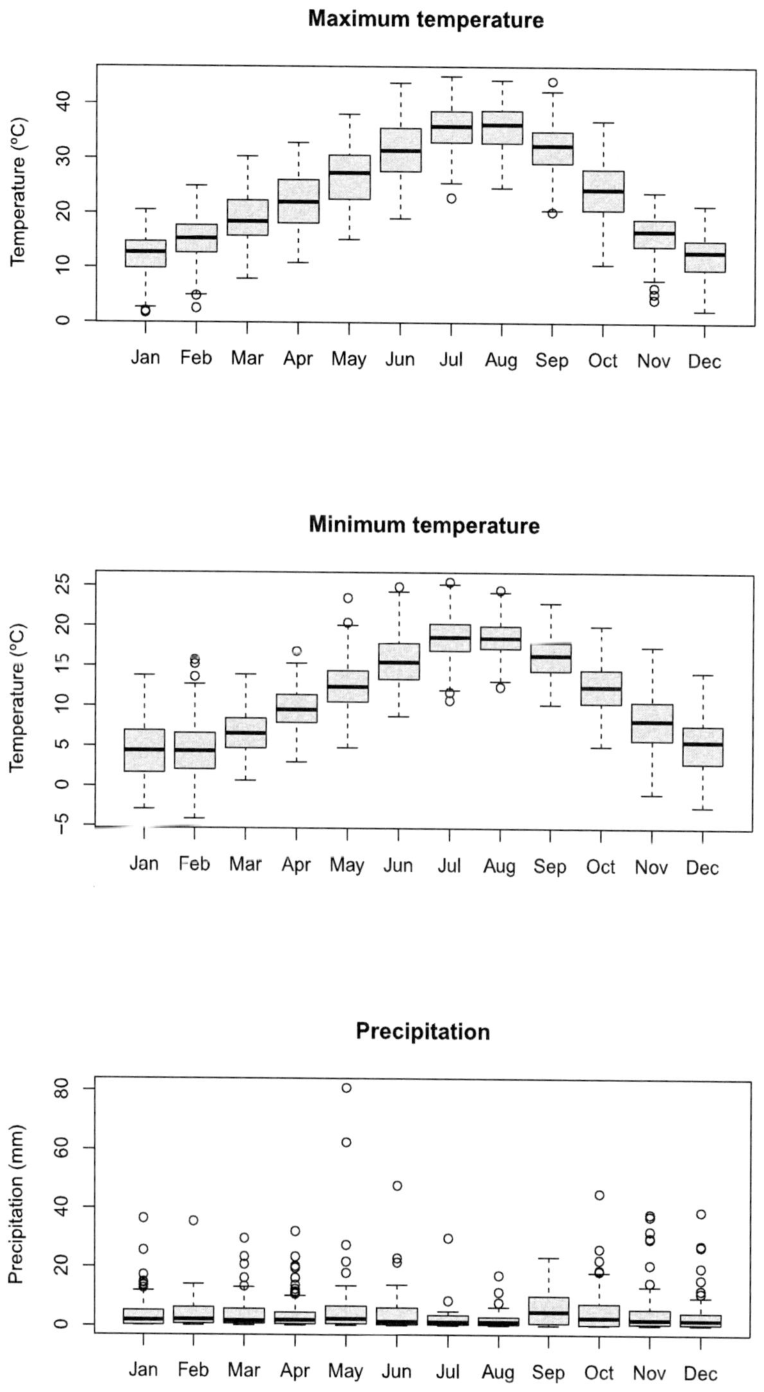

Fig. 3. Boxplots of maximum air temperature, minimum air temperature, and total precipitation by month.

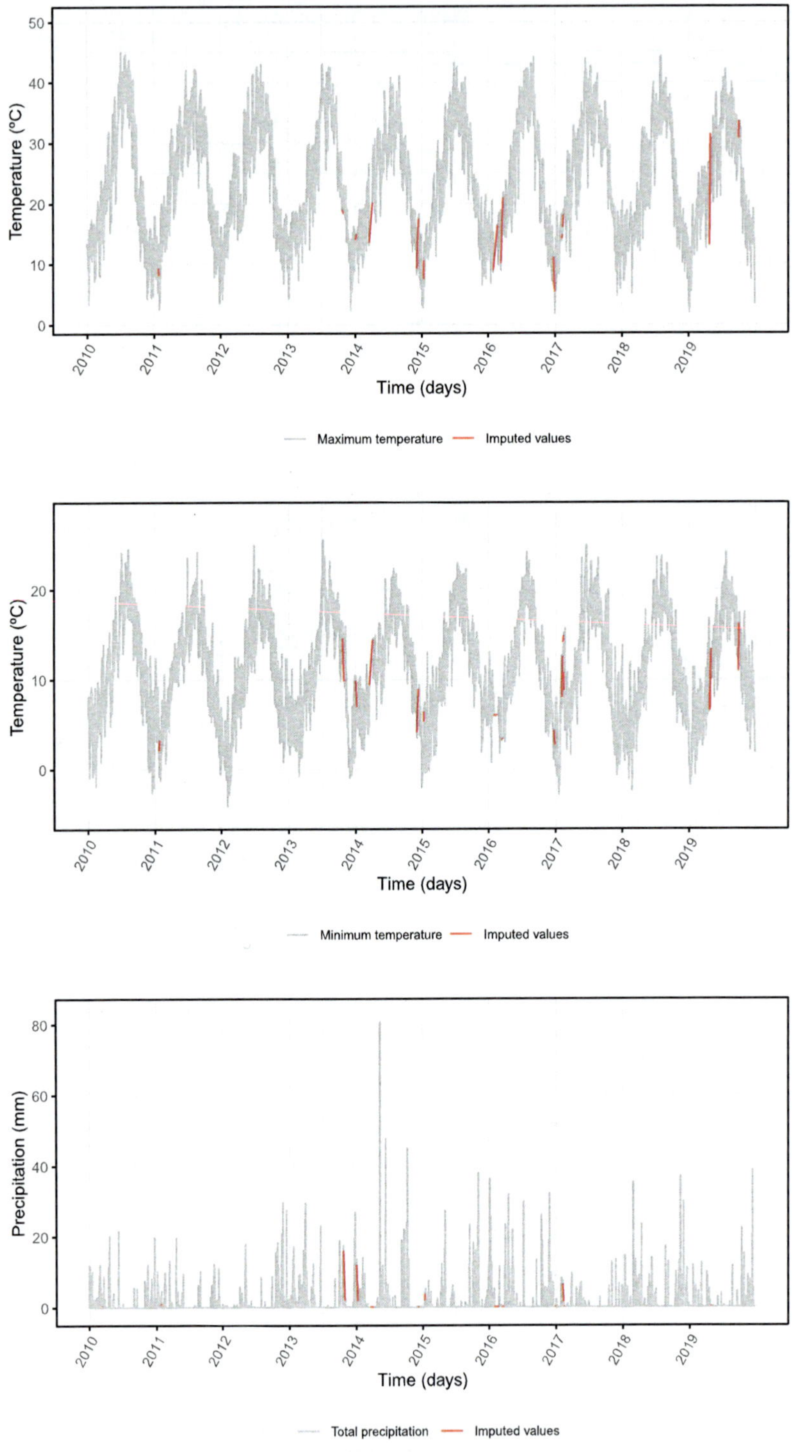

Fig. 4. Graphical representation of daily meteorological time series (in black), and imputed values (in red).

3 Discussion and Future Work

In this work, we address the daily irrigation problem of minimizing water consumption. For this, we selected three weather variables: maximum air temperature, minimum air temperature, and total precipitation. This type of preliminary data analysis performed in this work is fundamental in any time series analysis because it allows the study and identification of data behavior that will provide relevant information for the application of appropriate models, with the ultimate goal of forecasting, allowing for better modeling and forecasting.

After the exploratory analysis, the modeling and forecasting step follows. Given the behavior of meteorological time series, two types of models stand out: TBATS (Trigonometric Seasonal, Box-Cox Transformation, ARMA errors, Trend and Seasonal Components) and state-space models. These models have successfully been applied to modeling and foresight scenarios of environmental and weather variables.

TBATS models are suitable for modeling time series with strong trend and complex seasonal patterns (multiple seasonal patterns), which are predominant features in weather series. For example, in [11], they modeled the minimum air temperature series (recorded on the same farm) using TBATS models, which proved to be very appropriate for the daily forecasting approach.

State space models are a very flexible class of models as they allow several features of the data to be integrated explicitly into the model structure itself, and they also allow predictions to be made and updated in real time whenever a new observation is integrated into the model [12–14]. In addition, they allow incorporating covariates that are important to explain the process. In [15], they proposed a model with a state-space representation that establishes a stochastic linear relationship between the maximum temperature observed at a farm (the same database) and the h-days ahead forecast ($h = 1, \ldots, 6$) produced from a weather website (as a covariate).

In order to have more precise results, for future work it is suggested to extend the study to other regions of Portugal.

Acknowledgements. This research was partially supported by The Center for Research and Development in Mathematics and Applications (CIDMA) through the Portuguese Foundation for Science and Technology (FCT - Fundação para a Ciência e a Tecnologia), references UIDB/04106/2020 and UIDP/04106/2020. This research was partially financed by Portuguese funds through Portuguese Foundation for Science and Technology ("Fundaçào para a Ciência e a Tecnologia" - FCT), within the Projects UIDB/00013/2020 and UIDP/00013/2020 of CMAT-UM. This research was financed by national funds through FCT (Fundação para a Ciência e a Tecnologia) through the individual PhD research grant UI/BD/150967/2021 of CMAT-UM. This work was also supported by the Portuguese Foundation for Science and Technology (FCT) in the framework of the Strategic Funding UID/FIS/04650/2019, UIDB/00013/2020 and UIDP/00013/2020 of CMAT-UM. The authors are granted for the access to the meteorological data provided by the Direção Regional de Agricultura e Pescas do Norte. This work has been supported by FCT - Fundação para a Ciência e Tecnologia within the R&D Units Project Scope: UIDB/00319/2020.

References

1. United Nations: The United Nations World Water Development Report 2021: Valuing water. UNESCO, Paris (2021)
2. Ghalavand, Y., Hatamipour, M.S., Rahimi, A.: A review on energy consumption of desalination processes. Desalination Water Treat. **54**(6), 1526–1541 (2015). https://doi.org/10.1080/19443994.2014.892837
3. Zarzo, D., Prats, D.: Desalination and energy consumption. What can we expect in the near future? Desalination **427**, 1–9 (2018). https://doi.org/10.1016/j.desal.2017.10.046
4. Costa, C., Gonçalves, A.M., Costa, M., Lopes, S.: Forecasting temperature time series for irrigation planning problems. In: Proceedings of the 34th International Workshop on Statistical Modelling (IWSM 2019) (2019). http://hdl.handle.net/10773/26437
5. Lopes, S., Pereira, R., Pereira, P., Caldeira, A., Fonte, V.: Optimal control applied to an irrigation planning problem: a real case study in Portugal. Int. J. Hydrol. Sci. Technol. **9**(2), 173–188 (2019). https://doi.org/10.1504/IJHST.2019.098161
6. Gonçalves, A.M., Costa, M.: Predicting seasonal and hydro-meteorological impact in environmental variables modelling via Kalman filtering. Stoch. Environ. Res. Risk Assess. **27**, 1021–1038 (2012). https://doi.org/10.1007/s00477-012-0640-7
7. Costa, M., Monteiro, M.: Statistical modelling of water quality time series - the river vouga basin case study. In: Research and Practices in Water Quality. IntechOpen (2015). https://doi.org/10.5772/59062
8. Hyndman, R.J., Athanasopoulos, G.: Forecasting: Principles and Practice, 2nd edn. OTexts, Melbourne (2018)
9. Di Franco, G.: An alternative procedure for imputing missing data based on principal components analysis. Qual. Quant. **48**(3), 1149–1163 (2013). https://doi.org/10.1007/s11135-013-9826-4
10. Moritz, S., Sardá, A., Bartz-Beielstein, T., Zaefferer, M., Stork, J.: Comparison of different Methods for Univariate Time Series Imputation in R (2015)
11. Gonçalves, A.M., Costa, C., Costa, M., Lopes, S.O., Pereira, R.: Temperature time series forecasting in the optimal challenges in irrigation (TO CHAIR). In: Gaspar-Cunha, A., Periaux, J., Giannakoglou, K.C., Gauger, N.R., Quagliarella, D., Greiner, D. (eds.) Advances in Evolutionary and Deterministic Methods for Design, Optimization and Control in Engineering and Sciences. CMAS, vol. 55, pp. 423–435. Springer, Cham (2021). https://doi.org/10.1007/978-3-030-57422-2_27
12. Harvey, A.C.: Forecasting, Structural Time Series Models and the Kalman Filter. Cambridge University Press, New York (2009)
13. Costa, M., Alpuim, T.: Parameter estimation of state space models for univariate observations. J. Stat. Plan. Inference **140**(7), 1889–1902 (2010)
14. Shumway, R.H., Stoffer, D.S.: Time Series Analysis and Its Applications: With R Examples, 4th edn. Springer, Cham (2017). https://doi.org/10.1007/978-3-319-52452-8
15. Costa, M., Pereira, F.C., Gonçalves, A.M.: Improving short-term forecasts of daily maximum temperature with the Kalman filter with GMM estimation. In: Gervasi, O., et al. (eds.) ICCSA 2021. LNCS, vol. 12952, pp. 552–562. Springer, Cham (2021). https://doi.org/10.1007/978-3-030-86973-1_39

Beneficiaries of Social Disability Pension in Small Municipalities in the Northern Region of Portugal: Application of Cluster Analysis in the Identification of Potential Causes

Cristina Torres[1(✉)], Isabel Vieira[1], Cristina Lopes[1], Rui Monteiro[2], Carla Ferreira[2], and Inês Bem-Haja[2]

[1] CEOS.PP, ISCAP, Polytechnic of Porto, Porto, Portugal
ctorres@iscap.ipp.pt
[2] ISCAP, Polytechnic of Porto, Porto, Portugal

Abstract. This study is focused on the northern region of Portugal and addresses the number of road accidents with casualties and the number of beneficiaries of the social disability pension. Municipalities with low population density are mostly located in the hinterland. The phenomenon of population aging entails the problem of people who are unable to work and become beneficiaries of the social disability pension. The hierarchical clustering method applied in this paper characterizes 46 municipalities in northern Portugal with a population of less than 20,000 inhabitants – designated small – according to economic activity, rate of road accidents with casualties and disability pensioners. These municipalities were divided into two homogeneous groups or clusters. It was possible to identify that Cluster 1 comprises the innermost municipalities featuring few road accidents with casualties, not many disability pensioners, and a small number of companies compared to Cluster 2. This study can contribute to helping local governments' decision-making process to improve the economic activity, reduce the region's human desertification, and improve municipal road conditions.

Keywords: Beneficiaries of social disability pension · Small municipalities · Northern region of Portugal · Cluster analysis · Hierarchical method

1 Introduction

A major problem affecting Portugal is the desertification of the country's hinterland and its impact on the quality of life and the future. This work focuses on the problem of beneficiaries of social disability pensions who, in result of accidents and/or illness, are unable to work, and consequently unable to benefit from their works earnings that would provide them with a better quality of life.

J. Machado et al. (Eds.): icieng 2022, LNME, pp. 209–219, 2023.
https://doi.org/10.1007/978-3-031-09360-9_18

This problem may present an added dimension in municipalities with low population density, where any increase in this indicator will have impactful statistic implications for the active population of the respective territories in result of the low number of working-age citizens. In the analysis of the case under study, the official statistical indicators included in the Statistical Yearbook of the Northern Region 2018 [1] were used, published by Statistics Portugal in 2019.

The northern region of Portugal comprises 86 municipalities, which in turn comprise a Metropolitan Area and seven Intermunicipal Communities (see Fig. 1).

Fig. 1. Municipalities in the northern region of Portugal [2]

In Fig. 2 we can observe the 86 municipalities of the northern region of Portugal. Among these 86 municipalities, 10 have a population of over 100,000 inhabitants (large), where 6 of them are located in the Porto Metropolitan Area: Gondomar, Maia, Matosinhos, Porto, Santa Maria da Feira, and Vila Nova de Gaia; and 4 are located in Minho: Barcelos, Braga, Guimarães, and Vila Nova de Famalicão. These 10 large population-sized municipalities represent 12% of the total number of municipalities in the northern region. There are also 30 municipalities with a population of over 20,000 and less than 100,000 inhabitants (medium), representing 35% of the municipalities in the region, and 46 municipalities with a population of less than 20,000 inhabitants (small), which represent 53% of the region's municipalities. As can be seen in Fig. 2, most are located in the region's hinterland. Our study is focused on this group of small municipalities.

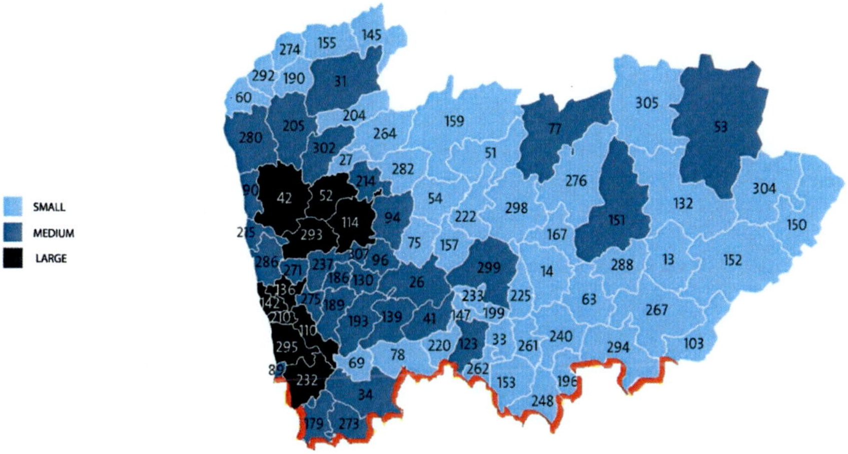

Fig. 2. Municipalities of the northern region, according to their inhabitants number [3]

In this context, if we add the Population Aging Index, according to Fig. 3, we find that the inland municipalities, which are also the smaller ones, are deeply affected by the reality of population aging. Any indicator that may have an influence on the working population of a municipality will contribute to exacerbating inequalities among the different territories, mainly from a socioeconomic perspective.

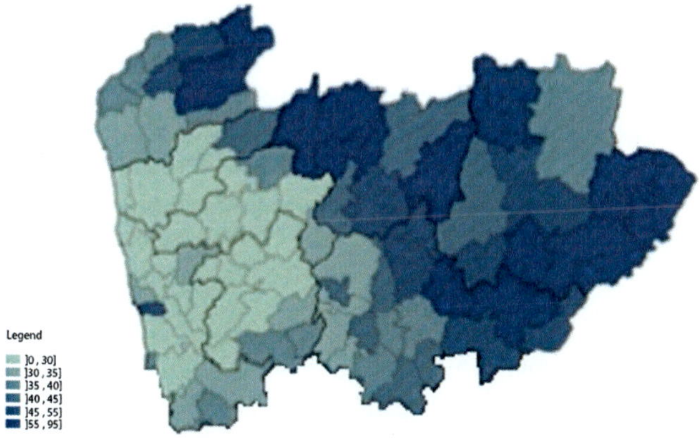

Fig. 3. Elderly dependency index: Number of people aged 65 and over, per 100 people of working age (15–64), by municipality in the northern region, in 2017 [4]

The main goal of the study is to understand the distribution of the number of beneficiaries of social disability pension to classify the municipalities into groups. For that, the search for other variables that could contribute to this dimension resulted in analyzing the high road traffic flow and the occurrence of road accidents with casualties, and also the economic activity of municipalities, proxied by the number of existing companies.

2 Brief Literature Review

Sarmento and Costa [5] define cluster analysis as the process of grouping variables accordingly to similarity criteria. Cluster analysis is important because it reduces the dimensionality of the problem, as it identifies the major patterns in the dataset. This statistic technique can be applied to a variety of research fields, such as health [6], water quality and consumption [7,8], electricity prices [9], finances [10], business and decision making [11,12], among others.

The objective of clustering is to make a partition of the dataset into distinct sets (clusters), according to a specific measure of similarity. The purpose of clustering is to minimize intra-cluster distance and maximize inter-cluster distance. To find the distance between two clusters, hierarchical or non-hierarchical techniques can be used.

In this paper, the hierarchical method is used. As Marôco [13] explains, it has to be done step by step. It is necessary to decide if we start by putting all the variables in a single cluster and then proceed to split into subgroups according to their distance measures, or the other way around. Depending on how the distances are measured, a different hierarchical method is obtained. Reis [14] suggests that cluster analysis can be done in five steps: starting by choosing the objects to be grouped; then defining a set of variables from which information is gathered to group the objects; afterwards it is necessary to define a measure of similarity between each of the two objects; the next step consists in selecting the criteria to aggregate or separate the objects; the final step is the validation of the results obtained.

3 Methodology

In order to collect a significant and credible sample of indicators from the municipalities in the northern region of Portugal, data was retrieved from the Statistical Yearbook of the Northern Region 2018 [1], which contains multiple indicators that characterize the region. As we intended to analyze only the small municipalities (with a population of less than 20,000 inhabitants), data from 46 municipalities were used. The chosen variables are the number of existing companies in each municipality, the number of road accidents with casualties that occurred in each municipality, and the number of beneficiaries of social disability pension existing in each municipality under study.

In the present analysis, IBM SPSS Statistics v26 software is used to apply hierarchical cluster analysis to these 46 municipalities. To perform the hierarchical cluster analysis, three steps were followed: Calculating the distances, linking the clusters, and finally selecting the "right" number of clusters.

The z-score standardization was used in the three variables. Several similarity measures and linkage methods were combined to group the municipalities into clusters. Finally, a validation step was performed, drawing boxplots and testing the differences between clusters.

4 Hierarchical Cluster Analysis

The combination of method and distance that produced the best results was the between-groups linkage method with squared Euclidean distance. The between-groups linkage method measures the distance between clusters as the average distance of all points within the clusters. The squared Euclidean distance increases the importance of large distances, while weakening the importance of small distances. To note that this is not the geographical distance between municipalities, but the distance between the numerical values of the three variables under study observed in each municipality in 2018. This combination of methods was the one that obtained a dendrogram that better differentiated the clusters thus formed.

In the dendrogram depicted in Fig. 4, the connected vertical lines designate joined municipalities. The municipalities are listed along the horizontal axis, and the vertical axis shows the distance between clusters when they are joined.

Dendrograms are a visual representation of the steps in a hierarchical clustering; they show how the clusters are merged and allow us to identify the appropriate number of clusters to keep. Single groups should be avoided, unless the single object in the group is clearly identified as an outlier. The dendrogram rescales the actual distances to numbers between 0 and 25, preserving the ratio of the distances between steps [7]. For the dendrogram in Fig. 4, the longest vertical lines between groups determine the cut line.

For the present study, the cut line could be set on the re-scaled distance 20, resulting in two groups. The results using other combinations of methods and similarity measures pointed to the same two clusters defined. This combination of methods was selected because it produced longer vertical lines and a balanced number of municipalities in each group. Table 1 shows the identification of the municipalities belonging to each cluster.

In Fig. 5, the geographic location of Cluster 1 (in yellow) and Cluster 2 (in green) can be verified. The remaining unmarked areas correspond to medium and large-sized municipalities, which were not included in this study. The analysis of the map in Fig. 5 allows confirming that the municipalities that are part of Cluster 1 are predominantly in the region's hinterland.

The descriptive statistics that characterize the clusters are presented in Table 2. In 2018, the municipalities in Cluster 1 had an average of approximately 1095 companies, 22 road accidents with casualties and 159 disability pension beneficiaries. The municipalities in Cluster 2 had, in 2018, an average

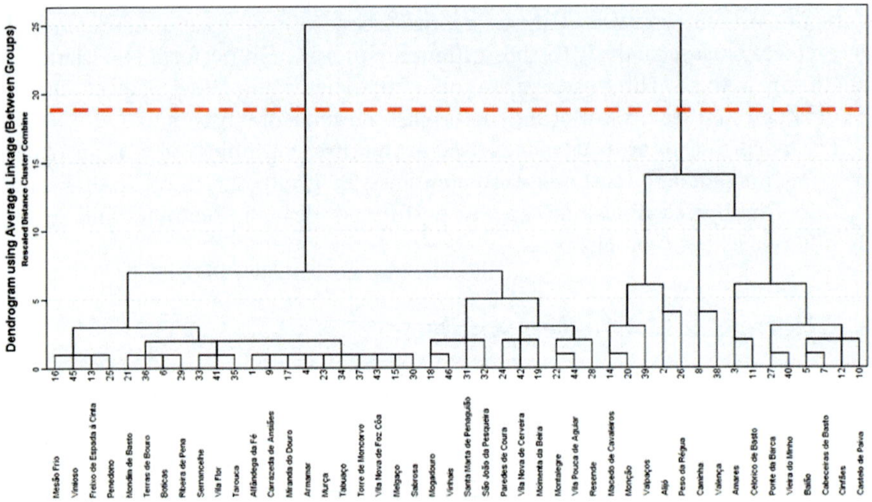

Fig. 4. Dendrogram for the cluster analysis (between-group linkage method with squared Euclidean distance)

of approximately 2012 companies, 49 road accidents with casualties and 402 disability pension beneficiaries. In general, Cluster 2 showed more variability regarding these variables than Cluster 1.

To validate the obtained cluster structure with a confirmatory analysis, boxplots were drawn separated by the created clusters, and tests of hypothesis were performed to verify if there were significant differences between the clusters for all the variables of interest.

By analyzing the boxplots for the three variables (Fig. 6), we found that Cluster 1 municipalities have a lower number of companies, road accidents with casualties and disability pensioners compared to Cluster 2 municipalities. 50% of Cluster 1 municipalities have 840 to 1334 companies, 16 to 27 road accidents with casualties and 112 to 213 beneficiaries of disability pension. Also, 50% of Cluster 2 municipalities have 1530 to 2461 companies, 40 to 58 road accidents with casualties and 329 to 466 beneficiaries of disability pension.

As can be seen in Table 3, for all clusters, the Shapiro-Wilk test resulted in all p-values > 0.05; therefore, for a 5% significance level, H_0 is not rejected, and normality is assumed, then parametric tests can be performed to compare the means for the variables under study. The differences between the means in each two clusters are analyzed using an independent samples T test.

As can be seen in Table 4, for all the variables under study, the Levene's test pointed to an equality of variances regarding the number of beneficiaries of social disability pension, as well as the number of road accidents with casualties. For the number of companies, the equality of variances could not be assumed, with a significance level of 5%. For all the variables under study, the T Test produced p-values < 0.05, so the null hypothesis, that assumes equality of means between

Table 1. Identification of the municipalities that comprise each cluster

Cluster 1		Cluster 2
Mesão Frio	Murça	Macedo de Cavaleiros
Vimioso	Tabuaço	Monção
Freixo de Espada á Cinta	Torre de Moncorvo	Valpaços
Penedono	Vila Nova de Foz Côa	Alijó
Mondim de Basto	Melgaço	Peso da Régua
Terras de Bouro	Sabrosa	Caminha
Boticas	Mogadouro	Valença
Ribeira de Pena	Vinhais	Amares
Sernancelhe	Santa Marta de Penaguião	Celorico de Basto
Vila Flor	São João da Pesqueira	Ponta da Barca
Tarouca	Paredes de Coura	Vieira do Minho
Alfândega da Fé	Vila Nova de Cerveira	Baião
Carrazeda de Ansiães	Moimenta da Beira	Cabeceiras de Basto
Miranda do Douro	Montalegre	Cinfães
Armamar	Vila Pouca de Aguiar	Castelo de Paiva
Resende		

Table 2. Descriptive statistics of the three variables, according to the two clusters

	Cluster	N	Mean	Std. Dev.	Min.	Percentiles			Max.
						25th	50th	75th	
Beneficiaries of social disability pension	1	31	159.16	61.150	54	112	140	213	286
	2	15	401.60	82.858	251	329	404	466	558
Number of road accidents with casualties	1	31	22.06	9.172	8	16	21	27	43
	2	15	48.53	11.344	32	40	47	58	72
Number of companies	1	31	1095.32	356.106	565	840	1065	1334	1777
	2	15	2012.40	678.819	1215	1530	1714	2461	3406

clusters, is rejected. Thus, with 5% of significance, we can conclude that there are significant differences between the means of the 2 clusters, for all the variables of interest.

The Pearson correlation matrix between the three variables is presented in Table 5. There is a high positive linear correlation ($r = 0.773$) between the number of road accidents with casualties and the number of beneficiaries of social disability pension, and a moderate positive linear correlation between the number of companies and the number of beneficiaries of social disability pension ($r = 0.590$), and between number of companies and the number of road accidents with casualties ($r = 0.640$).

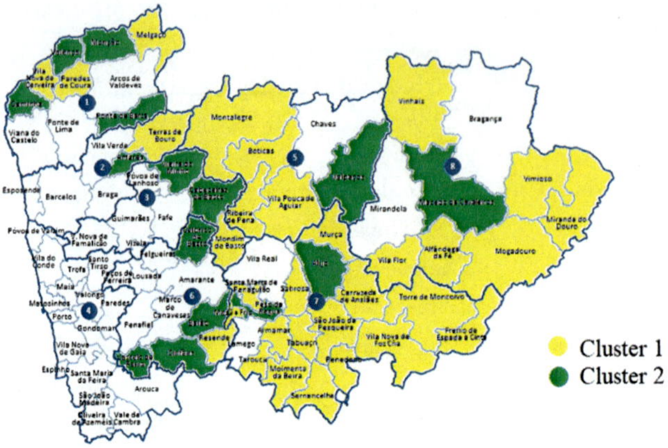

Fig. 5. Cluster 1 and Cluster 2 municipalities (Color figure online)

Table 3. Tests of normality

	Average linkage (between groups)	Shapiro-Wilk		
		Statistic	df	Sig.
Beneficiaries of social disability pension	Cluster 1	0.962	31	0.324
	Cluster 2	0.984	15	0.990
Number of road accidents with casualties	Cluster 1	0.934	31	0.055
	Cluster 2	0.960	15	0.701
Number of companies	Cluster 1	0.954	31	0.201
	Cluster 2	0.913	15	0.151

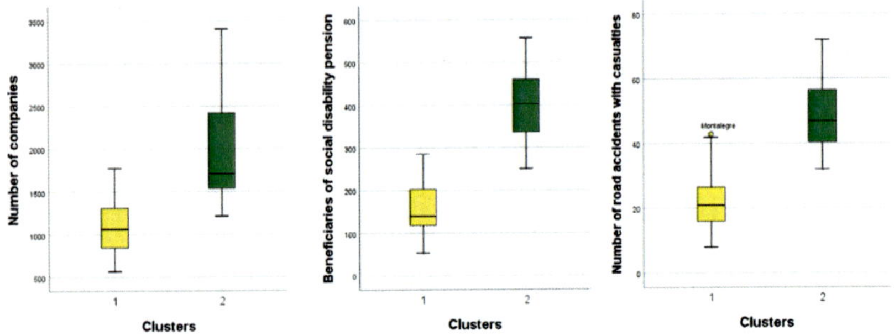

Fig. 6. Boxplot by clusters of variables Number of companies (*left*), Beneficiaries of social disability pension (*center*), Number of road accidents with casualties (*right*)

Table 4. Independent samples T Test

	Levene's test for equality of variances		Independent samples T Test	
	F	p-value	t	p-value
Beneficiaries of social disability pension	1.831	0.183	−11.203	<0.0005
Number of road accidents with casualties	1.418	0.240	−8.488	<0.0005
Number of companies	12.210	0.001	−4.915	<0.0005

Table 5. Correlation

	Beneficiaries of social disability pension	Number of road accidents with casualties
Number of road accidents with casualties	0.773	
Number of companies	0.590	0.640

5 Conclusions and Future Work

From the analysis of the results, we can conclude that 67.4% of the municipalities under study, with less than 20,000 inhabitants in the northern region, are part of Cluster 1, and 32.6% of the remaining same-sized municipalities in the same region are part of Cluster 2.

Cluster 1 municipalities are predominantly located in the region's hinterland. On average, Cluster 2 municipalities have approximately twice the average number of companies and the number of road accidents with casualties compared to Cluster 1 municipalities. Regarding beneficiaries of social disability pension, Cluster 2 municipalities have, on average, more than 2.5 times the number of beneficiaries of social disability pension compared to Cluster 1 municipalities. These differences are confirmed by the results of the T Test in Table 4, with a significance level of 5%.

As a final note, the number of beneficiaries of social disability pension is related to the number of companies and the number of road accidents with casualties. The greater the number of companies and the number of road accidents with casualties, the greater the number of social beneficiaries of disability pension. Therefore, we can conclude that the economic activity of a territory not only contributes to its valuation by creating jobs, but also brings with it a reality that impacts people's lives. The greater the number of companies, the greater the number of road traffic accidents, especially in low-density areas where the distances to be covered are greater; The existence of inefficient public transport networks consequently entail more expensive road transport of people and goods. This reality, expressed in the average number of road accidents with casualties, is certainly one of the factors that contribute to the number of beneficiaries of social disability pension.

The hierarchical clustering method applied in this paper characterizes the small municipalities of the northern region of Portugal according to economic activity, rate of road accidents with casualties, and number of beneficiaries of social disability pension. This study allowed differentiating two situations: on the one hand, it is important to increase both the number of companies of Cluster 1 municipalities and job opportunities. On the other hand, the number of companies is higher in Cluster 2 municipalities; however, it is important to improve road conditions to prevent road accidents, and consequently reduce the number of beneficiaries due to disability. This study can help local governments monitor the annual evolution of these indicators for improved decision-making.

To continue this research, a comparison with other regions of Portugal is already in progress. Another possibility to be developed is to use hierarchical methods that optimize the classification of municipalities in clusters.

Acknowledgments. This work is financed by Portuguese national funds through FCT-Fundação para a Ciência e Tecnologia, under the project UIDB/05422/2020.

References

1. Instituto Nacional de Estatística: Anuário Estatístico da Região Norte - 2018 (2019). https://www.ine.pt/xportal/xmain?xpid=INE&xpgid=ine_publicacoes&PUBLICACOESpub_boui=410498016&PUBLICACOESmodo=2
2. CCDR-N, Mapa Norte NUTS III. https://www.ccdr-n.pt/storage/app/media/files/ficheiros_ccdrn/institucional/mapa_norte.jpg
3. Fernandes, M.J., Camões, P., Jorge, S.: Anuário Financeiro dos Municípios Portugueses 2020. Ordem dos Contabilistas Certificados (2021). https://www.occ.pt/news/Anuarios/afmp2020.pdf
4. Observatório das migrações. Envelhecimento demográfico em Portugal. https://www.om.acm.gov.pt/-/3--8
5. Sarmento, R., Costa, V.: Clusters. In: Comparative Approaches to Using R and Python for Statistical Data Analysis, pp. 179–190. IGI Global (2017). https://doi.org/10.4018/978-1-68318-016-6.ch008
6. Liao, M., Li, Y., Kianifard, F., Obi, E., Arcona, S.: Cluster analysis and its application to healthcare claims data: a study of end-stage renal disease patients who initiated hemodialysis. BMC Nephrol. **17**(25) (2016). https://doi.org/10.1186/s12882-016-0238-2
7. Silva, E., Lopes, I.C., Correia, A., Gonçalves, A.M.: Hierarchical clusters of phytoplankton variables in dammed water bodies. In: AIP Conference Proceedings, vol. 1836, no. 1, pp. 020038-1-5. International Conference on Applied Mathematics and Computer Science (2017). https://doi.org/10.1063/1.4981978
8. Araújo, E., Ferreira, F., Silva, D., Bicho, E., Erlhangen, W.: Clustering domestic water consumption profiles: a Portuguese case study. In: XXVIII Meeting of the Portuguese Association for Classification and Data Analysis (JOCLAD 2021). Program and Book of Abstracts, pp. 77–78 (2021). http://www.joclad.ipt.pt/download/8/bookofabstracts_joclad2021.pdf

9. Cardoso, M., Martins, A., Lagarto, J.: Combining various dissimilarity measures for clustering electricity market prices. In: Milheiro, P., et al. (eds.) Estatística: Desafios Transversais ás Ciências dos Dados - Atas do XXIV Congresso da Sociedade Portuguesa de Estatística, pp. 197–212. Edições SPE (2021)
10. Castro, C., Lopes, C., Ferreira, F.A.: Ethical behaviour of firms and corruption - a cluster analysis. In: IX Conferência Internacional de Investigação e Intervenção em RH, no. 9 (2019). https://doi.org/10.34630/iirh.v0i9.2782
11. Fávero, L.P., Belfiore, P.: Cluster analysis. In: Fávero, L.P., Belfiore, P. (eds.) Data Science for Business and Decision Making, pp. 311–382. Academic Press (2019). https://doi.org/10.1016/B978-0-12-811216-8.00011-2
12. Silva, E., Correia, A., Borges, A.: Unveiling the dynamics of the European entrepreneurial framework conditions over the last two decades: a cluster analysis. Axioms **10**, 149 (2021). https://doi.org/10.3390/axioms10030149
13. Marôco, J.: Análise Estatística com o SPSS Statistics. Pero Pinheiro, Report Number (2014)
14. Reis, E.: Estatística Multivariada Aplicada. Edições Sílabo (2001)

Modelling and Analysing Product Development Processes in the Textile and Clothing Industry

Inês Xavier Fernandes[1], Filipe de Sá-Soares[2(✉)], and Anabela Tereso[3]

[1] Master's in Industrial Engineering, University of Minho, Campus de Azurém, Guimarães, Portugal
[2] Centre ALGORITMI, Department of Information Systems, University of Minho, Campus de Azurém, Guimarães, Portugal
fss@dsi.uminho.pt
[3] Centre ALGORITMI, Department of Production and Systems, University of Minho, Campus de Azurém, Guimarães, Portugal
anabelat@dps.uminho.pt

Abstract. The textile and clothing industry is one of the most important sectors in the Portuguese economy, operating in a market that is increasingly competitive, globalized and characterized by short lifecycles and consumer volatility. This leads to an increased need for agility and speed by manufacturers, in order to respond to the market demands. Thus, it becomes important to companies in the sector to understand and diagnose its product development processes in order to identify opportunities for improvement, potentially changing the organisation conventional routines This work adopted a case study research strategy to critically analyse and discuss the product development process of a major player in the fashion retail sector. For that goal, processes were mapped using business process modelling tools such as Business Process Model and Notation, highlighting the problems and opportunities in the development of new products in the textile and clothing industry, and providing context for the consideration of new measures to enhance its efficiency.

Keywords: Process modelling · Product development · Textile and fashion industry

1 Introduction

Headed to exportation, the Textile and Clothing Industry (TCI) remains as one of the most important Portuguese industrial sectors [1]. After a period of decline, boosted by the global economic crisis, the sector has been recovering, currently representing 10% of total national exports and 8% of volume of business with 19% of the employability of the manufacturing industry [2]. This is an industrial sector characterized by a high commitment to innovation, design and quality, in order to face a global and increasingly competitive market [3]. The fashion retail sector is characterized by very short life cycles, instability in consumer preferences and a high heterogeneity of supply, leading to increased competitiveness in the market. Therefore, this sector presents a volatile and

J. Machado et al. (Eds.): icieng 2022, LNME, pp. 220–233, 2023.
https://doi.org/10.1007/978-3-031-09360-9_19

risky market, with product life cycles planned to be short, in order to capture consumer interest [4]. Fashion apparel producers face the need to permanently deliver new products to fulfil the orders requested by retailers, boosted by consumer market trends. Due to the reduced life cycle of these products, their development must be fast and efficient, in order to develop a competitive advantage in the market. Characterized by the need for agility, the process of developing new products in the fashion clothing sector requires the adoption of practices that enhance efficiency. To pursue that excellence, the process of clothing development needs to be carefully understood and diagnosed, in order to find out opportunities for its improvement.

We argue that Business Process Management is a valuable tool for the modelling the Product Development Process (PDP) in the sector, assisting in the understanding of how the development of new products in the TCI is processed, and providing a basis for the projection of new measures that enhance its efficiency.

This paper is structured as follows: Sect. 2 presents the literature review, addressing relevant topics, namely Product Development and Business Process Management; Sect. 3 describes the research methodology followed; Sect. 4 explores the results of the study; finally, Sect. 5 concludes the paper, underlining some limitations and presenting relevant future research.

2 Literature Review

2.1 Product Development

The reference to a new product happens when a product or concept is totally new to the market, or totally new to the company, despite the existence of similar products on the market [5]. In an increasingly competitive market, the development of new or improved products and services is crucial for the survival and prosperity of organizations [6]. Product development is increasingly considered a critical business process for the competitiveness of companies, especially with the growing internationalization of the market, increasing diversity and variety and reducing the life cycle of products [7].

The design of a product is the process that takes place at the beginning of its life cycle, which includes moments of research, experimental development, and marketing [8]. Therefore, it consists in the sequence of steps or activities that an organization takes to conceive, design and market the product [9], in order to convert needs into technical and commercial solutions [10]. Product development activities start with the perception of a market opportunity and end with the production, sale, and distribution of a product. The precise and detailed definition of a PDP is useful and important because it [9]:

- Ensures higher quality of the process;
- Stimulates coordination between the elements involved in the process;
- Keeps the process planning updated;
- Allows managing process performance; and
- Promotes continuous improvement of the organization's PDP.

Product Development in Clothing and Fast-Fashion

The PDP is a divergent process, which varies depending on the context of the organization. Since this work is based on the TCI, and more specifically on the production of clothing for the fast-fashion market, it is important to know the characteristics of this market in order to understand its specific needs and the particularities of its product development process.

In the last 20 years, the textile and clothing industry has evolved significantly. Mass production began to disappear and products with a shorter life cycle appeared and the number of collections increased, forcing manufacturers to change their supply chain in order to be able to produce at a lower cost and with greater flexibility, quality and speed to market [11]. In part, this explains the instable nature of the clothing industry, due to the seasonality and subjectivity of the factors that lead to the purchase of products. Thus, to keep up with the complexity of the fashion apparel product, manufacturers need to understand and adopt the most efficient methods, in order to create and produce differentiated models in the shortest possible time [12]. The fast-fashion concept emerged to follow the trend of changing the paradigm of this sector, representing a market that is difficult to predict, with a high buying impulse, shorter life cycles and high volatility in demand [11]. This concept means that organizations must be able to react to new fashion trends in periods as short as 15 days to 1 month, in order to allow them to remain competitive in the market [13], unlike the traditional cycles, in which two annual collections were presented according to the seasons, increasing the number of cycles per year from 2 to approximately 50 [14]. Given these characteristics of the market, the product development process has to respond accordingly. Senanayake [15] identifies the main differentiating characteristics of the fast-fashion clothing PDP as being:

- **Pull strategy**, given that the products developed will be products based on what the market is looking for and asking for at the moment, instead of push production in which the manufacturer introduces products to the market based on history.
- Design and development of **small productions of multiple styles in short cycles**, in order to respond to the market, providing a variety of options in smaller quantities to increase purchase motivation and increase sales.
- **Speed in design approval**, in order to fulfil an efficient validation process and reduce the waiting time until delivery to the customer.
- **Prefabricated raw material**, to guarantee a flexible flow of material for the development of the part.
- **Close relationship between the elements of the supply chain**, since they will be elements that will work closely to develop the product and it is necessary that they are integrated and aligned, so that quickly, and in collaboration, they can respond to the dynamics of the market.
- **Multidisciplinary in the development teams**, since it is a process that goes through several stages, intervening in different areas of action, and thus will require the integration of several elements, with experience in different themes.

Thus, the PDP in the textile and clothing sector, particularly with production for fast-fashion, will have to follow the complexity of the product, creating and producing differentiated models, in a short period of time, while following the trends of the fashion market [16].

2.2 Business Process Management

Originating from the Latin processus, process represents a set of interrelated activities and tasks, which are initiated in response to an event, and which aim to achieve a specific result for the process consumer [17]. A business process is distinguished from other processes since, as its name implies, it is a process applied in a business. Hence, a business process is defined as a set of activities, logical and sequential, that allow the transformation of inputs into outputs, in order to deliver value to the customer [18]. Business processes are coordinated activities that create a final product or service, with inputs and outputs, as well as a defined beginning and end [17].

In a business, it is possible to classify its processes in three different ways [19]: primary processes (essential processes of the organization, responsible for adding value to the customer); support processes (support the organization's primary processes); and management processes (with the objective of measuring, analysing, monitoring, and coordinating all processes of an organization). Each one with its objective, all the processes of the organization are necessary for its proper functioning, and therefore its management is necessary, and hence the need for Business Process Management (BPM). Despite the different definitions of BPM, it is possible to reconcile them as the discipline that involves any combination of modelling, automation, execution, control, measurement, and optimization of an organization's process flow [17]. As a result, there is a set of methods, techniques and tools for the design, interaction, control, and operational analysis of business processes, involving people, organizations, applications, documents, and other sources of information.

To represent business processes, visual tools are commonly used which consist of diagrams representing the process, namely, Business Process Model and Notation (BPMN), that allows to illustrate, document, and shape the way an organization develops its business across all its levels, i.e. at the strategic, tactical and operational level [17].

Business Process Model and Notation (BPMN)

BPMN was introduced in 2004 as a standard language for business process modelling. It is a visual representation that serves to model and represent the operational flows of an organization [20]. This tool has the main objective of providing a notation that is understandable by any user of the process, being simple to use and allowing to create flow diagrams that are independent of the type of business they represent [21].

The BPMN language is grouped into four major categories of elements [20]:

Flow Objects

These elements allow to build the business processes, and include the activities (squares), events (circles) and gateways (rhombus), as shown in Fig. 1 [22]:

Fig. 1. Flow objects at BPMN

- **Activities –** Work units to be carried out during the business process. Among the activities we can have processes, subprocesses and tasks. An activity is a discrete action with a well-defined beginning and end, that is, a function that is performed continuously is not an activity in the BPMN context. Among tasks, one may find several types, depending on their purpose, and they can be distinguished by the use of icons in the upper left corner of the activity element.
- **Events –** Something that occurs in the process and how the process responds to this event (in the case of being a "receive" event), or how the process generates the signal that something has happened (in the case of being a "send" event). Events may be initial, intermediate, or final.
- **Gateways –** Decisions in a process, being used to demonstrate the convergence or divergence of the flow of control between activities, events, among others. They are graphically represented by a rhombus and again, depending on their type, take different graphical representations.

Connection Objects
Elements used to connect the flow objects and represent the relationships between them. There are three types of connection objects, depending on their function [21] (Table 1).

Table 1. Types of connection objects

Type	Function	Representation
Sequence flow	Defines the order in which activities are performed within a process.	
Message flow	Represent the flow of messages between two entities, which can be senders or receivers.	
Association	Represent the links between artefacts and other elements in BPMN.	

Swimlanes
Elements used to denote a participant in a process and group all the activities carried out by that same participant. Subdividing the pool into lanes allows the activities to be organized and categorized by each entity that performs them (Fig. 2). The concepts pool and lane, serve exactly, as their names indicate, to replicate the operation of a pool, in which each swimmer stays in his lane, just like each process contained in its element.

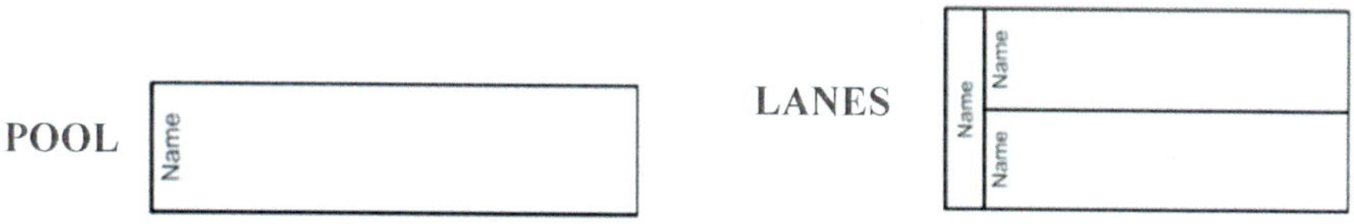

Fig. 2. Pool and lanes in BPMN

Artefacts
Elements with the purpose of providing complementary information about the represented process. They can be data objects, groups or annotations and each one has a specific purpose, as illustrated in Table 2.

Table 2. Types of artefacts in BPMN

Type	Function	Representation
Data object	Represents the data requested or produced by a process.	
Group	Visual aid that allows to aggregate various elements for documentation and analysis.	
Annotation	To add additional process specific information in text format.	

3 Research Methodology

In order to understand the PDP in TCI, this paper reports the application of a case study research strategy. This strategy consists of an empirical method that investigates in depth a topic or phenomenon of interest, within its real-life context [23]. Thus, the case study aims to explain, describe and explore a specific case in the context in which it occurs, thus revealing the multifaceted nature of the subject of interest [24]. In case study research, the word 'case' refers to the units of analysis such as people, groups, organizations, change processes, among others [25].

In this context, the case study strategy was applied in a group of companies that stands out in the TCI sector as one of the largest Portuguese textile groups. This group is one of the few complete vertical units worldwide, controlling its production from spinning to confection, and following the entire transformation process of the raw material, from the cotton wool to the finished garment. It is one of the largest suppliers of the *Inditex Group* worldwide, and therefore its production needs to fit the requirements of the fast-fashion market, with speed and agility being the watchwords in its operation.

To support the findings in this case study, the technique of participant observation was employed. Participant observation is based on collecting data from the point of view of user and researcher active in the work. This technique made it possible to monitor the processes of key users. To enhance the relevance of the data collected, users were selected, taking into account their role in the processes under study. This selection was made together with the company's production director, based on her extensive knowledge of the process.

4 Results

4.1 Resources

To carry out the PDP, the company makes use of various resources, namely human and material. In terms of human resources, the PDP relies on its employees and external resources. Regarding internal employees, Table 3 provides a summary of the relationship between their functions and quantities.

Table 3. Human resources involved in the PDP

Function	Description	N° of users
Fashion designer	Search, creation, and development of the design of the piece	9
Graphic designer	Search, creation, and development of graphics for adornment of the piece	7
Modeller	Development of moulds for the production of the garment	8
Seamstresses	Cutting, sewing, and making of the sample	14
Head of seamstresses	Management and organization of the sample confection	1
Salespeople	Communication with the client, negotiation of the sale and accompaniment of the development process	19
Warehouse operator	Organization and operation of the sample warehouse; includes knitwear, fabric, and accessories sample warehouse	7
Sourcing	Search of knitwear/textile/accessories from external suppliers	1
R&D	Research and creation of new knits and fabrics for in-house production	4
Production director	Management of the entire process and resolution of disagreements that may arise during the process	1
Administrative staff	Collaboration in the administrative process of product development (budgeting, technical sheets, among others)	4

The process is organized into teams, with each team being assigned to departments according to the customers with whom the group works. Each team has at least one salesperson, one fashion designer, one graphic designer and one modeller. Eventually, when there is a department with a greater flow of work, there may be a reinforcement of employees in the functions described. In collaboration with the sales teams, there are elements that are transversal, collaborating with all the development teams, namely the R&D team – responsible for searching and creating new knit and woven fabrics, in order to produce the bases for the development of new pieces (new compositions, weights, mixtures, among others); the sourcing team – responsible for sourcing raw materials when they do not exist internally, be it knitwear, fabrics or accessories; the operators of the samples warehouse, both for fabric and knitwear – responsible for organizing the materials and preparing the quantities requested by the developers; and the seamstresses of the samples production – responsible for making the samples.

In what concerns material resources involved in the process, the raw material for garment manufacture and the operational resources that will support the employees involved in the process must be addressed. In relation to raw materials, the general components handled in the PDP are knitwear, fabrics, and accessories. The operational resources to make the garments, such as sewing machines, will also be used, as well as computers running creative development software.

Regarding the periodicity of the process, it is characterized by its variability, following the volatility of the market in which it is inserted. Thus, it is difficult to predict or estimate the time interval in which a new development will be necessary. The process was monitored over the time. Development requests were formalized in the form of collections of 15 to 30 models, with an interval of approximately 15 to 20 days. According to the employees involved, this is an adequate estimate of their periodicity, but it should not be taken as stationary, given its intrinsic instability.

4.2 General Process

Maintaining its core business in the garment trade, the creative development of new designs and products is essential for the company production activity, and consequently, its survivability. This process consists of a preliminary stage to production, in which the creative development teams, together with the sales teams and all other stakeholders, create and negotiate the piece with the customer, in order to result in a purchase and therefore start the group's production process. Throughout this process, the design of the garment, its patterns, its textile material, its accessories, and the details of its manufacture will be defined, according to the customer's wishes, following a cyclical process until final approval, as shown in Fig. 3.

Given the complexity of the PDP, it was important to categorize the activities in three sub processes Product Development and Creation Process, Negotiation Process and Raw Material Search and Supply Process. As mentioned, the process begins with a contact with the customer and the identification of their needs. Given that contact, the creative development process is triggered – Development and Creation Process, namely design and modelling; as well as the search for materials to compose the piece – Raw Material Search and Supply Process. This process culminates in the confection of the sample, which will be sent to the customer. The next step depends on the customer's decision:

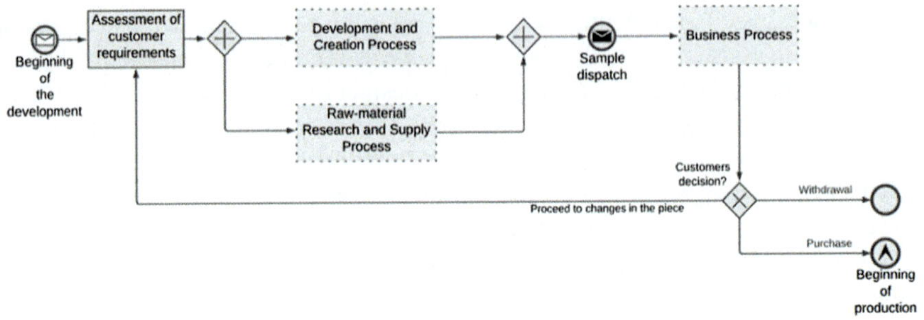

Fig. 3. General overview of PDP

to make changes to the garment (and thus start a new development cycle); to buy the garment (and start the production process) or to give up the development (and terminate the process). Each of these processes were modelled using BPMN, highlighting the dynamics between the different participants in the process.

Development and Creation Process

This is the most creative part of the process, which relies on the support of various collaborators who aim to create the concept of the garment and explore their creative abilities to do so. It involves fashion designers, graphic designers, modellers, and confection. Figure 4 depicts the process at a general level.

Negotiation Process

Regarding the negotiation process, its main focus is the role of the salesperson. It is through him that customer's wishes are discussed and worked on, until the developed item is purchased. Figure 5 displays the BPMN structure of this activity.

Raw Material Search and Supply Process

Finally, it becomes essential to address the Raw Material Search and Supply Process (Fig. 6) where the activities required to supply the components necessary for the part are represented, with a highlight to the users of sample warehouses, as well as to the sourcing and the R&D department.

4.3 Discussion

In this section, we identify and discuss the main lessons learned from the development of this work, which aimed at understanding the working environment in the PDP, in order to identify its key processes, diagnose its main problems, and find out opportunities for improvement.

The PDP analysed in the case of study is weakly formalized and controlled, being in an embryonic state regarding its maturity in efficiency. Thereafter, there is a need for action in order to spot possible opportunities for improvement. Indeed, the monitoring of the agents of the DPD and the mapping of their roles and activities, made possible the assimilation of crucial problems whose mitigation was required. Among these problems,

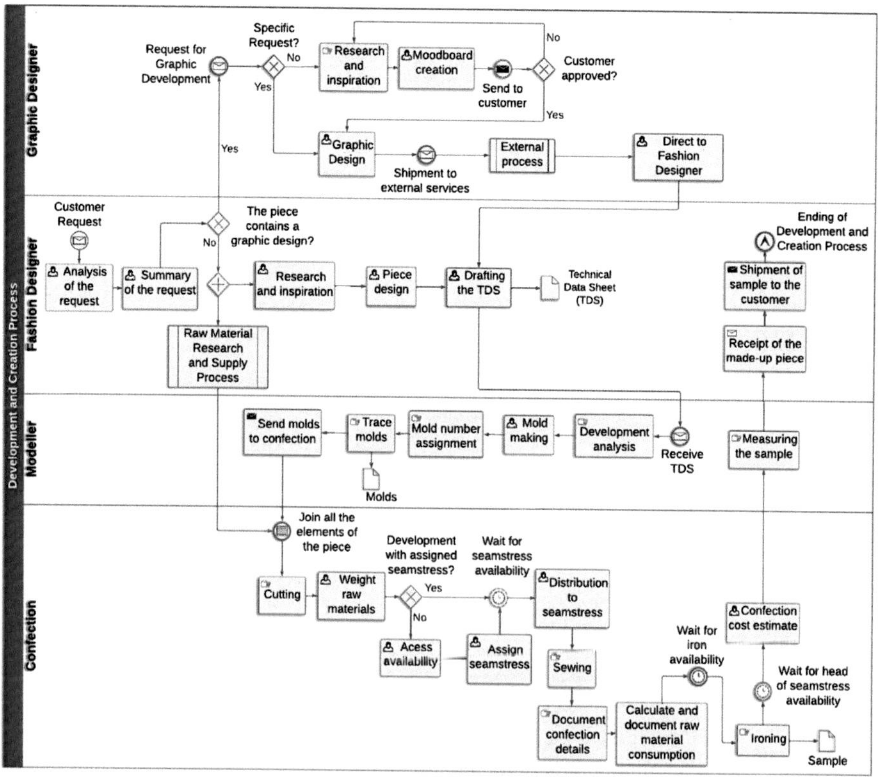

Fig. 4. BPMN - development and creation process

the dispersion of information in the process, the informal working method and procedure, and the lack of visibility of the process stand out. These characteristics can be observed through the BPMN diagrams where, for example, the various documents that circulate in the process are represented, showing the dependence on paper and the possibility of misinterpreted information, as well as the difficulty of sharing information between different contributors to the process, represented by the diagram's lanes.

With regard to the dispersion of information, this is due to the existence of multiple communication channels, often even in the employees' personal communication channels, making it impossible to maintain a communications history and to record information. Additionally, the use of paper, as a method of transmitting information, increases the dispersion of information and the transmission of incorrect information due to misinterpretation. The informal working method, with no standard working procedures, is also a critical problem in the process. This problem is verified, for example, in the inexistence of task management processes and the lack of process control points. Finally, the lack of visibility of the process status is also a major problem in PDP. Both from the perspective of visibility of the state of a development, and from the perspective of visibility of all the developments occurring at the moment and historically. As for the status of a development, the lack of visibility leads to a constant need to query other users

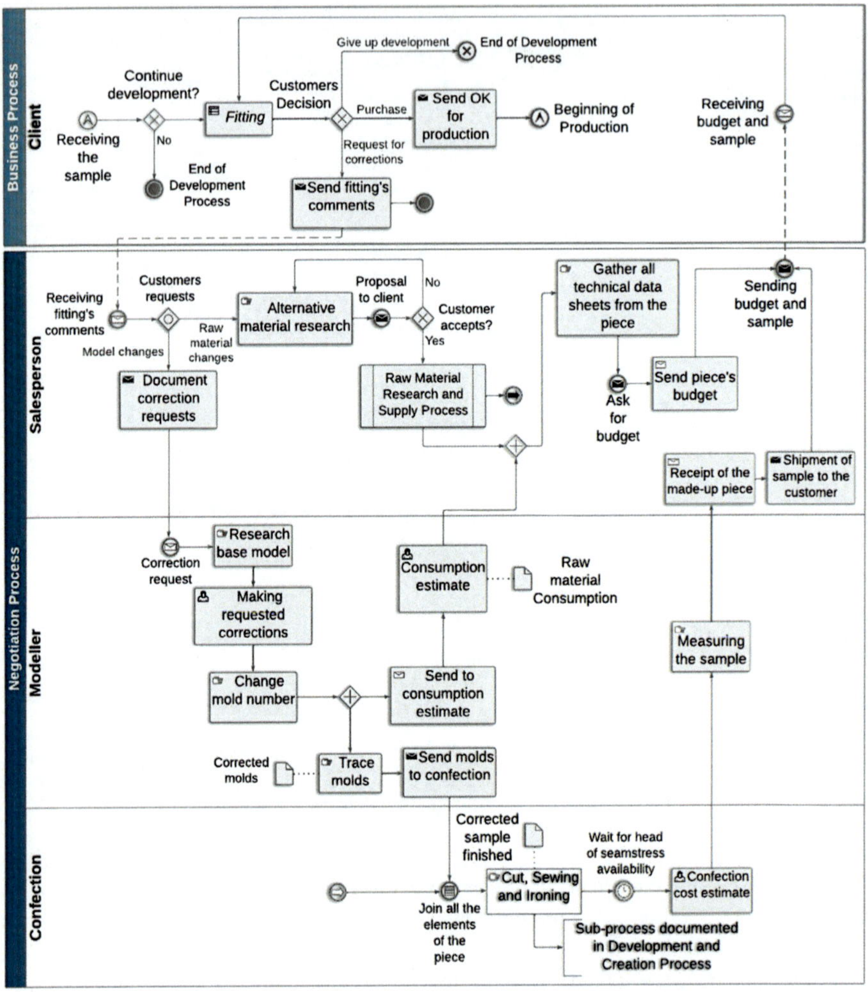

Fig. 5. BPMN - negotiation process

to understand at what stage the development is, interrupting their work and consequently decreasing productivity. Additionally, the lack of visibility of all the work is worrying in two ways: the first being the lack of understanding of all the work being developed at the moment, preventing the management of priorities; and the second being the lack of a record of all the work already developed, hindering data analysis and consequently the management and performance control of the process.

The PDP was performed without standardization, leading to disperse working methods. However, the need and the benefits of introducing new practices in PDP were recognized by employees and management. Regarding the life cycle of a PDP, the process scrutinized in this study was limited to the actual product development, excluding any

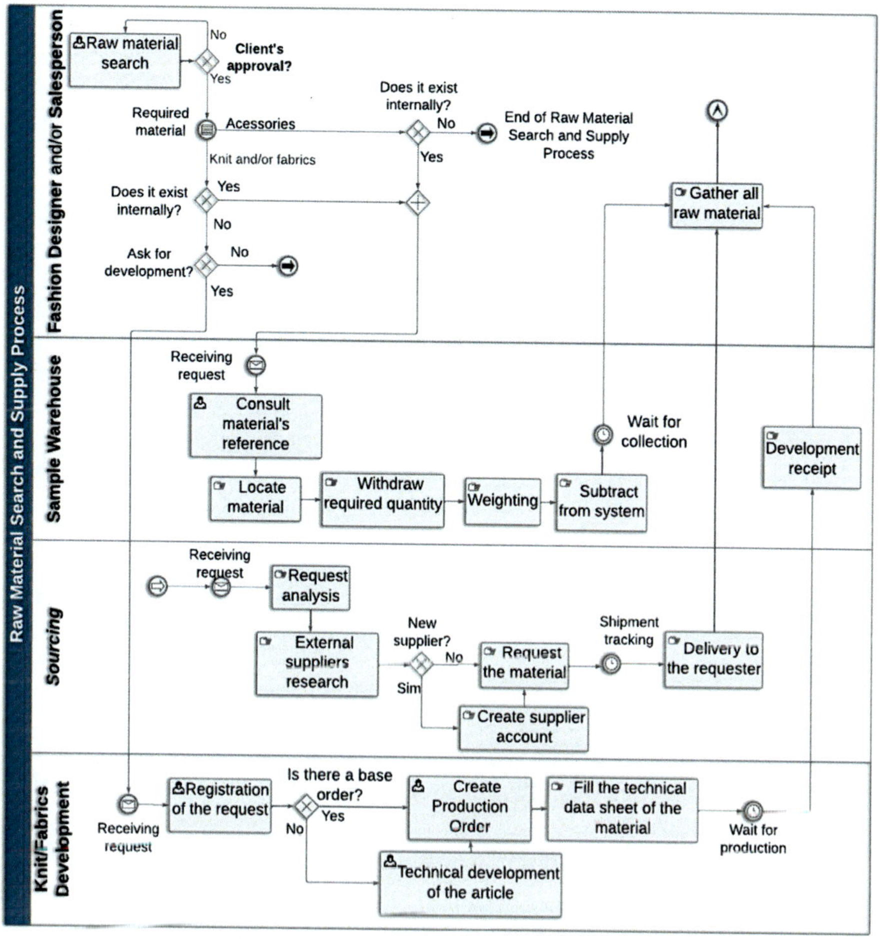

Fig. 6. BPMN - raw material search and supply process

pre-development and post-development phases, corresponding respectively to the initiation/planning, and closing of the process, therefore disregarding the potential benefits of each of those phases.

5 Conclusion

The purpose of this work was to diagnose and analyse PDP practices in the TCI, modelling the PDP of the case of study, in order to identify improvement opportunities and provide context to adopt practices that improve its functioning. The participant observation of the employees' routines, with the objective of mapping the PDP, enabled a closer look of those involved in the process, which facilitated the understanding and documentation of their practices and concerns.

As known, case studies intend to provide insight in a specific case, so it has its limitations since the findings of this paper may not correspond to the reality of PDP processes in all CTI and therefore this point should be emphasized. Accordingly, the next step would be to perform the same study in other manufacturers, making it possible to assess comparative lines between the various organizations and be able to create a management model adapted to the sector where they operate, and then create common metrics and performance indicators that allow the performance of the process to be critically evaluated.

At this case study, product development took place without method and standardization. According to what was mapped throughout this work, it is noted the division of new product development into three macro processes: the Development and Creation Process (where the graphic and fashion designers develop their work), the Negotiation Process (mainly explored by the salespeople with the help of administrative technicians) and the Raw Material Search and Supply Process (carried out by all internal and external raw material suppliers). These processes are carried out, in general, in a manual way and with dependence on paper, with no standardized processes throughout its development.

This work contributes to the adoption of practices that enhance the efficiency of the PDP in the TCI, increasing the knowledge about its activities, so that improvement opportunities might be identified and seized upon.

Acknowledgement. This work has been supported by FCT – *Fundação para a Ciência e Tecnologia* within the R&D Units Project Scope: UIDB/00319/2020.

References

1. CITEVE: Indústria Têxtil e do Vestuário: Roadmap para a Inovação 2012–2020. Vila Nova de Famalicão, Portugal (2012)
2. ATP: Portuguese Textile Indicators. https://atp.pt/wp-content/uploads/2019/06/Indicadores-de-Gestão-Atualiz.-Relatorio-19-06-2019.pdf. Accessed 09 Jan 2021
3. Cardoso, C., Quelhas, V.: Indústria Têxtil e de Vestuário: Uma referência a nível mundial. Portugal Global, pp. 12–23 (2018)
4. Dapiran, P.: Benetton: global logistics in action. Int. J. Phys. Distrib. Logist. Manag. **22**(6), 7–11 (1992)
5. Panwar, J., Bapat, D.: New product launch strategies: insights from distributors' survey. South Asian J. Manag. **14**(2), 82 (2007)
6. Kahn, K.B., Kay, S.E., Slotegraaf, R.J., Uban, S. (eds.): The PDMA Handbook of New Product Development, 3rd edn. Wiley, Hoboken (2013)
7. Rosenfeld, H., et al.: Gestão de Desenvolvimento de Produtos: Uma referência para a melhoria do processo, 1st edn. Editora Saraiva, São Paulo (2006)
8. OECD/Eurostat: Oslo Manual: Guidelines for Collecting and Interpreting Innovation Data, 3rd edn. OECD Publishing, Paris (2005)
9. Ulrich, K.T., Eppinger, S.D.: Product Design and Development, 6th edn. McGraw-Hill Education, New York (2016)
10. Smith, R.P., Morrow, J.A.: Product development process modeling. Des. Stud. **20**(3), 237–261 (1999)
11. Bhardwaj, V., Fairhurst, A.: Fast fashion: response to changes in the fashion industry. Int. Rev. Retail Distrib. Consum. Res. **20**(1), 165–173 (2010)
12. Parker-Strak, R., Barnes, L., Studd, R., Doyle, S.: Disruptive product development for online fast fashion retailers. J. Fash. Mark. Manag. **24**(3), 517–532 (2020)

13. Tran, Y., Hsuan, J., Mahnke, V.: How do innovation intermediaries add value? Insight from new product development in fashion markets. R&D Manag. **41**(1), 80–91 (2011)
14. WRI: The Apparel Industry's Environmental Impact in 6 Graphics. World Resources Institute. https://www.wri.org/insights/apparel-industrys-environmental-impact-6-graphics. Accessed 27 May 2021
15. Senanayake, M.: Product development in the apparel industry. In: Nayak, R., Padhye, R. (eds.) Garment Manufacturing Technology, pp. 21–57. Woodhead Publishing, Sawston (2015)
16. Moretti, I.C., Braghini Junior, A.: Reference model for apparel product development. Indep. J. Manag. Prod. **8**(1), 232–262 (2017)
17. von Rosing, M., von Scheel, H., Scheer, A.W.: The Complete Business Process Handbook: Body of Knowledge from Process Modeling to BPM. Morgan Kaufmann, Waltham (2014)
18. Harrington, J.H., Essling, E.K.C., van Nimwegen, H.: Business Process Improvement Workbook: Documentation, Analysis, Design, and Management of Business Process Improvement. McGraw-Hill, New York (1997)
19. ABPMP: BPM CBOK Versão 3.0: Guia para o Gerenciamento de Processos de Negócio, 1st edn (2013)
20. vom Brocke, J., Rosemann, M.: Handbook on Business Process Management, 2nd edn. Springer, Heidelberg (2014)
21. OMG: Business Process Model and Notation (BPMN). Object Management Group (2013)
22. Ciaramella, A., Cimino, M.G.C.A., Lazzerini, B., Marcelloni, F.: Using BPMN and tracing for rapid business process prototyping environments. In: ICEIS 2009 - 11th International Conference on Enterprise Information Systems (2009)
23. Yin, R.K.: Case Study Research: Design and Methods, 5th edn. SAGE Publications, Thousand Oaks (2018)
24. Crowe, S., Cresswell, K., Robertson, A., Huby, G., Avery, A., Sheikh, A.: The case study approach. BMC Med. Res. Methodol. **11**(1), 1–9 (2011). https://doi.org/10.1186/1471-2288-11-100
25. Saunders, M., Lewis, P., Thornhill, A.: Research Methods for Business Students, 5th edn. Pearson Education Limited, Edinburgh (2008)

Improvement of Internal Transport with the Use of Milk Run Concept

Joanna Kochańska, Anna Burduk(✉), and Aleksandra Małysiak

Wroclaw University of Science and Technology, Wybrzeże Wyspiańskiego 27, 50-370 Wrocław, Poland

{joanna.kochanska,anna.burduk}@pwr.edu.pl

Abstract. The proper functioning of production enterprises depends on the regularity of the flow of production factors. That should be ensured by proper transport management, not only outside but also inside the plant. In accordance with the principles of Lean Logistics, there was developed a project of improvements for selected logistics processes of one of the leading manufacturers of the household appliances industry. The aim of the project was to improve the internal transport within 4 stages: stage 0 - current state analysis, stage 1 - implementation of Milk Run concept and Kanban system, stage 2 - improvement of implemented solutions, stage 3 - development of proposals for further improvements. There were characterized transport problems in the form of low frequency of deliveries, resulting in overstock and downtime. Therefore, Kanban system and Milk Run concept were implemented - there were prepared: appropriate packaging and component storage locations, new transport equipment and the required documentation. The introduced changes significantly improved the transport of components to the analyzed area, however, there was noted a problem of employee overload. In response to identified problem sources, there was developed a new division of responsibilities together with some facilitations of the process. As a result of the project, there were developed proposals for changes in the logistics process and its documentation. Their implementation allowed the company to improve the production flow. Finally, there were presented proposals for further improvements, including changes of: warehouse supply logistics, component storage locations and packaging of delivered components.

Keywords: Internal transport · Lean logistics · Milk run concept

1 Introduction

Regularity of flows, required for correct operation of manufacturing companies, manifests itself by uniform supplies of production factors and by uniform outflow of the manufactured goods [1], thus reducing wastage in form of excess transport, inventories and awaiting. An important role in reorganization of transport plays a branch of the Lean Manufacturing philosophy, popularized by prominent industry analysts J.P. Womack and D.T. Jones [2], namely the Lean Logistics [3]. Since the time of the first publication in that the term "lean" was used in relation to manufacturing systems [4], companies in

J. Machado et al. (Eds.): icieng 2022, LNME, pp. 234–247, 2023.
https://doi.org/10.1007/978-3-031-09360-9_20

the whole world use the principles determined by this term to improve their processes, both the manufacturing processes and the production-related ones, gaining this way significant profits [5–8].

The necessity to improve logistic processes was noticed also at one of the leading manufacturers of household goods. As a response to the existing problems in form of variable wastages, a project was developed, aimed at improving the in-house transport by means of the methods and techniques of the Lean philosophy, with particular reference to the Milk Run concept. Figure 1 shows schematically execution of the project.

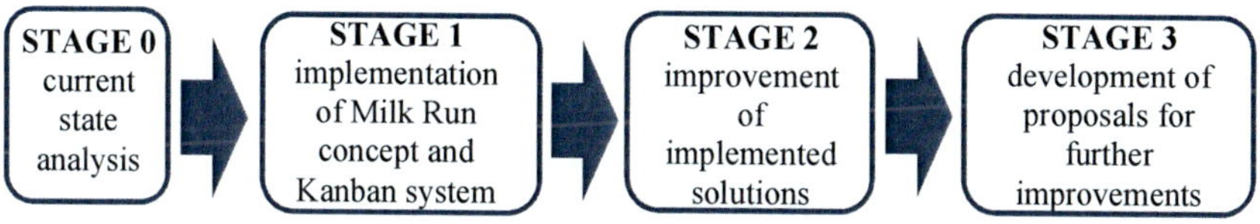

Fig. 1. Layout of execution of the improvement project

The project was designed in 4 major stages, starting from analysis of the present state, through implementation and improvement of the Lean Logistics solutions (the improvements in stages 1 and 2 have actually been implemented in the company), up to development of a plan of future innovations on the ground of the results obtained for the first 3 stages.

The aim of the study was to analyze the current state of the internal transport in the company, to propose organizational soliutions that can reduce negative results of the implemented solutions (improving the internal transport in the company) and then to investigate the impact of the introduced methods on the results within the company. The analysis was carried out based on the data from the production documentation (i.e. production plans, technological processes, BOMs, standardized work sheets) and with the use of the following methods: Gemba Walk, photographs of the working days, production time measurements.

2 Lean Logistics

A well-organized in-house transport should guarantee displacement of a given load through the shortest routes, with possibly effective use of suitable transport means and with possibly short times of loading and unloading, as well as at maximum transport speed with safety rules maintained. The Lean Logistics assumes supporting the continuous material flow and execution of deliveries to customers in proper times and locations, with maintained proper levels of quality and costs [9]. It aims at frequent deliveries in small lots based on a pull system, resulting in accelerated material flow, balanced production line or reduced inventory, as well as in elimination of production downtimes caused by shortages of components [10, 11].

In the in-house logistics, the pull system is used for feeding continuous-flow production lines and centers. Both the pull system and the continuous flow are basic components of the Just-In-Time (JIT) system [12]. In practice, it assumes supplying components in small containers fit on the storage racks in their point-of-use. To that end, manually pushed carts or so-called tugger trains are used, to resign deliveries on pallets. Storage is perceived not as a security measure but as a loss and deliveries are executed directly to the line, not to the warehouse. The Kanban system is used for effective management of material flow. The Transport Kanban system controls displacement of products to the processes located down the value stream. The in-house transport Kanban indicates the necessity to collect a component from its storage location and to transport it to the receiver process. In a company practice, the Kanban signals take various forms, from empty containers or storage areas, through colored balls or triangle tables up to electronic signals. However, Kanban cards are most often used [13].

3 Milk Run Concept

The Milk Run is a delivery method in that a means of transport executes, during one course, many loadings and unloadings of various materials, according to the previously accepted route. The transport route is fixed, determined by connecting the target points in the so-called loop [14, 15] (Fig. 2).

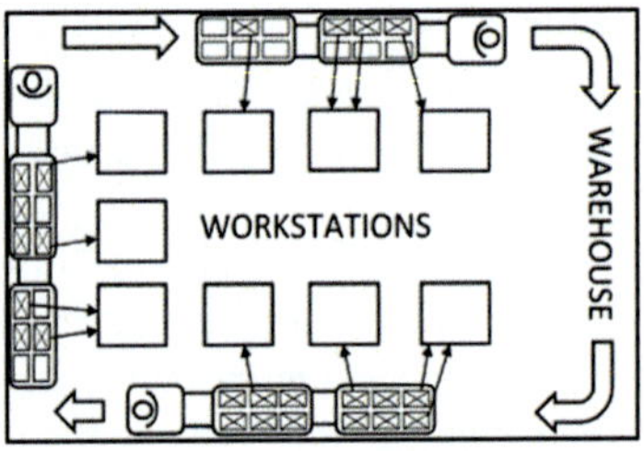

Fig. 2. Milk Run concept [16]

The idea is based on the old system of delivering milk to households [17]. The delivery is executed according to the JIT concept [18]. Cyclic deliveries are realized on the grounds of the Kanban system and sequential deliveries are realized according to the accepted schedule [19, 20]. The idea finds its application in both external and internal transport [9, 21]. In practice, a logistic train and multiple use containers are used for execution of deliveries in close transport [20]. A characteristic feature of the Milk Run concept is that, as a result of increased frequency of deliveries, quantity of inventories in the workstations is reduced. Fluidity of feeding the production line increases as a result of its higher throughput and productivity by using a minimum number of transport facilities thanks to minimization of empty courses, reduction of transport distances and improvement of work ergonomics [14, 22–26].

4 Case Study

4.1 The Characteristics of the Studied Area

The case study was realized within cooperation with one of the world-leading manufacturers of household goods. The concepts of in-house transport improvements were developed on the grounds of measurements and analyses performed within one of the production sub-processes. The sub-process includes 2 manufacturing areas of the base component (W1 and W2) and 2 sub-assembly areas (P1 and P2) feeding 2 different production lines (1 and 2). The process parameters for the above-mentioned areas accept the values given in Table 1.

Table 1. Basic parameters of the analyzed areas

Production lines 1 and 2; Areas W1 and W2	Takt time	51 s
	Size of production lot	18 pcs.
	No. of shifts per day	3
	No. of work days per week	6
Area P1	No. of work shifts per week	2
Area P2	No. of work shifts per day	1

With regard to work times of individual areas, the basic zones served by the logistic worker are W1 and W2, where components are transported from 4 different places of the production floor: 2 preliminary processes, a small component warehouse and a high-bay store.

Analyzed was the entire logistic process, covering transport of the stored components performed on the grounds of sequential delivery (based on the so-called shopping list) and of the delivery according to consumption (demand). Figure 3 shows layout of a fragment of the production floor, where the areas involved in the considered logistic process are distinguished.

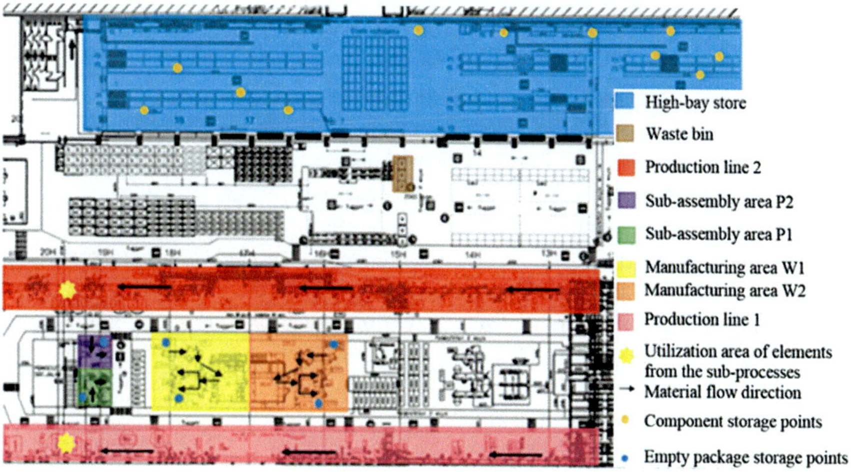

Fig. 3. Layout of a fragment of the production floor with distinguished selected areas

The analyzed logistic process includes:

- collecting components according to the shopping list and demand from the high-bay store,
- delivering components in the designated places of the areas W1, W2, P1 and P2,
- collecting empty packages of components from the areas W1, W2, P1 and P2 and discarding them to suitable bins in front of the high-bay store.

The analyzed logistic process does not include the transport of finished products because they are transported with the automatic conveyor.

4.2 The Course of the Implementation of Improvement Activities

The analyzed production lines together with their related processes and the entire infrastructure were moved from another factory, so initially the so-far applied practices and solutions were accepted, adopting them only to the new conditions. As the implemented processes were recognized, the following problems became apprehensible:

- irregularities in deliveries of components,
- missing standardization of work of the employee feeding the area,
- high time-consumption of the delivery process,
- frequent stoppages caused by shortages of components,
- large stocks of components at the workstations,
- large number of the components subjected to damages,
- non-ergonomic works of the operators.

These problems resulted from the practices, like:

- too late receipt of the signal about a transport need,
- infrequent and sizeable deliveries,
- storage of components in the operators' working space,
- incorrect way of storing components at a workstation.

Average time of the stoppages resulting from shortages of components was 1096 min and 1153 min per month for W1 and W2, respectively (on the grounds of data from 6 months), but average quantity of the work in progress was 9 cartons of a given component for each of the areas. As a priority, reduction of the work in progress and elimination of stoppages caused by shortages of components were accepted. In response to the identified problems, within the stage 1, the following measures were introduced in the company:

- the Kanban system, including the related actions:
  - a container was accepted as the information carrier,
  - numbers of the Kanban containers were determined for each of the components,
  - the containers were marked with Kanban labels,

  - capacities of shelf-racks adjacent to workstations were calculated,
  - new shelf-racks were designed and manufactured,
  - new shelf-racks were properly marked;

- the electronics supermarket in the P1 area, storing components in the quantity adapted to the sequence,
- the logistic train (in place of pallet trucks or forklifts),
- the Milk Run concept, including the related actions:
- transit times of the tugger were calculated,

  - duration times of individual actions executed by the milkman were estimated,
  - transit routes were designed,
  - a standardized work sheet was prepared, including the transit map.

To determine numbers of the containers, the function of delivery volume, circulation time and size of one container was used:

$$N = \frac{DT(1+\alpha)}{n_p} \quad (1)$$

where:

N – number of containers (pcs.),
D – average demand for a given component per time unit (pcs.),
T – average time of one complete circulation of a container; time from placing the order to its execution (s.),
α – possible safety margin (pcs.),
n_p – capacity of a container in pieces (pcs.).

Capacity of the shelf-racks adapted to the components was calculated from the formula:

$$p = 2n + 1 \quad (2)$$

where:

p – capacity of the shelf-rack (pcs.),
n – number of containers of the given component used per hour (pcs.).

For the workstations, where the components are transported according to the sequence, it was accepted that the shelf-rack should accommodate 3 cartons of a given component, on the assumption that one sequence is always in use and the other two make a time buffer for the route operator for execution of the next delivery. The same assumption was made for the possible safety margin (α) for which 2 sequences were determined.

For the designed Milk Run, it was assumed that consumption time of one sequence of the components in the areas W1 and W2 was 15.3 min. On the grounds of known speed of the logistic train and the route length from the hall layout, measurements of times of the actions executed by the milkman made it possible to calculate average total transit time of one loop, amounting to ca. 4.5 min. The obtained result was increased by the time required for mounting and dismounting the tugger, to obtain in total ca. 5.5 min.

$$t = \frac{s}{v'} + (t_s \times n_s) \quad (3)$$

where:

t – loop time (s),
s – distance covered by the logistic train (m),
v′ – average speed of the logistic train (m/s),
t_s – time of mounting/dismounting the tugger (s),
n_s – number of stops (pcs.).

In the case when the milkman would execute, in the loop, a delivery of one sequence only, the time remaining for collecting and unloading the components would be ca. 9.8 min. In the company, the milkman course functioned already successively in different lines, so it was found on the grounds of the possessed experience that the time is too short. So, it was decided that, in one loop, the milkman will feed the areas W1 and W2 with the components for the next two sequences. This meant that the time of one course had to be close to 30.6 min. Moreover, the employee's load could exceed 95%, so the one-loop execution time would not exceed 29 min. The individual transit times were re-calculated and, after estimating the action times, a milkman's standardized work sheet was developed, as well as a map on that the route with consecutive stops were marked, see Fig. 4.

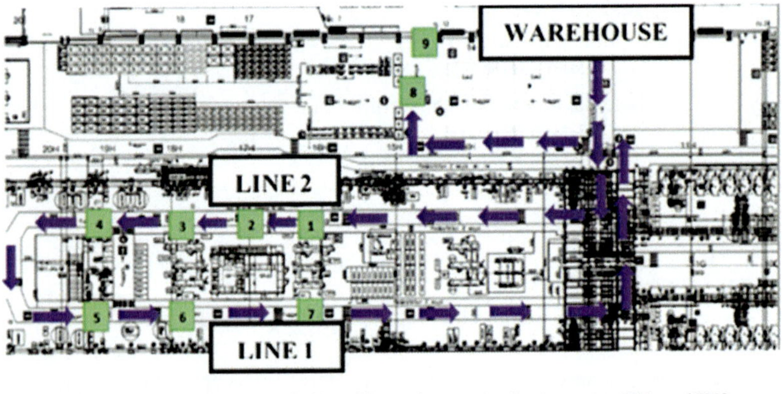

Fig. 4. A part of a map of the milkman's transits in the areas W1 and W2

Table 2 shows the basic data related to the process after implementation of the above-descripted solutions.

Fluidity of the process and quantity of the work-in-progress was reduced by 50% in comparison to the previous state, but the time of stoppages caused by shortages of the components was reduced by ca. 80%. However, it was impossible to reach the

Table 2. Basic process parameters after implementation of the Kanban process and the Milk Run concept

Number of milkmen	1
Milkman's loop time	29.0 min
Milkman's load	95.0%
Number of stops	9
Average quantity of the work-in-progress [cartons in a container]	4.5
Average monthly downtime caused by missing components in W1	218.4 min
Average monthly downtime caused by missing components in W2	233.2 min

intended goal, i.e. complete elimination of downtimes resulted by delayed deliveries of the components.

In order to identify further possibilities of improvements, photographs of the milkman's working days were taken. The examination was performed during one 8-h work shift. The obtained data were introduced to a worksheet and the times of individual actions were calculated, as well as total execution times of the sequent loops. The milkman's actions and their percentages in the entire working work are shown in Fig. 5.

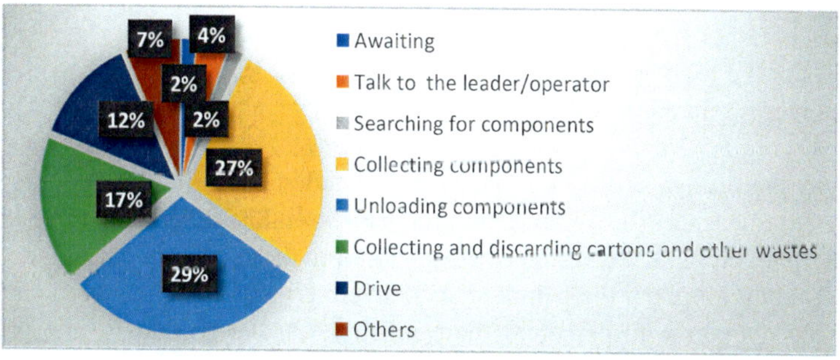

Fig. 5. Percentage distribution of duration of the milkman's actions

The worker spent the largest part of his/her time on unloading the components and their collecting from the warehouse, which is understandable with regard to the nature of the executed work. As much as 16.8% of his time, the milkman used for cutting-open the cartons and discarding them with other wastes to the bins, but 12% of the worktime was taken by transits. After deducting the breaks, the milkman's total work time was 7 h and 22 min. During that time, the employee made 14 loops, of that over 60% were longer than 29 min. The maximum transit time was 45.2 min and the average result was 30.1 min.

After the performed analyses, observations and talks with the employees, further problems existing after implementation of the first stage of improvements were identified in the following forms:

– exceeding the assumed execution time of one loop, resulting, among others, from:

  - long times of unloading cartons with components and cutting-out empty packages,
  - long distances between locations of individual components in the warehouse,
  - searching and awaiting the components missing in the warehouse or finding them in another warehouse,
  - a necessity to transport additionally incomplete sequences due to damages of components,
  - large part of the time used for talks with the co-workers in order to determine the component to be additionally delivered,
  - imprecise standardized work sheets.

- irregular utilization of components by W1 and W2,
- delivering components in larger quantities than those accepted (gathering excessive stocks).

Within the stage 2, in relation to the appearing problems, mainly those of excessing the limit value 29 min during execution of one line, it was decided to employ an additional worker and several actions aimed at improving the work were undertaken:

- the previous one milkman's duties were divided to two employees,
- new, detailed standardized sheets were developed, together with exact transit maps (Figs. 6 and 7); tables with stop numbers of both milkmen were used in the W1 and W2 areas, as well as in the warehouse,
- libraries of small components delivered in the Kanban system were created on carriages of the logistic train, to reduce times of some transits,
- Kanban tables related to additional demands were installed in the W1 and W2 areas.

The previous duties of one milkman were divided in the following way: one worker was assigned to each of the W areas, but the sub-assembly areas were to be thereafter served alternatively. The sequential components from these areas occur in the shopping list every second loop, but the components delivered in the Kanban system are replenished by that worker who perceives the need first. On the grounds of the previously performed observations completed with additional measurements, average execution times for each action executed by the milkmen were determined and the loop for each of them was suggested.

The introduced changes affected the process positively, which was evidenced by values of the indices shown in Table 3. The accepted goal was successively attained, the inventories were reduced to a minimum and the stoppages caused by shortages of components were completely eliminated.

The obtained results were most strongly affected by separation of the previous duties of one milkman between two employees. However, this solution is not fully satisfactory, since employment of an additional person involves the costs including, among others, remuneration of the employee or refurbishment and maintenance of the logistic train. It would not make any problem, if the milkmen were maximally loaded, but this condition was not met here. Times of the loops obtained on the grounds of the standardized work sheets every time consider collection and unloading the components in the sub-assembly areas 1 and 2. Actually, the production process in the sub-assembly areas 1

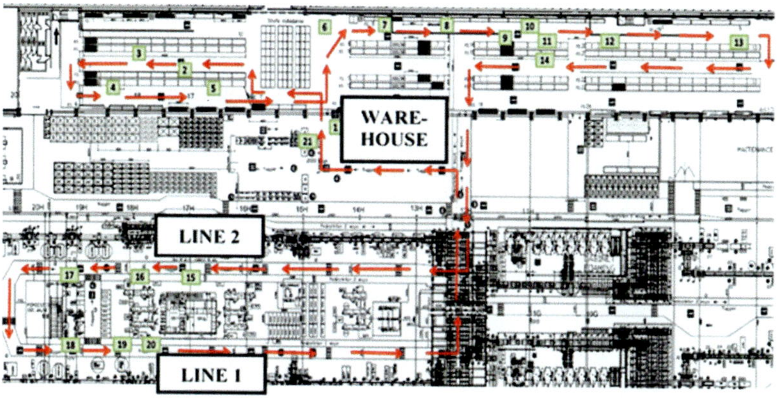

Fig. 6. Map of the milkman's transits in the area W1

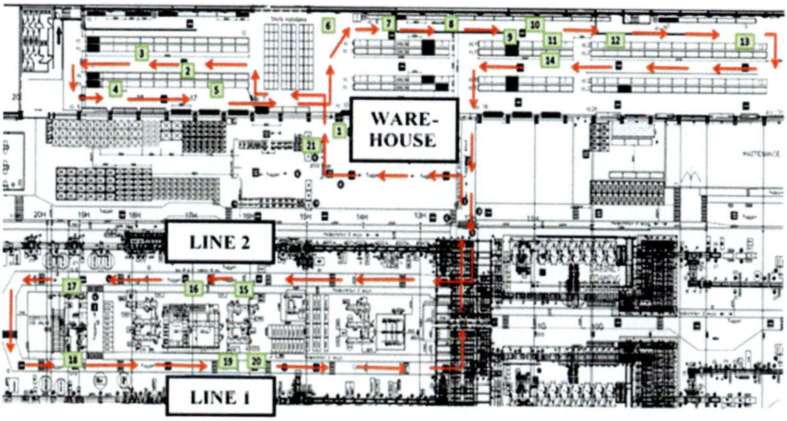

Fig. 7. Map of the milkman's transits in the area W2

Table 3. Basic process parameters after improving the implemented solutions

	W1	W2
Number of milkmen	1	1
Milkman's loop time	27.5 min	27.4 min
Milkman's load	90.2%	89.8%
Number of stops	21	21
Average quantity of the work-in-progress [cartons in a container]	2.7	3.0
Average monthly downtime caused by missing components	1.0 min	0 min

and 2 is started each week, on the average, for 34% and 11% of the worktime of the respective manufacturing area W. This means that, in most cases, times of the milkmen's loops are reduced by ca. 5.5 min (the result obtained by adding-up duration times of the actions related to serving the sub-assemblies, taken from the standardized work sheets). In addition, most components delivered in the Kanban system are replenished one or two times per shift only, which makes it possible to save the next 3.4 min per loop of each worker. Considering these two cases, it can be estimated that the shortest transit times will be respectively 18.6 and 18.5 min for the milkmen from the areas W1 and W2, which corresponds to the loads of 60.9% and 60.7%.

4.3 The Discussion of the Results

Basic indices of the process before and after implementation of the subsequent stages of the improvements are shown in Table 4.

Table 4. Comparison of basic process parameters at three stages of the implemented improvements

	Stage 0		Stage 1		Stage 2	
	W1	W2	W1	W2	W1	W2
Number of milkmen	-		1		1	1
Milkman's loop time	-		29.0 min		27.5 min	27.4 min
Milkman's load	-		95.0%		90.2%	89.8%
Number of stops	-		9		21	21
Average quantity of the work-in-progress [cartons in a container]	9.2	9.0	4.4	4.5	2.7	3.0
Average monthly downtime caused by missing components	1096.0 min	1153.0 min	218.4 min	233.4 min	1.0 min	0 min

Times of the milkmen's loops and their loads were determined on the grounds of the data acquired from the standardized work sheets. The other parameters were estimated on the grounds of the data from six months for the stages 0 and 1, and from three months for the stage 2. After the first stage of the improvements, the number of excessive components was reduced on the average by 50% for both W areas and the time of stoppages caused by shortages of components was reduced by ca. 80%. After the second stage of the improvements, values of these parameters decreased respectively on the average by 35% and 100% in relation to the previous state. The number of stops in the loop increased by 12, which is related to making the transit route more detailed. The average loop time decreased by ca. 1.5 min for both milkmen.

5 Conclusion, Limitations and Directions for Further Research

The actions applied within the project of improvement of the in-house transport made it possible to reach the goal accepted by the company, i.e. to eliminate the stoppages caused by shortages of components and to reduce the inventory level to a necessary minimum.

The presented transit times are theoretical. In fact, after examining the results of the second stage of the improvements, the times are much shorter, which results in low load degree of the workers. Therefore, the third stage of further improvements was suggested, within two possible solutions of the problem of the relatively low milkmen's load were suggested:

1. extending the workers' responsibilities by additional actions, e.g. servicing the adjoining W3 (not only to increase the degree of effective use of the milkmen's worktime, but also to make it possible to discharge the worker who currently feeds W3 with the components collected from the high-bay store, loaded in ca. 52%);
2. introducing the next improvements in order to reduce duration times of individual actions executed by the milkmen and, in consequence, reducing the number of workers to one, among others by:

   – changing locations of components in the warehouse (arranging them closer to each other, as well as reducing the transit time and the number of stops, see Fig. 8),

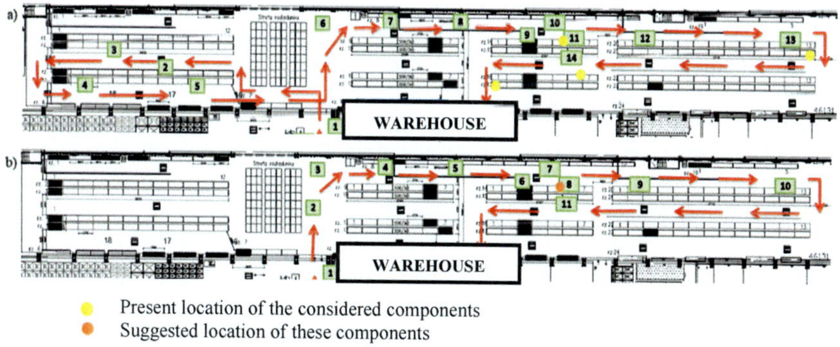

Fig. 8. Map of the milkman's transit: a) present state, b) suggested changes

– eliminating the times of awaiting for delivery of missing components to the warehouse (irregularities in the deliveries can indicate too high load of the operators or a necessity to improve functioning of the inventory management system),
– reducing the number of components delivered in cartons (to reduce the time for opening the cartons in the workstation and for cutting them out before discarding to the bin), by putting them previously to special containers.

It is possible that much wider analysis of the situation before introducing the solution would bring more satisfactory results with the first implementation, however the company decided to focus only on the most important issues. It is possible to optimize the number and location of the stops (to make the transport process faster and more ergonomic), taking into account the location of the components in the logistic train, as well as the parameters of the logistic train and the tugger. It is also possible to minimalize the risk of potential collisions. Such an optimization is possible with the use of intelligent algorithms [16, 27]. The main limitation for the improvements is not only the course of the production process but also the layout of the production hall together with the possible transport roads and possible component storage areas.

References

1. Biniasz, D.: The role and functions of the internal transport of small manufacturing companies - a case study. Logistyka (Logistics) **3**, 533–542 (2014)
2. Womack, J.P., Jones, D.T.: Machine that Changed the World. Rawson Associates, New York (1990)
3. Fynes, B.: From lean production to lean logistics: the case of microsoft Ireland. Eur. Manag. J. **12**(3), 322–331 (1994)
4. Graban, M.: The term "Lean Production" is 25 years old – some thoughts on the original John Krafcik article (2019). https://www.leanblog.org
5. Jylhä, T., Junnila, S.: Learning from lean management - going beyond input-output thinking. Facilities **31**(11/12), 454–467 (2013)
6. Stadnicka, D., Antosz, K.: Lean in large enterprises: study results. Int. J. Soc. Educ. Econ. Manag. Eng. **7**(10), 1376–1382 (2013)
7. Jasiulewicz-Kaczmarek, M., Saniuk, A.: How to make maintenance processes more efficient using lean tools? In: Goossens, R.H.M. (ed.) AHFE 2017. AISC, vol. 605, pp. 9–20. Springer, Cham (2018). https://doi.org/10.1007/978-3-319-60828-0_2
8. Burduk, A., Kotowska, J.: Production line with embedded Lean approach. Res. Logist. Prod. **7**(3), 221–235 (2017)
9. Baudin, M.: Lean Logistics: The Nuts and Bolts of Delivering Materials and Goods. Productivity Press, New York (2004)
10. Wronka, A.: Lean logistics. J. Posit. Manag. **7**(2), 55–63 (2016)
11. Karlin, J.N.: Defining the lean logistics learning enterprise: examples from Toyota's North American supply chain. Dissertation, University of Michigan (2004)
12. Bozer, Y., Ciemnoczolowski, D.: Performance evaluation of small-batch container delivery systems used in lean manufacturing – Part 1: system stability and distribution of container starts. Int. J. Prod. Res. **51**(2), 555–567 (2013)
13. Dimitrescu, A., Babiş, C., Niculae, E., Chivu, O., Dascălu, L.: Efficiency of a production line by application of the Kanban method. JRISS **1**(1), 29–34 (2019)
14. Alnahhal, M., Ridwan, A., Noche, B.: In-plant milk run decision problems. In: 2014 International Conference on Logistics Operations Management, pp. 85–92 (2014)
15. Kovács, A.: Optimizing the storage assignment in a warehouse served by milkrun logistics. Int. J. Prod. Econ. **133**(1), 312–318 (2011)
16. Kotowska, J., Markowski, M., Burduk, A.: Optimization of the supply of components for mass production with the use of the ant colony algorithm. In: Burduk, A., Mazurkiewicz, D. (eds.) ISPEM 2017. AISC, vol. 637, pp. 347–357. Springer, Cham (2018). https://doi.org/10.1007/978-3-319-64465-3_34

17. Zhang, X., Zou, L., Zhuang, H.: The studdy in supply chain of auto parts based on milk-run. OJBM **4**(4), 778–783 (2016)
18. Satoglu, S.I., Sahin, I.E.: Design of a just-in-time periodic material supply system for the assembly lines and an application in electronics industry. Int. J. Adv. Manuf. Tech. **65**, 319–332 (2013)
19. Ciemnoczolowski, D., Bozer, Y.: Performance evaluation of small-batch container delivery systems used in lean manufacturing – Part 2: number of Kanban and workstation starvation. Int. J. Prod. Res. **51**(2), 568–581 (2013)
20. Gyulaiac, D., Pfeiffera, E., Sobottka, T., Váncza, J.: Milkrun vehicle routing approach for shop-floor logistics. Procedia CIRP **7**, 127–132 (2013)
21. Kilic, H.S., Durmusoglum, M.B., Baskak, M.: Classification and modeling for in-plant milk-run distribution systems. Int. J. Adv. Manuf. Tech. **62**, 1135–1146 (2012)
22. Bocewicz, G., Nielsen, I., Gola, A., Banaszak, Z.: Reference model of milk-run traffic systems prototyping. Int. J. Prod. Res. **59**(15), 4495–4512 (2021)
23. Lean Enterprise Institute. http://lean.org.pl
24. De Moura, D.A., Botter, R.C.: Delivery and pick-up problem transportation – milk run or conventional systems. IJM&P **7**(3), 746–749 (2016)
25. Patel, D., Patel, M.B., Vadher, J.A.: Implementation of milk run material supply system in vehicle routing problem with simultaneous pickup and delivery. IJAIEM **3**(11), 122–123 (2014)
26. LeanCenter. http://www.leancenter.pl
27. Kłosowski, G., Gola, A., Amila, T.: Computational intelligence in control of AGV multimodal systems. IFAC-PapersOnLine **51**(11), 1421–1427 (2018)

Reducing Environmental Impact Using Vehicle Route Planning

Lia Coelho de Oliveira[1,2](✉), Oleksandr Pavlenko[1], and Jorge Esparteiro Garcia[1,2]

[1] ADiT-LAB, Instituto Politécnico de Viana do Castelo, Rua Escola Industrial e Comercial Nun' Álvares, 4900-347 Viana do Castelo, Portugal
liaoliveira@esce.ipvc.pt
[2] INESCT TEC, Porto, Portugal

Abstract. Companies focus on achieving high service levels and need to combine short service times with the dynamics between cost and quality. Their transportation systems are therefore a fundamental part; they must be reliable and efficient. This study was implemented in a company of the marine industry, and its final product has special characteristics that require special transportation, i.e., they need a truck with a special structure to be able to transport the boats. This situation causes the vehicle to return empty to the company, a route that the company must support economically. The company has already approached several options with logistic service providers (3PL) without obtaining positive solutions. It is in this sense that the present project arises, which aims to develop a tool for the creation of round-trip circuits, given that in the current context the company intends to acquire a vehicle with reduced environmental impact. In a first phase we analyze the company's needs based on the unique characteristics of the final product, then we study the existing options on the market. Culminating in the proposal of a vehicle that allows performing a circuit in round trip (distribute the final product and return with raw material and not empty) powered by renewable energy.

Keywords: Round trip · CO_2 emissions · Load factor · Transportation

1 Introduction

The competitiveness of a company in today's markets has gone from a mere dynamic between costs and quality, to a scenario that requires constant innovation providing the need for integration of new technologies focused on process optimization. Product life cycles are becoming shorter and shorter, given the increasing differentiation and customization required by the markets, translating into more complex and tense flows, with a higher risk of disruption, since inventory is increasingly considered a risk.

This scenario leads to the need for frequent orders, in small lots, associated with increasingly shorter delivery times. To remain competitive, supply chains must rely on high service levels and increasingly on short response times, replacing forecasts with knowledge of demand. Thus, it is crucial that companies can develop their production and

J. Machado et al. (Eds.): icieng 2022, LNME, pp. 248–256, 2023.
https://doi.org/10.1007/978-3-031-09360-9_21

distribution systems, as well as their strategies, to be able to meet customer expectations at the lowest possible cost [1].

In association with competitiveness, the European Commission has combined environmental sustainability as a priority in European policies, launching a set of measures in the "White Paper of 2011". In this it outlined goals for the transport sector to be achieved by 2050 published in the European document "Roadmap to a Single European Transport Area – Towards a competitive and resource efficient transport system" [2, 3]. This document has been widely analyzed by the scientific and industrial community. And it was with this in mind that the present project was born.

The North American multinational company operating in Portugal is a producer of recreational boats and intends to align its constant transport needs in Europe with the goals set by the EC for 2050. To this end, it has defined that the integration of CO_2-free vehicles is a priority in the first phase.

Their final product has peculiar characteristics that require specific transportation, i.e., they need a truck with a custom-designed trailer to transport the recreational boats (typically 2 boats per truck) to allow the stability of the product, but also due to their size. This leads to the empty return of the vehicle to the company, a route that the company will have to bear economically, and the service is outsourced. The company has already approached several options with logistic service providers without obtaining more advantageous solutions.

The shipping, under analysis, is done by road to all Europe, and there are markets with a predictable cadence. The business group to which it belongs has more companies in Europe that interact with each other, either by sending the final product (pleasure boats with dimensions from 5 m to 10 m), or by sending raw materials (palatized fiberglass coils, appliances, decorative products, electrical cables, steering wheels, among others). In this context, the company proposed the study of route planning, structuring circuits with integration of final product distribution and raw material collection, considering the particularities of the case study and using vehicles without fossil fuel consumption [4].

This paper is divided as follows. Section 2 outlines the Vehicle Routing Problem. Section 3 presents the methodological approach used in this research. Section 4 introduces the proposed solution to the identified problem. Section 5 presents the main findings. Finally, the last section presents the conclusions and suggestions for future lines of research.

2 Vehicle Routing Problem

Within the context of operational research, the Vehicle Routing Problem (VRP) has been widely studied. This is a classic and important operational research problem. In a simplistic way, it consists in assigning a fleet of vehicles to serve clients respecting a given set of constraints to optimize the objective. It has several variants in which changes are verified in its constraints, some are added, others are eliminated or relaxed [5].

Today the objective of VRP does not only contemplate time and costs (for example: minimizing the number of vehicles, traveled distances), but also environmental issues, thus responding to global policies that aim to combat climate change. The transport of goods is responsible for the consumption of a large part of the world's oil reserves, and the activity is expected to increase in the coming years because of increasing globalization [6]. Data from 2015 for Europe attributes 23.5% of total CO_2 emissions to road transport, and of these 72.9% are attributed to road transport, which in turn is divided into two more significant groups, emissions attributed to cars (61% of the 72.9%) and freight transport which represents a total of 37.7% [7]. These values, contrary to what is desirable, have increased. Highlighting other negative effects associated with the increase in transport (not only goods), such as congestion, noise pollution, air pollution and accidents, showing direct impact on the quality of life of citizens [8, 9].

The targets set by the European Commission to reduce carbon emissions [2, 3] have deserved the attention of the scientific community. Accelerated the development of VRP variants aimed at reducing fuel consumption and CO_2 emissions by establishing relationships between distances, load volumes, and speeds [10, 11]. The Green Vehicle Routing Problem (GVRP), an extension of VRP, appears in the literature with the goal of minimizing travel distance and carbon emissions [12].

The growing innovations in fuels and in vehicle propulsion systems have allowed significant CO_2 reductions and the appearance of vehicles with high load capacity free of diesel consumption. In the market, manufacturers of vehicles for the transportation of goods already offer attractive solutions for long distances, allowing companies to consider the transition of their fleets to vehicles powered by Liquefied Natural Gas. This is the alternative to diesel that currently has a supply network and the necessary autonomy. However, the industry sees hydrogen fuel cells as the future [13]. Electric Vehicles (EV) are also available, but their reduced autonomy (around 100 km) becomes an advantage only when integrated into short distance, typically urban, routes [14].

We can find several works in the literature that address EV adoption by focusing beyond range, on restrictions imposed by the still limited charging network, battery backup option, and on fleet management options.

Lin, Zhou, & Wolfson (2016), developed an Electric Vehicle Routing Problem (EVRP) considering a limited capacity of EV batteries as well as the recharging activity. In the model, the authors associated battery consumption with vehicle speed and load, comparing diesel trucks (VRP) and electric trucks (EVRP) on the same route with the goal of minimizing total costs. They concluded that EV charging times significantly increased labor costs.

The European Union in the goals set for 2050 contemplates that the traffic of goods up to 300 km will continue to be mainly done by truck, however, in addition to the incentives to decarbonize the fleet promotes the dissemination of tracking and tracing technologies (GPS, RFID) and the profitability of loading capacity [15].

In long distance (routes over 300 km), the White Paper mentions as a goal for 2050 a 50% reduction in road freight transport. The strategy involves multimodality dependent on the development of railway (for example, the Portuguese program Ferrovia2020) and maritime infrastructures. Betting on technological development and the profitability of cargo capacity [16].

3 Methodology Approach

To solve the problem, we looked at the most classical variation of the VRP model, the capacitated vehicle routing problem (CVRP). It was necessary to reflect the distinct capacity constraints (only one of the trailers allows the closing and return of raw material) for each of the trailers, distinguishing collection and delivery points considering the perspective of minimizing the distance traveled. To calculate costs all variable and fixed components were considered with the objective of allowing the most accurate result possible for comparison with current costs, considering a 6-year vehicle amortization period.

The data used regarding vehicles, consumption, range, and distribution networks were collected from manufacturers and distributors enabling a realistic and operational scenario.

This consisted in the development of a heuristic model based on a variant of the classical Vehicle Routing Problem with capacity restriction for raw material collection with minimized road transport costs, the so-called "Traveling Salesman Problem". Its use was integrated in a two-stage heuristic, that is, in a first stage by flow type, and in a second stage using TSP the collection (route or routes) was defined. The heuristic was developed in Excel VBA allowing the use by the company and evaluation using performance indicators of the routes in operation.

4 Proposed Solution

Before developing the tool, it was necessary to define the type of vehicle to be used to consider its characteristics in the definition of routes (Sect. 4.1). Subsequently, we proceeded to build a tool in Excel VBA which has as input the data extracted from the company's ERP (Sect. 4.2), structured in a two-step heuristic as previously mentioned.

4.1 Vehicle Type

To identify a vehicle with reduced environmental impact, several manufacturers/distributors were consulted. MAN was excluded because it did not present a solution that would allow inbound flow (restriction that one of the trailers had to close). DAF did not submit a bid. Mercedes does not have a vehicle with the desired requirements at the moment (under development at the time of the study). IVECO presented a solution, being a liquefied gas-powered vehicle - IVECO STAILES NP 460. The brand ensures the installation of a loading structure that is adaptable to the project under study allowing inbound flow.

The costs presented in Table 1 and Table 2 allow the comparison between the two types of vehicles. To calculate the cost per kilometer were considered fixed costs (licensing, taxes, insurance, human resources, amortization) and variable costs (maintenance, fuel, tires). Based on the company's activity history we estimated the average monthly kilometers to be driven (12,727 km per month or 140,000 km per year, considering 11 months of activity).

Table 1. Comparative cost of vehicle acquisition

Heading level	Standard diesel vehicle	IVECO NP460
Equipment	95436 €	171 261 €
Interest 48-months	9 544€	17 126€
10-year residual value	0€	0€
Total cost	104 979 €	188 387€

Table 2. Cost structure cost comparison

Types of costs	Standard diesel vehicle	IVECO NP460
Km's per month	12727	12727
Km's per year	140 000	140 000
Service life (years) 10 years	6	6
Depreciation/km	0.119€	0.214€
Depreciation/year	16 701,21€	29 970,68€
Average cost of maintenance	5868.96€	4560€
Average cost of maintenance/km	0.038€	0.030€
Tires cost/km	0.0155	0.0155 €
Fuel cost/km	0.23€	0.18€
Driver cost/km	0.207	0.207
Other fixed costs/km	0.0705€	0.0535€
Total cost/km	0.68€	0.70€

By analyzing the total cost per kilometer driven it is possible to notice the difference between fossil fuel and liquefied gas-powered truck. As the acquisition value of the gas vehicle is very high compared to the fossil fuel vehicle, the amortization is higher. On the other hand, we save on fuel and vehicle maintenance, which is cheaper compared to a fossil fuel vehicle. The cost of tires and drivers is unchanged. A strong advantage of the gas truck is that it reduces the level of CO_2 pollution and in this way, we protect our environment. According to information from the IVECO brand it is possible to reduce the emission of CO_2 by up to 95%.

"...And reducing CO_2, contributing to shrink the carbon footprint of the transport industry, by reducing CO_2 emissions by as much as 95% with compressed or liquified biometh..."[1]

But it is not only the CO_2 reduction that led to the choice of the gas vehicle, but it is also the efficiency of the vehicle. Each vehicle with this structure will do the work of two vehicles with standard features, that is, the normal vehicle used today since it will have a double trailer. The maintenance interval is every 90,000 km and the useful life of the truck is one million kilometers. This makes the vehicle much more efficient compared to Diesel vehicles.

4.2 Tool

For the development of the tool the following data was used as input:

- Final product characteristics: dimensions; Load factor.
- Transportation: vehicle dimension; autonomy; CO2 emissions; consumption; costs/km; fuel type.
- Raw-Materials: characteristics; dimensions; supplier.
- Suppliers: raw- material; name; address; coordinates; Incoterm.
- Customer: product; name; address; coordinates; Incoterm.
- Gas supply network: coordinates.

Phase 1: Clusters Definition
Every week the shipping department receives the plan of deliveries and collections that it must execute. Based on this data, in the first phase, the tool allocates each elemen to one of two groups, group S if it is a supplier or group C if it is a customer. Afterwards, it eliminates the elements that do not have an INCOTERM that defines the responsibility for collection/distribution overland to the company.

Phase 2: Route Construction
Based on the information of the clusters, we start the elaboration of the routes based on the VRP heuristic and obeying a set of restrictions. Thus, the starting point is the company's facilities that will establish a connection with the nearest point, not yet visited, included in cluster C (outbound flow).

If there are elements in cluster C (deliveries to be made) the closest point not yet visited should be added. If it obeys two restrictions: the occupancy rate cannot exceed 100%; the distance between two consecutive points cannot exceed 750 km. This procedure is repeated until all elements of the cluster under analysis are visited.

Of the outbound routes prepared, those with an occupancy rate greater than 70% are automatically validated.

When each outbound route ends, the construction of the inbound route begins. Thus, the tool analyzes the S group with the objective of establishing a connection with the

[1] IVECO Brochure – Page 6 (29/05/2019) - https://www.iveco.com/Common/Documents/Brochures/new_stralisNP.pdf.

closest point not yet visited. However, it will only be validated if the distance is not greater than 500 km. If there are elements in cluster S (raw material collections) the closest point not yet visited should be added. If it obeys two restrictions: the occupancy rate cannot be greater than 24 Ton or 90 m^3; the distance between two consecutive points cannot exceed 200 km. This procedure is repeated until it visits all the elements of the cluster under analysis.

Since there is only one vehicle of this type, it is necessary to ensure the optimization of its use, so routes with occupancy rates below 95% are discarded. In constructing the routes, the following constraints were considered:

- Distance between suppliers equal to or less than 500 km;
- Load Capacity;
- Load Factor outbound (equal to 100%).

Given the information entered, the system will analyze if the load factor corresponds to 100%. When the load factor does not reach 100% the system will warn and advise to go back to the previous step to add more goods to be transported to reach full load. If we want to get better efficiency from the circuit, we must ensure that the load factor is equal to 100%, but it will also be allowed to send the truck with load factor lower than 100% in which case the circuit is no longer more advantageous for the company.

Then when we get the load factor the system will analyze the distance between the last unloading point and the first loading point to ensure that the distance the truck will travel empty cannot exceed 500 km and simultaneously the system will analyze the distance between all unloading points and between loading points to present several types of economic and fast circuits considering the supply points and the driver's driving time.

5 Main Findings

As an example, shown in Fig. 1 the result of a circuit to be executed in Italy, the vehicle leaves the company base loaded with final product, with a load factor higher than 95%, which is equivalent to 3 non-motorized pleasure boats to be delivered in France and Italy.

After unloading all the boats, the distance traveled empty cannot exceed 500 km and between suppliers 200 km in return the load factor of the vehicle cannot be less than 95% (on the trailer destined for raw material). As there is a company's factory in Italy in the suppliers' concerted zone, on the return circuit the vehicle can take up to two boats for example to Spain or France.

The major concern in planning the circuit is the refueling points for the gas vehicles. Due to the autonomy of this type of vehicle of 1600 km and the maximum distance between refueling points cannot exceed 700 km.

The circuit concerning Italy has several refueling points whose distance between them on average corresponds to 400 km. At this moment it is an ideal route to create a circuit that will unite outbound and inbound transportation costs, and the company will have a single transportation cost, and at the same time will facilitate the collection of raw materials from Italian suppliers.

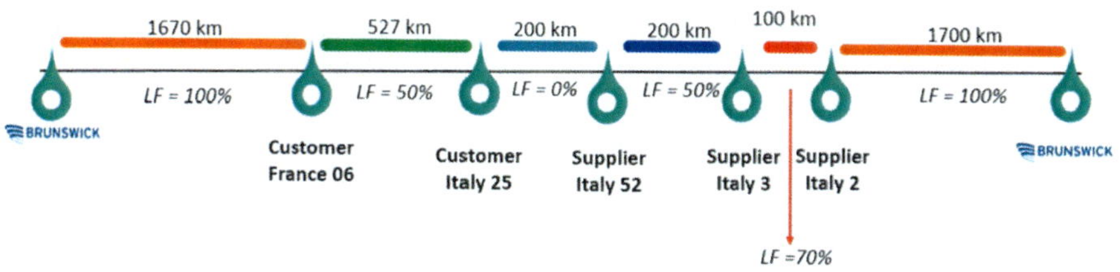

Fig. 1. Circuit - Italy

The circuit can always be adjustable to company's need if any points need to be added on the route by adjusting the following criteria: Supply points; Distance traveled in empty.

In the various simulations performed with data from 6 consecutive months, other circuits appeared, but with rates in Inbound lower than 75%. By analyzing the concentration distribution of flows, it was possible to observe a large discrepancy between Inbound and Outbound. For example, the same country receives 15% and is responsible for sending 85% of the Inbound flows via land. This scenario makes it difficult to implement more circuits.

Respecting all the criteria the operator will obtain the new circuit plan and this circuit can serve as instructions for the driver. It should be noted that in the historical flow analysis of the 12 months prior to this study this rate would not be at risk, and given the seasonality of the business, higher values would be admissible.

6 Conclusions and Future Research

The company that was the object of this study, produces boats of considerable size requiring vehicles classified as special transport.

As expected, the associated costs are high, a factor that originated the present work, focused on the integration of the inbound and outbound flows, analyzing the possibility of round-trip route planning (using a model of vehicle capacity routing problem) to be executed by vehicles without fossil fuel consumption with recourse to customized double trailer. This was done using one of the variants of VRP based on capacity constraints (CVRP). As expected, the environmental and economic impact is significant.

The current phase is the development of a Risk Sharing model with logistic operators (3PL) with the objective of assigning the operational management to a 3PL, as well as the initial investment.

Recently, during the pandemic period, the production volume increased, and it is of interest to develop the study to the remaining supply chain and proceed with the implementation.

References

1. Ertogal, K., Darwish, M., Ben-Daya, M.: Production and shipment lot sizing in a vendor–buyer supply chain with transportation cost. Eur. J. Oper. Res. **176**(3), 1592–1606 (2007)

2. Kallas, S.: White Paper on Transport: Roadmap to a Single European Transport Area–towards a Competitive and Resource-efficient Transport System. Office for Official Publications of the European Communities (2011)
3. Margaritis, D., Anagnostopoulou, A., Tromaras, A., Boile, M.: Electric commercial vehicles: practical perspectives and future research directions. Res. Transp. Bus. Manag. **18**, 4–10 (2016)
4. Brar, G.S., Saini, G.: Milk run logistics: literature review and directions. In: Proceedings of the world Congress on Engineering. WCE (2011)
5. Golden, B., Raghavan, S., Wasil, E.: The Vehicle Routing Problem: Latest Advances and New Challenges, vol. 43. Springer, New York (2008). https://doi.org/10.1007/978-0-387-77778-8
6. Islam, D., Ricci, S., Nelldal, B.: How to make modal shift from road to rail possible in the European transport market, as aspired to in the EU Transport White Paper 2011. Eur. Transp. Res. Rev. 8–18 (2016)
7. European Comission: EU (2017) EU Transport in figures – Statistical Pocketbook 2017 (2017). https://ec.europa.eu/transport/facts-fundings/statistics/pocketbook-2017_en
8. Nocera, S., Cavallaro, F., Irranca Galati, O.: Options for reducing external costs from freight transport along the Brenner corridor. Eur. Transp. Res. Rev. **10**, 1–18 (2018). https://doi.org/10.1186/s12544-018-0323-7
9. EU: White paper on transport (2011)
10. Zhang, J., Zhao, Y., Xue, W., Li, J.: Vehicle routing problem with fuel consumption and carbon emission. Int. J. Prod. Econ. **170**, 234–242 (2015)
11. Bektas, T., Laporte, G.: The pollution-roting problem. Transp. Res. Part B: Methodol. **45**(8), 1232–1250 (2011)
12. Wu, C., Visutarrom, T., Chiang, T.C.: Green vehicle routing problem: the tradeoff between travel distance and carbon emissions. In: 2018 15th International Conference on Control, Automation, Robotics and Vision (ICARCV), pp. 1659–1664 (2018)
13. EuroTransporte: Estratégia da IVECO apoiada pela Comissão Europeia (2019). http://www.eurotransporte.pt/noticia/2/629/estrategia-da-iveco-apoiada-pela-comissao-europeia/
14. Cano, Z., et al.: Batteries and fuel cells for emerging electric vehicle markets. Nat. Energy **3**(4), 279–289 (2018)
15. Speranza, M.G.: Trends in transportation and logistics. Eur. J. Oper. Res. **264**(3), 830–836 (2018)
16. Chen, M., Milenkovic, M., Prosen, M., van Meijeren, J., Val, S.: Smart-rail - smart supply chain oriented rail freight service. In: Proceedings of 7th Transport Research Arena TRA 2018, 16–19 April 2018, Vienna, Austria (2018)

Check for updates

Benefits of University-Industry R&D Collaborations: A Systematic Literature Review

João Barbosa[1], Gabriela Fernandes[2(✉)], and Anabela Tereso[3]

[1] Master's in Engineering Project Management, University of Minho, Campus de Azurém, 4800-058 Guimarães, Portugal
pg41128@alunos.uminho.pt

[2] CEMMPRE, Department of Mechanical Engineering, University of Coimbra, Pólo II, Rua Luis Reis Santos, 3030-788 Coimbra, Portugal
gabriela.fernandes@dem.uc.pt

[3] Centre ALGORITMI, Department of Production and Systems, University of Minho, Campus de Azurém, 4800-058 Guimarães, Portugal
anabelat@dps.uminho.pt

Abstract. University-Industry R&D collaboration (UIC) project is a temporary organization with a collaborative work environment, with heterogeneous partners who have collective responsibilities, and seek to grant benefits to their different stakeholders. This paper aims to gather and bring understanding into what these benefits are, and, through a systematic literature review (SLR) develops a catalog of these benefits in the specific context of UIC projects. The SLR revealed 213 initial papers that after the application of the exclusion criteria was reduced to 125 and of those after the analysis only 54 papers were selected. From these, 45 benefits of UIC projects were identified with varying frequency values, such as 'Patents', 'Publications' and 'New Knowledge Creation'. The collection of benefits was then divided into three categories – Outputs, Outcomes, and Impacts – where it was found that the number of Outputs mentioned in literature was residual when compared to the identified Outcomes and Impacts, highlighting that the benefits of UIC projects go much further beyond the direct Outputs of the project. These results can be used as a source of information for benefits management of UIC projects and for supporting the development of a methodology to evaluate the benefits and societal impact of UIC projects.

Keywords: University-Industry Collaboration · Benefits · R&D projects

1 Introduction

University-Industry research and development Collaboration (UIC) projects play a vital role in stimulating open innovation that leads to new products, processes, and services that create value for customers and broader societal impact in employment and economic growth [1]. The Covid-19 pandemic currently demands an unprecedented number of UICs, and their success has never been more important for society. But, while there is

J. Machado et al. (Eds.): icieng 2022, LNME, pp. 257–280, 2023.
https://doi.org/10.1007/978-3-031-09360-9_22

an increasing prevalence of UICs [2], they often fail to meet stakeholders' expectations [3].

UIC projects lead to new innovations, strategy implementation, or competitive advantages created [4]. For organizations, the development of their strategy is established based on its vision, mission, politics, and internal and external factors, regardless of the scope of the work done. The desired results or outcomes, for the project outputs, are defined as benefits, which are sought in project management, taking into account the time and resource constraints among other critical success factors [5, 6].

It is long recognized that productive research teams require a balance between challenge and security, arrange of 'creative tensions' whose dimensions include the relationships between science and its application, and between individual independence and organizational coordination [7].

The success of a project cannot be dissociated from the evaluation of the direct benefits (outcomes) and indirect benefits (impacts), which will define how a project will be perceived by all its stakeholders. This goes beyond the responsibility of assuring that a project is concluded respecting the triple constraints of time, cost, and scope (quality) [8]. For project success, the value of its output should be considered as well as the satisfaction of its stakeholders [9] and the reach of strategic objectives [10].

A benefit can be defined as "a measurable improvement that derives from the results obtained" [11]. As a product of apparent mutations, it presents itself as an improvement for the project stakeholders [12]. So, a benefit is considered a result whose nature and value can assume numerous forms and, ultimately, be something that brings advantages to an organization [13].

A robust definition of project success can be understood as the conclusion, respecting the triple constraints, that results in outputs, like products or services, that produce outcomes and impacts, with net positive value to the project stakeholders [14]. The successive iterations of this definition lead to a multidimensional framework with the elements from the triple constraint, stakeholder satisfaction, and the achievement of the expected functions [15].

Projects involve multiple internal or external stakeholders, where a stakeholder is "an individual, group, or organization who may affect, be affected by, or perceive itself to be affected by a decision, activity, or outcome of a project" [16], each of whom may have specific ideas about what constitutes value [17]. The evaluation of project value will always be directly connected to what is the frame for evaluating and accessing its true value [18].

It can be considered that for the key stakeholders affected, these benefits are perceive in different degrees, as such, UIC projects bring not only considerable economic benefits to both universities and industries, but also promote the integration of education, science, technology, and economy, i.e., it should also result in social benefits [19]. But a question remains, what are the benefits perceived by the stakeholders of these projects, and how they can be classified? This paper aims to answer this question by providing a Systematic Literature Review (SLR) on the perceived and identified benefits in the literature.

This paper is divided into four sections. First, it gives a brief background on the UIC project benefits and its categorization. Secondly, it presents the procedure used for the SLR. Third, the analysis of the results is presented, and finally, the discussion and conclusions are outlined.

2 Background

2.1 University-Industry R&D Collaboration Benefits

Companies are more open to innovation and investment in research and development (R&D), which positively influences UIC projects, leading to more significant investment in R&D. Companies increasingly recognize the value of collaborating with universities to enhance their innovation capabilities. Collaboration transfers existing knowledge among organizations facilitates the creation of new knowledge and produces synergistic solutions. R&D collaboration is viewed as an essential input to innovation. The European Union has made innovation the central theme of the Horizon 2020 research funding program. The collaboration between universities and industries is encouraged by governments to enhance national competitiveness and wealth creation. Therefore, when developing UIC projects, the focus should be the societal impact, i.e., the demonstrable contribution that the research makes to society and the economy, benefiting individuals, organizations, and nations. UIC projects need to address systematically the various ways people can benefit from their research. This is even more important than ever because major funding bodies worldwide now consider impacts a fundamental aspect of almost all research programs.

UIC projects improve access to science and technology. It enhances knowledge sharing on mutually agreed terms by promoting multi-stakeholder partnerships that mobilize and share knowledge, expertise, technology, and financial resources to support the achievement of the Sustainable Development Goals in all countries, especially in developing countries.

The stakeholders in the UIC have different primary objectives and motives: new ideas and knowledge creation in the case of university versus profit orientation of companies [20]. For collaboration to emerge and be effectively sustained, each of the stakeholders needs to have identified at least one benefit that is desired and deemed impossible to achieve by other means, otherwise the possibility of having an isolated decision by one party to go their own way is almost certain [7]. As such, to ensure the success of a UIC project, one must consider the challenges that come with deferred and incompatible expectations regarding benefits [21].

It is clear that the identification and classification of these benefits should be seen as an important exercise as a means to improve the tools available for benefits management - understood as a continuous process that aims to identify, plan, measure, and assure a follow-up or control of the benefits from the beginning of the project until the realization of last remaining benefit that was initially accounted for [22] and ultimately project success. This same facet is seen in literature as the benefits focus leads to an improved project success [23] and its relevance is evident by the dedication given by the 'Standard of Program Management', from the Project Management Institute, which highlights that the prospective benefits should be registered, analyzed, classified and planned in detail, pursued and transferred using a five-stage process: (1) benefits identification;

(2) benefits analysis and planning; (3) benefits delivery; (4) benefits transition; and (5) benefits sustainment [16].

2.2 Benefits Categorization

From the literature, there are surprisingly different interpretations of what a benefit is. Therefore, to reach an understanding of the benefits identified in the literature, it is important to categorize them and thus understand their purpose and importance.

The UIC project benefits can be categorized into five classes: benefit type; benefit nature; benefit perpetuity; benefit agent; and benefit scope [13]. These are described in the following paragraphs.

- Benefit type
  - Strategic: for industry perspective, these are the most important as they correlate the project objectives with organization strategy [24];
  - Operational: these directly influence the daily activities of an organization by changing its nature and possibly improving it [20];
  - Economic: benefits that impact market value assets and resources regulation [13, 25];
  - Social: perceived as the knowledge transfer between university, industry, and also society, it can be identified as social capital [26].
- Benefit nature
  - Tangible: benefits measurable in a quantifiable, objective way, e.g. creation of products and processes [10];
  - Intangible: only valued in subjective terms, they are, for example, related to the improvement of innovation capabilities [27];
- Benefit perpetuity
  - Long term: outcomes are only perceived to be achieved after the project lifecycle [28];
  - Short term: benefits that cause immediate effects and that can be realized even before the end of the project lifecycle [28];
- Benefit agent
  - University: these are benefits that mainly or solely impact university stakeholders [21];
  - Industry: these are benefits that mainly or solely impact industry stakeholders [21];
  - Society: Social benefits can also be obtained at the regional and national levels, such as the generation of jobs, the development of new products or the training of human resources, all of which contribute to improving the economy [29];

- Benefit scope
  - Value creation: relates to every benefit that is deemed as a net positive value (which in fact could be, at first, a broader term that encompasses all benefits) as such focus on those that are non-fungible [30];
  - Strategy: those that are in line with the organization strategy and contribute directly to the Key Performance Indicators (KPIs) [16];
  - Resource: those related to tangible assets or goods that are used as a result of the project [13];
  - Quality/performance: the improvements achieved regarding the functional aspects of products and services by improving their capabilities/competencies and the actual or perceived quality [29];
  - Employability: benefits that contribute to or are directly responsible for the creation or maintenance of jobs [31];
  - Knowledge: benefits directly or indirectly responsible for the improvement of knowledge transfer and absorption capacity [32];
  - Inter-relational: these will mostly be indirect benefits related to triggering informal relationships between stakeholders [5].

Another possible categorization is to group the benefits into outputs, outcomes, and impacts. This categorization is integrated into most of the UIC generalized frameworks, however there is no consensus on how to classify them [14]. Based on the framework ultimately developed by Perkman [28], which breaks down the UIC project lifecycle into four stages: inputs, in-process activities, outputs, and finally impact, Plewa et al. [24] furthered the development of the framework that led to the collaboration perspective subdivision of outputs as more tangible and direct or 'hard' results and outcomes, and less tangible ones, with 'softer' collaboration effects, coming possibly later on time [20]. Muros et al. [14], based on the proposal from Plewa et al. [24], rearranged the definition for the UIC activities results as follows:

- Outputs: are the direct products, services, or other properties, which are delivered as a result of the UIC project. They are typically tangible and countable and are delivered to individual and institutional stakeholders in the short term [14, 43], for example, a scientific publication or patent [24].
- Outcomes: are the direct benefits or detriments of the UIC project by individuals and institutions [14]. UIC outcomes derive from outputs, their effects might be perceived as positive or negative for the stakeholder, they also can be tangible or intangible and be experienced directly or indirectly over a wide time range. For example, gaining access to equipment or resources [34, 49].
- Impacts: are the UIC results experienced indirectly by individuals, institutions, and societies [14], thus the directness of the effect is the main difference with outcomes [20]. For example, it could be the increase in organization and individual reputation or status.

For the classification of the benefits found, the researchers of this study agreed that a minor change should be made to the above definition of impact. From literature, it was established that this definition refers to indirect experiences, but there are also contradictory definitions that consider impacts as the change that can be credibly attributed to an intervention, an effect of an intervention, or contribution of an intervention that results in UIC agents receiving indirect impacts from UIC in the medium to long term, including individuals, institutions, and societies [35]. Therefore, impact on this study is understood as an indirect or unintended benefit or detriment that affects stakeholders (irrelevant from their degree of influence or how they are affected), which may or may not contribute to broader project objectives.

3 Research Methodology

The Systematic Literature Review (SLR) frames an entire investigation process, which goes beyond a mere review of what was written on the investigation subject. A SLR answers a question or questions to be investigated and consists of a research methodology that locates active studies (valid and relevant), selects and evaluates contributions, and then analyses and synthesizes the information previously obtained, so that reasonably clear answers can be obtained as to what is and is not known about the object under study [36].

3.1 Research Design

The biggest obstacle to developing a SLR in a management environment is the lack of repositories that excel in rigor and organization, observed for example in the medical field. "There is greater emphasis on the application of reviews of the "classical" literature" [38], which can be divided into eight categories: (i) systematic or synthesis by evidence; (ii) narrative; (iii) conceptual; (iv) speed; (v) realistic; (vi) criticism; (vii) expertise; (viii) state of the art [37], the latter being the most prevalent in the fields of engineering and management. However, the objective of this study is to conduct a systematic review of the literature or SLR, which excels in applying strict criteria to identify, evaluate and sum-up the literature, which includes studies published in a peer journal, as well as "gray" or difficult to identify literature [36]. This research study, although initially based on the process proposed by Thomé et al. [38], changed its process to a more direct and simplified form, nevertheless, to compare both the base process is as described.

- First step - Planning and formulation of the problem:
  - Definition of the scope of the theme - applied with the theme itself;
  - Confirmation of the investigation issue;
  - Conceptualization of the topic, the scope of research, and keywords.
- Second step - Literature review:
  - Selection of repositories and databases;
  - Application of the chosen keywords;

  - Reading abstracts and applying exclusion criteria;
  - Full review of selected texts;
  - Review of the search protocol on the basis of the revised texts.

- Third step - Qualitative and quantitative analysis:

  - Check the framework of the objective of the study and its idealization and application;
  - Scrutinize methodology is taken into account:

    - Publication bias – verify the inadvertent exclusion of potentially relevant articles as a consequence of the selection of databases [39];
    - Search methodology skewed by external/internal factors outside the control of the research team;
    - Statistical analysis of the appearance of each factor/benefit/impact;
    - Quantitative analysis of the preponderance of each key element in the model.

- Four step - Determination of the relationship between studies:

  - Summary of the information collected;
  - Analysis and interpretation of information;
  - Establish common points and differences.

- Fifth step - Analysis and synthesis of information:

  - Comparison of key concepts and metaphors;
  - Creation of conductive thread between studies;
  - Elaboration of comprehensive and integrative narrative.

- Sixth step - Interpretation:

  - Synthesis of the relationships obtained in the fifth point;
  - Diagram creation.

- Seventh - Presentation of the results:

  - Preparation of a systematic review article of the literature;
  - Preparation of the report.

3.2 Data Collection

The terms for the search were initially based on a seed document [11]. These were selected and categorized using as a base the PICOC structure (Population, Intervention, Comparison, Outcome, and Context), unlike the recommendation by Cruz et al. [40], the term 'Comparison' was not applied as this answers the question 'compared to what?' and in this case, such comparison point does not add value to the main research question.

The main level called "Population" defines the search context (UIC, University-Industry collaboration) – see Table 1.

The context section reflects the work done to determine what was the scope for the documents presented. In this case, the expectation was to gather all material related to benefits in the context of project/benefits evaluation.

The combinations were all developed using SCOPUS as the database for all searches. To reach the final set, a process of trial and error was used, with several different combinations of either logic connectors ('AND'; 'OR') or keywords where a set of variables (Terms, search terms, or connectors) were blocked, and iteratively changed the free variables to create a list of the best results for each combination. This was an iterative process that resulted in 14213 results or 0 results in the limits for the combinations set – see Table 2.

Table 1. Search terms.

Terms	Concept	Search terms
Population	Industry/university partnerships - application of partnerships between universities, polytechnics, or other educational institutions and small, medium, or large companies	"University-Industry" OR "Industry-University" OR "UIC" OR "IUC"
Intervention	Research and development projects	"R&D" OR "Innovation" OR "Collaborative R&D"
Outcome	Benefits and social impact - included but not limited to economic benefits, market, knowledge, academic relevance	"Project success" OR "Success factors" OR "Program and project benefits" OR "Social impact" OR "Project benefits"
Context	Development of evaluation methods	"Performance measures" OR "Weighted scoring approach" OR "Project evaluation" OR "KPI"

The most acceptable result before any additional filtering, the one with the combination on Table 2, generated 213 documents. This result was further filtered by the nature of the documents. The research was carried out only in academic or research publications, conference papers, or conference reviews, only in English language. The year filter was also used, and only publications from 2011–2021 were included. The cut-off at 2011 was done since this is a topic with a large number of publications, and as it is well-recognized in literature, earlier works are generally included in the latest research. The final total number of publications was 159.

Next, these 159 publications were analyzed, a brief study of abstract, introduction and conclusion was done, from which 125 publications were selected and to which a more thorough analysis was undertaken. This whittled down the number to a total of 54 publications, 35 of which directly mentioned the benefits. The remaining 19 publications indirectly mention benefits or provide information regarding benefits categorization.

Table 2. Example of search strings used.

Search string	Results
("university-industry" OR "industry-university" OR "UIC" OR "IUC") AND ("R&D" OR "Innovation" OR "Collaborative R&D") AND ("social impact" OR "success factors" OR "project success" OR "benefits" OR "KPI" OR "performance measures" OR "project evaluation")	Selected with 213 hits
'IUC' OR 'R&D' OR 'Project success' OR 'Performance measures'	String with higher number of hits: 14213
'IUC' AND 'R&D' AND 'Project success' AND 'Performance measures'	String with lower number of hits: 0

3.3 Data Analysis

Following the initial document filtering process, a content analysis was undertaken, more specifically a conventional form of content analysis, which can provide a more comprehensive image of the subject that is being studied [40].

A sequential cumulative process for identification of the relevant data and concurrently categorizing and eliminating duplicated information [41, 42], the analysis of the literature was done by a specific order, in this case, age of the paper analyzed from the oldest to the most recent.

A conventional content analysis dictates that future reanalysis of the literature might be needed to confirm the findings [41]. This was done to confirm, add or redefine the categories of benefits identified previously. At the moment that a final set of benefits had been defined, the information regarding those was analyzed in order to obtain a consolidated description of each benefit.

The result of the process described before is presented next. According to affiliation, as Fig. 1 shows, there is a short number of institutions that contributed with more than one

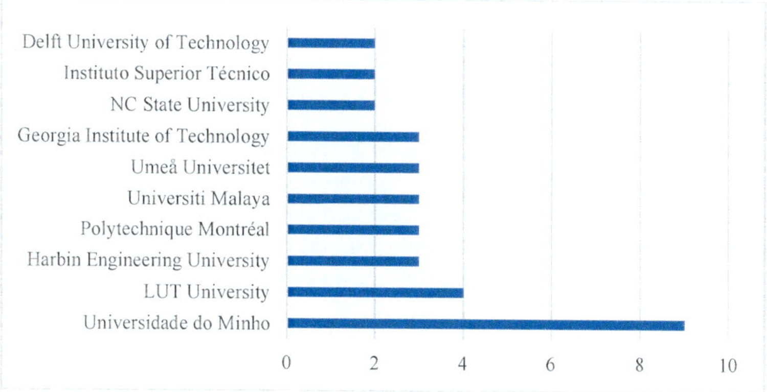

Fig. 1. Documents by affiliation.

paper with new knowledge, from the total 125 papers 34 come from the 10 institutions in Fig. 1 and the remaining 91 papers associated with 91 different institutions.

According to the document subject (area of interest), Business, Engineering, and Social Sciences account for most of areas of interest, as shown in Fig. 2.

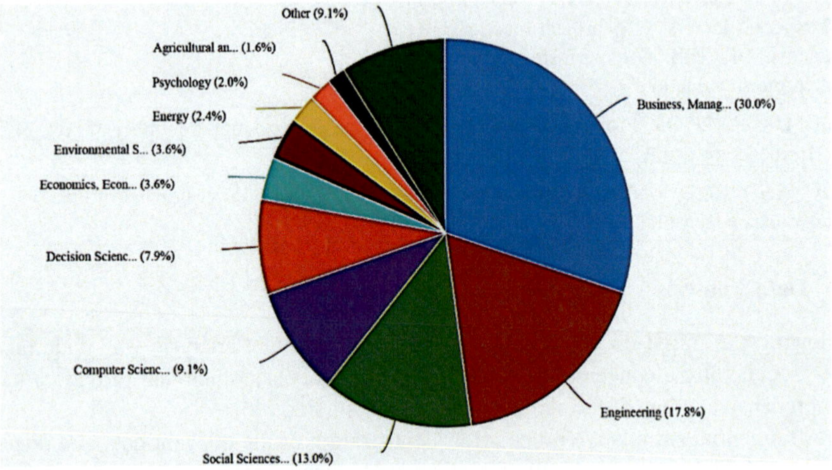

Fig. 2. Documents by subject.

The great majority of the papers found are articles, conference papers and only a single review found, regarding this specific review, the scope does not crossover with the one for this research.

From this overview, some broad conclusions can be taken. The main contributors for the knowledge are few which might indicate a concentration of information or perhaps motivation for the development of the subject. This is also supported by the areas of interest that saw some development in this subject.

4 Results

4.1 Benefits of University-Industry R&D Collaborations

Table 3 summarizes the results from the SLR. A general label that gathers different naming for the same concept was defined. Along with each label, there is a summarized definition for further clarification. It is worth mentioning that although no specific SLR was found, there were researchers that gathered a critical amount of information, to such a degree that a few articles can account for most of the identified benefits.

Table 3. Benefits of University-Industry R&D collaborations.

ID	Benefit	Description	References
B.1	Patents	Number of patents/inventions submitted and accepted, with a strong number of forwarding citations or patents in productions	[5, 8, 11, 20, 25, 26, 28, 29, 33, 43–51, 54, 59]
B.2	Publications	Number of publications submitted and accepted. Increase the capacity of scientific production (industry and academic joint publications) with high volume of citations	[5, 8, 11, 20, 25, 26, 28, 29, 43, 45, 48, 49, 52, 53]
B.3	Influence in public governance	Number of allocations in steering committees, and other public positions of public governance by members of academia and industry	[5, 7, 9, 11, 20, 25, 26, 29, 52, 56]
B.4	Industry experience for students	Successful apprenticeships undertaken during the project by students leading to meaningful industry skills acquired	[5, 10, 11, 20, 25, 31, 46, 49, 53, 58]
B.5	Start-up/spin-off creation	Number of Start-ups/spin-offs created as a result of the UIC project	[5, 10, 20, 29, 31, 44, 53, 56]
B.6	Improve market position	Sales growth, business market position, market share	[5, 10, 11, 20, 23, 27, 47, 51]
B.7	Re-position current product/process	Change of products market target or objectives, change production or methodology process to fit new objectives	[5, 9–11, 20, 29, 48, 53]
B.8	Novel technology	Development of new technologies that will not necessarily be distributed but serve as the basis for new real-world applications	[5, 9, 10, 20, 23, 25, 29, 46, 47, 58]

(*continued*)

Table 3. *(continued)*

ID	Benefit	Description	References
B.9	Skills development	Training collaborators on specific knowledge areas or helping them achieve higher education degrees. Have implemented and normalized the process of continuous development and workforce training - focus on highly specialized workforce	[5, 10, 11, 23, 25, 26, 28, 51, 60]
B.10	Market access	Diminishing the barriers to entry either by creating superior products or services or by having a paved ground for those markets	[5, 20, 44, 48, 56, 61]
B.11	Product or process improvement	Improving products regarding the form or fit function, and or optimizing industrial process	[9, 11, 20, 29, 46, 48]
B.12	Disruptive products	Development of new technologies that can replace an entire market segment. Achieve a technological breakthrough as an indirect consequence of the project objective	[9, 11, 20, 23, 46, 48, 56]
B.13	Integration of qualified resources	Possibility of having integrated both in industry and academia physical and intellectual resources from both sides	[7, 9, 21, 23, 25, 27, 49, 52, 53, 56, 57, 60]
B.14	New knowledge creation	Ability to generate more abstract knowledge and optimize knowledge sources	[9, 11, 23, 25–28, 44, 48, 49, 53, 58, 60]
B.15	Knowledge absorptive capacity	Ability to generate more knowledge sources as well as the improvement of the knowledge absorption capability. Reinforcement of scientific and technological knowledge transfer to industry	[9, 11, 23, 27, 28, 44, 48, 53]

(continued)

Table 3. (*continued*)

ID	Benefit	Description	References
B.16	New ideas for future collaboration	New ideas or initiatives that are formed during or because of the project outputs	[9, 44, 47, 52]
B.17	Create new channels of collaboration	Services regarding industry consulting or the possibility of supporting new projects. Promoting the mobility of resources and access to specialized consultancy	[9, 20, 25, 28, 29, 50, 53]
B.18	Financial resources	Access to public and/or private funding regarding new project development. Financial incentives in form of rewards or benefits for successful project completion that are awarded to researchers, also royalties/equity from intellectual property license or sell	[9–11, 20, 21, 23, 46, 52, 56]
B.19	Facilitated access to human resources	Ability to integrate and attract new skilled students and collaborators	[11, 23, 28, 29, 44, 46, 49, 50, 53]
B.20	Cost reduction	Reducing project development cost - cost sharing; risk sharing	[11, 21, 23, 29, 44, 52, 56, 62]
B.21	Reputation increasing	Increased credibility of individuals and institutions resultant from project success	[11, 26, 29, 52, 60]
B.22	Relationships/trust bonds	Relationships/trust, and other bonds that connect individuals and institutions	[7, 11, 20, 21, 26, 27, 44, 49, 56]
B.23	Access to new knowledge and technology	Lasting informal access to novel information - channels developed by both academics and professionals for information exchange	[11, 23, 25, 26, 46, 53, 56]
B.24	Job creation	Creation of new opportunities for increasing the number of collaborators by either increasing the production needs or creating new functions	[7, 25, 29, 31]

(*continued*)

Table 3. (*continued*)

ID	Benefit	Description	References
B.25	Economic improvement	Economic diversification and growth by means of services and product development	[11, 23, 31, 33, 48, 52]
B.26	Wage improvement	Wage improvement by enhancing the effectiveness of the production cycle or by increasing the level of specialization needed	[23, 31]
B.27	Improve curriculum	Teachable knowledge; industry experience; theoretical knowledge	[20, 46, 52, 53, 57]
B.28	New products developed	Development of new alternative products for existing markets and problems	[23, 28, 29, 33, 48, 50, 53, 55, 58, 63]
B.29	Certification achievement	Certifications for specific faculties or regulations as a direct benefit from the project results	[20, 29]
B.30	Increase production	Improvement in services or products chain value; increase process efficiency	[11, 23, 53]
B.31	Organizational development	Organizational and inter-organizational development as a result of an increase in efficiency provided by the developed solutions	[11, 53, 56]
B.32	Sustainable development	Sustainable development for new products and services with a focus on environmental protection while striving to maintain or improve quality of life	[56]
B.33	Improve safety workspace	Improving working conditions with the objective of implementing HSE (health and safety executive) departments	[56]

(*continued*)

Table 3. (*continued*)

ID	Benefit	Description	References
B.34	Improve time management	Increase effectiveness of the time usage, by gathering new tools to maximize the productivity of each task – reducing risks of delays and accomplishing project schedule	[7, 51]
B.35	Ability to recognize scope	Improve the capacity to understand the project scope during the lifecycle, increasing the effectiveness of the work done – increasing the probability of accomplishing project scope	[7]
B.36	Management and organizational quality	Improvement of process and procedures as well as revisions of organizational structure that increases continuous improvement predisposition	[49, 53]
B.37	Improve product quality	Increase the functional, visual, physical characteristics of the product result – accomplishing project quality	[11, 23, 51]
B.38	PhD, MSc degrees	Number of master's and doctoral theses obtained under the UIC project context	[7, 11, 20, 51, 52]
B.39	Improve innovation process	Increasing the speed of the innovation process, to improve lead time and optimize the steps taken to develop new knowledge and technologies. Improvement of innovation capacity; continuous monitoring of technological changes	[11, 52, 53]

(*continued*)

Table 3. (*continued*)

ID	Benefit	Description	References
B.40	Portfolio diversification	Create multiple differentiated products or services that generate new revenue streams improving organization performance and reducing the overall risks	[11, 20]
B.41	Improve KPIs	Creating a framework for systematic improvement of KPIs by furthering the goals	[9, 11]
B.42	Increase exports	Improvement in services or products chain value that results in cost reduction and consequently increased exports	[7, 11, 27]
B.43	Access to state-of-art resources	Access to up-to-date technology and knowledge	[52, 62]
B.44	Research feedback	Gained insights into one's own academic research. Affiliation with a safe environment to receive feedback on ideas/results/theories. Field-tested one's own theory and research	[46]
B.45	Reorientation of the research agenda	Creation of new research programs, influence on the university research path	[23]

4.2 Analysis

Using first a quantitative perspective to analyze the results of the UIC project benefits referred to in the literature, it is clear that the most frequently mentioned project benefits are 'Patents' (B.1) and 'Publications' (B.2), 20 and 15 references, respectively, while contrasting to the group average of 5.61 references/papers mentioning each benefit – see Fig. 3. There is also a group of benefits 'Sustainable development' (B.32), 'Improve safety workspace' (B.33), 'Ability to recognize scope' (B.35), 'Access to state-of-art resources (B.43)', 'Research feedback' (B.44), and 'Reorientation of the research/development agenda' (B.45) were only mentioned in one paper.

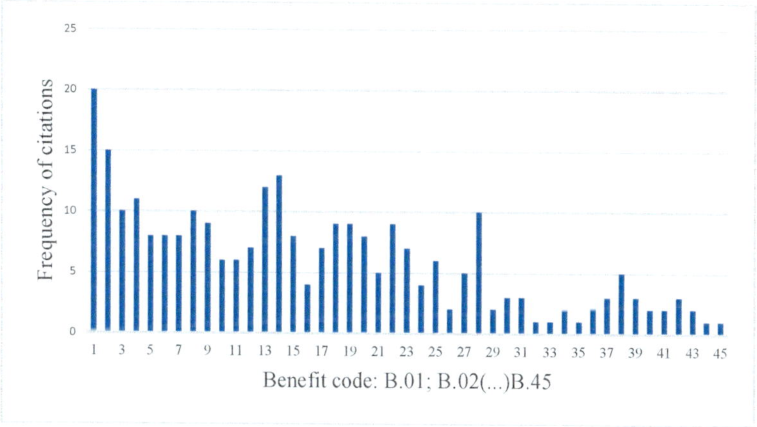

Fig. 3. Quantitative analysis of the identified benefits.

Figure 4 shows that the frequency of benefits mentioned by each paper is heterogeneous at best with a standard deviation of 4.87, with an average of 7.6 benefits and a median of 8 benefits by paper. Regarding the limits of the range of benefits mentioned, observing above Fig. 4, it is possible to identify the papers with more benefits pointed out. These were Fernandes et al. [11] with 26 benefits mentioned, and Feller et al. [53] and Albats et al. [20] with 18, closing the top three and on the other end of the spectrum, it is possible to observe a total of three papers with a single benefit mention [54, 61, 63].

What is possible to confirm regarding the division of citations by paper, and is reflected by the median value, is a presentation of several sources with few inputs, punctuated by a small number of papers that present an appreciable concentration of knowledge on the benefits of UIC projects.

Finally, it is possible to analyze the division of benefits between their classes, these are, as mentioned before, Output (product of a project), Outcome (direct benefit or hindrance), and Impact (indirect benefit or risk realized), all of these are presented in the Fig. 5 bellow.

Most benefits are divided between Impacts and Outcomes with the sum of the latter two classes representing 86.7% of all benefits identified. There is no statistical difference in relative weight concerning representativeness – Outcomes category has 20 benefits attributed and Impacts category has 20 benefits attributed.

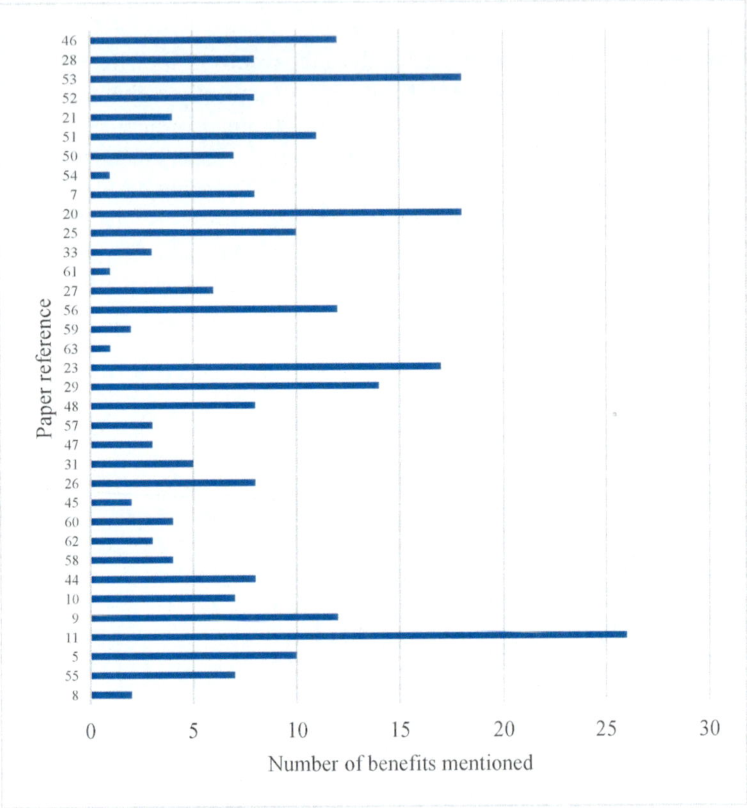

Fig. 4. Frequency of benefits mentioned by paper.

Regarding Outputs, it is possible to access that only 13.7% of the total benefits, corresponding to five benefits, are found to be direct products from the projects. These are: 'Patents' (B.1), 'Publications' (B.2), 'Novel technology' (B.8), 'New knowledge creation' (B.14) and 'PhD, MSc degrees' (B.38). It is possible to access that the benefits considered as Outputs are also the most mentioned on average with the 'Patents' and 'Publications' being the most pointed out benefits overall and having the minimum references of five – 'PhD, MSc degrees'.

5 Discussion and Conclusions

The research reported in this paper has a couple of main contributions to make. The first is the identification and gathering of benefits that are cited as expected, perceived, or experienced as a result of a R&D project in UIC context. The concentration for information related to this area of knowledge within the benefits management field is low [11, 63] having found several papers whose work is restrained manly to no more than three or four UIC benefits [8, 21, 33, 44, 46, 51, 52, 56, 57, 60, 62, 64] and the reason for this might be associated with the goal for each individual study that, unlike

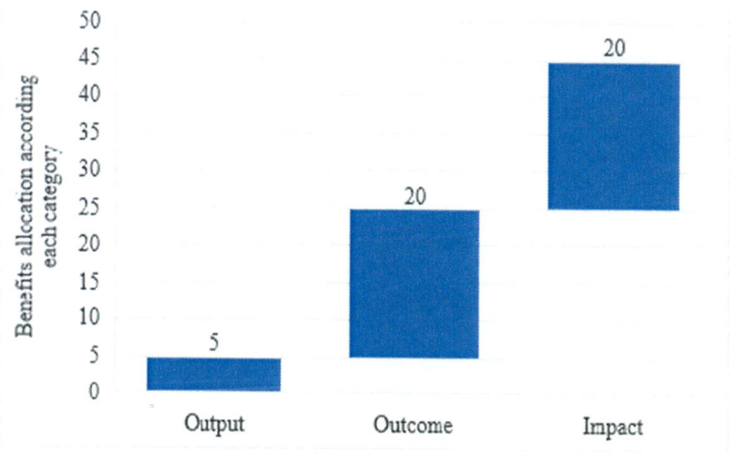

Fig. 5. Division of benefits according to their category.

any other project where the benefits are the final objective, in this case, these are merely variables or effects to the underlying subject which is being studied.

In line with the trend observed before, upon observing the initial pull of papers available before final selection, although there is a concentration of body of work within a select number of academic institutions (see Fig. 1), which represents 34 papers out of a total of 125 results, means that the remaining 95 documents are affiliated within a single institution.

Finally, regarding the provenance of entries, when classifying them by the areas of study in which they are undertaken, the same level of heterogeneous results is found with several disciplines covering the subject, while for the benefits citations by paper and benefits citations by institution it is possible to associate the dispersion with growing interest in the subject of evaluation of project success [58, 65], for the area of study, as well as for the issues in general, is it possible to conclude that "science is becoming trans-disciplinary, public funding is increasingly scarce, scientific results are focused on knowledge utility in society and science is becoming industrialized as academics forge closer links with industry" [36].

The use of an SLR method to gather all information pertaining on project benefits was the most efficient approach as it is a well-known research methodology that has been used by many authors to deep their knowledge [40, 66].

The results produced by this research might lead to questioning if it was effective in gathering all benefits referred in literature. In this respect an overall result of 45 benefits in total can appear as a low number – even taking into account that for several entries a common designation was applied when it was concluded that authors were classifying differently benefits with the same definition, for example, 'Organizational development' (B.31) with different authors referring to the same benefit in different ways [11, 51, 54]. The reasoning for the high frequency of the benefits 'Patents' (B.1), 'Publications' (B.2), and 'New knowledge creation' (B.14) might be related to the common ground associated with these three benefits, which are so generalized that are cited in several studies and researches as a means (a variable) to reach other conclusions on different subjects [50, 67].

At last, an attempt to divide the benefits found into a set of categories was done, this was based not only on the name for each benefit but on an analysis of the remarks made in each paper regarding each single one of them. The categorization of output, impact and outcome comes as mentioned on the background section from a merge of similar concepts that in some sense complete one another, mainly regarding the category 'Impact' which bears a more simplified and yet possibly a more inclusive definition.

What is possible to conclude by this division is that the results from a project matter more when talking about Outcomes or Impacts than Outputs, as it is defended by some authors [68–70] and is now reflected by the exercise done in this paper which demonstrates a greater connection of benefits with Outcomes and Impacts, meaning that the benefits which have a direct connotation with value have greater importance.

Moving from the frequency of benefits identified, one question that remains is the intrinsic value of these to the stakeholders, in a sense of studying what benefits are most valued by each of the stakeholders. There are some references in the literature that point to these tendencies or preferences, but a comprehensive analysis of the literature on this subject should be undertaken.

Additionally, another question that follows is how should these be prioritized? For this possibly the study of multiple criteria decision-making (MCDM) research might be conducted. This has become an important area of research for dealing with complex decision problems that could be a possible avenue [71, 72].

Acknowledgement. This work has been supported by FCT – Fundação para a Ciência e Tecnologia within the R&D Units Project Scope: UIDB/00285/2020 and UIDB/00319/2020.

References

1. Faria, J.R., Ferreira, J.J., Johnson, K.H., Mixon, F.G., Wanke, P.F.: Agglomeration economies and university program creation in the knowledge economy. Socioecon. Plann. Sci. **72**, 100800 (2020)
2. Figueiredo, N.L., Ferreira, J.J.M., Figueiredo, N.L., Ferreira, J.J.M.: More than meets the partner: a systematic review and agenda for University-Industry cooperation. Manag. Rev. Q. **72**(1), 231–273 (2022)
3. Nsanzumuhire, S.U., Groot, W.: Context perspective on university-industry collaboration processes: a systematic review of literature. J. Clean. Prod. **258**, 120861 (2020)
4. Shenhar, A.J., Dvir, D.: Reinventing Project Management: The Diamond Approach to Successful Growth and Innovation, 1st edn. Harvard Business School Publishing (2007)
5. Cross, S.E., Mcconnell, D.P.: How a research university supports university-industry collaboration. In: 2017 IEEE Technology and Engineering Management Society Conference, TEMSCON 2017, pp. 25–31 (2017)
6. Kerzner, H.: Project Management: A Systems Approach to Planning, Scheduling, and Controlling, 12o edição. Wiley, Hoboken (2017)
7. Garrett-Jones, S., Turpin, T., Diment, K.: Managing competition between individual and organizational goals in cross-sector research and development centres. J. Technol. Transf. **35**(5), 527–546 (2010)

8. Soh, P.H., Subramanian, A.M.: When do firms benefit from university-industry R&D collaborations? The implications of firm R&D focus on scientific research and technological recombination. J. Bus. Ventur. **29**(6), 807–821 (2014)
9. De Fuentes, C., Dutrénit, G.: Best channels of academia-industry interaction for long-term benefit. Res. Policy. **41**(9), 1666–1682 (2012)
10. Bellini, E., Piroli, G., Pennacchio, L.: Collaborative know-how and trust in university–industry collaborations: empirical evidence from ICT firms. J. Technol. Transf. **44**(6), 1939–1963 (2018). https://doi.org/10.1007/s10961-018-9655-7
11. Fernandes, G., Pinto, E.B., Araújo, M., Machado, R.J.: Planning benefits realization in a collaborative university-industry R&D funded program. In: 2017 International Conference on Engineering, Technology and Innovation (ICE/ITMC), pp. 1037–1045 (2017)
12. Sandberg, J., Holmström, J., Napier, N., Levén, P.: Balancing diversity in innovation networks: trading zones in university-industry R&D collaboration. Eur. J. Innov. Manag. **18**(1), 44–69 (2015)
13. Andrade, R., Fernandes, G., Tereso, A.: Benefits management in university-industry R&D collaborative projects: a review on benefits and success factors. Procedia Comput. Sci. **100**, 921–927 (2016)
14. Galan-Muros, V., Davey, T.: The UBC ecosystem: putting together a comprehensive framework for university-business cooperation. J. Technol. Transf. **44**(4), 1311–1346 (2017). https://doi.org/10.1007/s10961-017-9562-3
15. Joslin, R., Müller, R.: Relationships between a project management methodology and project success in different project governance contexts. Int. J. Proj. Manag. **33**(7), 14–32 (2015)
16. Project Management Institute: A Guide to the Project Management Body of Knowledge, 6a edição. Project Management Institute, Inc. (2017)
17. Martinsuo, M.: The management of values in project business: adjusting beliefs to transform project practices and outcomes. Proj. Manag. J. **51**(4), 389–399 (2020)
18. Fernandes, G., O'Sullivan, D., Pinto, E.B., Araújo, M., Machado, R.J.: Value of project management in university-industry R&D collaborations. Int. J. Manag. Proj. Bus. **13**(4), 819–843 (2020)
19. Jing, C., Ming, S., De-Cheng, F.: The game analysis on interests distribution of industry-university-research cooperative technology innovation. In: International Conference on Management Science and Engineering - Annual Conference Proceedings 200802171023, pp. 1081–1087 (2011)
20. Albats, E., Fiegenbaum, I., Cunningham, J.A.: A micro level study of university industry collaborative lifecycle key performance indicators. J. Technol. Transf. **43**(2), 389–431 (2017). https://doi.org/10.1007/s10961-017-9555-2
21. Tartari, V., Breschi, S.: Set them free: scientists' evaluations of the benefits and costs of university-industry research collaboration. SSRN Electron. J. 1–41 (2012)
22. Newton, P.: Principles of Project Management (2015). www.free-management-ebooks.com
23. Fernandes, G., O'Sullivan, D.: Benefits management in university-industry collaboration programs. Int. J. Proj. Manag. **39**(1), 71–84 (2021)
24. Plewa, C., Galan-Muros, V., Davey, T.: Engaging business in curriculum design and delivery: a higher education institution perspective. J. Higher Educ. **70**(1), 35–53 (2015)
25. Lucia, Ó., Burdio, J.M., Acero, J., Barragán, L.A., Garcia, J.R.: Educational opportunities based on the university-industry synergies in an open innovation framework. Eur. J. Eng. Educ. **37**(1), 15–28 (2012)
26. Wang, J., Shapira, P.: Partnering with universities: a good choice for nanotechnology start-up firms? Small Bus. Econ. **38**(2), 197–215 (2012)
27. Rantala, T., Ukko, J.: Performance measurement in university–industry innovation networks: implementation practices and challenges of industrial organisations. J. Educ. Work. **31**(3), 247–261 (2018)

28. Perkmann, M., Neely, A., Walsh, K.: How should firms evaluate success in university-industry alliances? A performance measurement system. RD Manag. **41**(2), 202–216 (2011)
29. Isiordia-Lachica, P.C., Valenzuela, A., Rodríguez-Carvajal, R.A., Hernández-Ruiz, J., Romero-Hidalgo, J.A.: Identification and analysis of technology and knowledge transfer experiences for the agro-food sector in Mexico. J. Open Innov. Technol. Mark. Complex. **6**(3), 59 (2020)
30. Al-Tabbaa, O., Ankrah, S.: Social capital to facilitate "engineered" university-industry collaboration for technology transfer: a dynamic perspective. Technol. Forecast. Soc. Change **104**, 1–15 (2016)
31. Ford, C.M., O'Neal, T., Sullivan, D.M.: Promoting regional entrepreneurship through university, government, and industry alliances: initiatives from Florida's high tech corridor. J. Small Bus. Entrep. **23**(1), 691–708 (2010)
32. Cao, X., Li, C.: Evolutionary game simulation of knowledge transfer in industry-university-research cooperative innovation network under different network scales. Sci. Rep. **10**(1), 1–1 (2020)
33. Shi, X., Wu, Y., Fu, D.: Does university-industry collaboration improve innovation efficiency? Evidence from Chinese Firms. Econ. Model. **86**, 39–53 (2020)
34. Locatelli, G., Greco, M., Invernizzi, D.C., Grimaldi, M., Malizia, S.: What about the people? Micro-foundations of open innovation in megaprojects. Int. J. Proj. Manag. **39**(2), 115–127 (2021)
35. European Commission. Directorate-General for Regional and Urban Policy: Guidance document on monitoring and evaluation: European Cohesion Fund, European Regional Development Fund: concepts and recommendations: the programming period 2014–2020. Commission Services (2018)
36. Smart, P., et al.: Open science and open innovation in a socio-political context: knowledge production for societal impact in an age of post-truth populism. R&D Manag. **49**(3), 279–297 (2019)
37. Petticrew, M., Roberts, H.: Systematic Reviews in the Social Sciences: A Practical Guide. Blackwell Publishing, Hoboken (2008)
38. Thomé, A.M.T., Scavarda, L.F., Scavarda, A.J.: Conducting systematic literature review in operations management. Prod. Plan. Control **27**(5), 408–420 (2016)
39. Cooper, H., Hedges, L.V., Valentine, J.C.: The Handbook of Research Synthesis and Meta-Analysis, 2nd edn (2009)
40. Cruz, A., Tereso, A., Alves, A.C.: Traditional, project management agile and lean a systematic literature review. J. Modern Proj. Manag. 1–9 (2020)
41. Leung, D.Y., Chung, B.P.M.: Content analysis: using critical realism to extend its utility. In: Handbook of Research Methods in Health Social Sciences, pp. 827–841 (2019)
42. Hsieh, H.F., Shannon, S.E.: Three approaches to qualitative content analysis. Qual. Health Res. **15**(9), 1277–1288 (2005)
43. Boehm, D.N., Hogan, T.: Science-to-business collaborations: a science-to-business marketing perspective on scientific knowledge commercialization. Ind. Mark. Manag. **42**(4), 564–579 (2013)
44. Bishop, K., D'Este, P., Neely, A.: Gaining from interactions with universities: multiple methods for nurturing absorptive capacity. Res. Policy **40**(1), 30–40 (2011)
45. Rupika, A.U., Singh, V.K.: Measuring the university-industry-government collaboration in Indian research output. Curr. Sci. **110**(10), 1904–1909 (2016)
46. Lee, Y.S.: The sustainability of university-industry research collaboration: an empirical assessment. J. Technol. Transf. **25**(2), 111–133 (2000)
47. Szücs, F.: Research subsidies, industry–university cooperation and innovation. Res. Policy. **47**(7), 1256–1266 (2018)

48. Hu, X., Tang, Y., Motohashi, K.: Varied university-industry knowledge transfer channels and product innovation performance in Guangdong manufacturing firms. Knowl. Manag. Res. Pract. **19**(2), 197–207 (2021)
49. Cui, H., Zhu, X., Wang, H.: Collaborative innovation of low-carbon technology from the triple helix perspective: exploring critical success factors based on DEMATEL-ISM. Polish J. Environ. Stud. **29**(2), 1579–1592 (2020)
50. Jones, J., Corral de Zubielqui, G.: Doing well by doing good: a study of university-industry interactions, innovationess and firm performance in sustainability-oriented Australian SMEs. Technol. Forecast. Soc. Change. **123**, 262–270 (2017)
51. Fernandes, G., Barbosa, J., Pinto, E.B., Araújo, M., Machado, R.J.: Applying a method for measuring the performance of university-industry R&D collaborations: case study analysis. Procedia Comput. Sci. **164**, 424–432 (2019)
52. Dan, M.-C.: Why should university and business cooperate? A discussion of advantages and disadvantages. Int. J. Econ. Pract. Theor. **3**(1), 2247–7225 (2013)
53. Feller, I., Ailes, C.P., Roessner, J.D.: Impacts of research universities on technological innovation in industry: evidence from engineering research centers. Res. Policy **31**(3), 457–474 (2002)
54. Messeni Petruzzelli, A., Rotolo, D.: Institutional diversity, internal search behaviour, and joint-innovations: evidence from the US biotechnology industry. Manag. Decis. **53**(9), 2088–2106 (2015)
55. Franco, M., Haase, H.: University-industry cooperation: researchers' motivations and interaction channels. J. Eng. Technol. Manag. JET-M **36**, 1–51 (2015)
56. Philippi, D.A., Maccari, E.A., Cirani, C.B.S.: Benefits of university-industry cooperation for innovations of sustainable biological control. J. Technol. Manag. Innov. **10**(1), 17–28 (2015)
57. Sebolao, R., Ntshoe, I.: Work-integrated practices in a technology education setting. J. Psychol. Africa **27**(1), 97–100 (2017)
58. Zhang, J., Lu, Y., Xie, Z., Haile, D., Williamson, K.: Game-theory approach on a university-industry collaboration model. In: ASEE Annual Conference and Exposition Conference Proceedings (2016)
59. Wang, X., Fang, H., Zhang, F., Fang, S.: The spatial analysis of regional innovation performance and industry-university-research institution collaborative innovation-an empirical study of Chinese provincial data. Sustainability **10**(4), 1243 (2018)
60. Höjer, M., Larsen, K., Wintzell, H.: Sustainable communications and innovation: different types of effects from collaborative research including university and companies in the ICT-sector. In: Hercheui, M.D., Whitehouse, D., McIver, W., Phahlamohlaka, J. (eds.) ICT Critical Infrastructures and Society, vol. 386, pp. 170–182. Springer, Heidelberg (2012). https://doi.org/10.1007/978-3-642-33332-3_16
61. Fudickar, R., Hottenrott, H.: Public research and the innovation performance of new technology based firms. J. Technol. Transf. **44**(2), 326–358 (2018). https://doi.org/10.1007/s10961-018-9695-z
62. Hewitt-Dundas, N.: Research intensity and knowledge transfer activity in UK universities. Res. Policy **41**(2), 262–275 (2012)
63. Gray, D., McGowen, L., Michaelis, T.L., Leonchuk, O., Rivers, D.: A policy mix experiment to promote start-up success: exploratory evaluation of the NSF Small Business Innovation Research (SBIR)/Industry University Cooperative Research Center (IUCRC) membership supplement. J. Technol. Transf. **47**, 176–212 (2020). https://doi.org/10.1007/s10961-020-09794-6
64. Caloghirou, Y., Giotopoulos, I., Kontolaimou, A., Korra, E., Tsakanikas, A.: Industry-university knowledge flows and product innovation: how do knowledge stocks and crisis matter? Res. Policy **50**(3), 104–195 (2021)

65. Rossi, F., Rosli, A.: Indicators of university–industry knowledge transfer performance and their implications for universities: evidence from the United Kingdom. Stud. High. Educ. **40**(10), 1970–1991 (2014)
66. Mengist, W., Soromessa, T., Legese, G.: Method for conducting systematic literature review and meta-analysis for environmental science research. MethodsX **7**, 100777 (2020)
67. Parreiras, R.O., et al.: A flexible multicriteria decision-making methodology to support the strategic management of science, technology and innovation research funding programs. Eur. J. Oper. Res. **272**(2), 725–739 (2019)
68. Aldabbas, H., Pinnington, A., Lahrech, A.: The role of innovation in the relationship between university–industry collaboration in R&D and ISO 9001. Int. J. Innov. Sci. **12**(4), 365–383 (2020)
69. Shah, S.K., Pahnke, E.C.: Parting the ivory curtain: understanding how universities support a diverse set of startups. J. Technol. Transf. **39**(5), 780–792 (2014)
70. Triulzi, G., Pyka, A., Scholz, R.: R&D and knowledge dynamics in university-industry relationships in biotech and pharmaceuticals: an agent-based model. Int. J. Biotechnol. **13**, 137–179 (2014)
71. Sun, C.C.: A performance evaluation model by integrating fuzzy AHP and fuzzy TOPSIS methods. Expert Syst. Appl. **37**(12), 7745–7754 (2010)
72. Nelson, R.G., Azaron, A., Aref, S.: The use of a GERT based method to model concurrent product development processes. Eur. J. Oper. Res. **250**(2), 566–578 (2016)

Facility Layout Design in Textile MSMEs. Literature Review of Resilient Indicators

José Albán-Palango[1], Franklin Tigre[1], Freddy Lema[1], Pablo Flores-Siguenza[2], and Lorena Siguenza-Guzman[2](✉)

[1] Carrera de Ingeniería Industrial, Facultad de Ingeniería en Sistemas, Electrónica e Industrial, Universidad Técnica de Ambato, Ambato, Ecuador

[2] Department of Computer Sciences, Universidad de Cuenca, Cuenca, Ecuador

lorena.siguenza@ucuenca.edu.ec

Abstract. The capacity to respond and adapt to risks and problems in an organization is critical for business success. Any type of weakness causes inefficient use of resources. On the contrary, flexible facilities can ensure the continuity of operations in the face of disruptive events, which significantly harm the company. However, flexibility is not achieved only with the optimization of facilities, as resilient approaches can enhance it. This research synthesizes the variables and indicators with greater use in three different areas, business resilience, textile industry, and the facility layout problem (FLP). A systematic literature review was conducted, considering 99 studies published in 2010–2021. The documents were analyzed using the Atlas.ti software, a 4W (When, Who, What, and Where) analysis, and three research questions posed through the PICO strategy. The findings indicate that the scientific interest regarding resilience has been increasing in the last six years, specifically in evaluation methods and approaches to identify resilience factors and indicators in the industry through fuzzy mathematical models. It is also highlighted the scarcity of resilient FLP studies the importance of indicators for an accurate model. Finally, a set of possible indicators for the facility layout design with a resilient approach based on FLP factors is proposed.

Keywords: Facility layout problem · Resilience · Textile MSMEs · Indicators

1 Introduction

The need for companies to adapt to an uncertain future represents a challenge, especially in pandemic times [1]. It has generated abrupt changes, resulting in a challenging competitive environment in the areas of price, quality, time, and innovation [2]. This is most evident in micro, small and medium-sized enterprises (MSMEs), whose response to the sanitary crisis has been affected by their limited resources [1, 3]. MSMEs are one of the leading integrators of economic growth, creating job opportunities and becoming contributors to large companies such as suppliers of goods and services. They represent over 70% of world production; thus, a considerable drop in their output drastically impacts the economy. Hence, developing and implementing change strategies is essential to remain in a highly competitive environment [4]. In turn, sectors like textile and footwear present

J. Machado et al. (Eds.): icieng 2022, LNME, pp. 281–293, 2023.
https://doi.org/10.1007/978-3-031-09360-9_23

adaptability problems due to the very dynamic nature of their processes. Thus, organizations must use different methodologies or tools to provide optimal solutions to the issues and risks generated [5].

One of these necessary tools is the facility layout. This tool optimizes the physical space, following a series of criteria under certain restrictions, such as shape, size, orientation, or availability [6]. Moreover, flexibility in the industries is another crucial point to improve their performance and provide adequate responses to different adversities. In this sense, the facility layout problem (FLP) fits somewhat into these flexibility requirements [7]. However, plant designs do not always react adequately under unfavorable circumstances, generating instabilities and failures in activities. In situations like this, resilience plays a crucial role. Its application allows organizations to anticipate their responses to perturbations and react before, during, and after a problem [8]. Several authors have studied the concept of resilience from different perspectives. However, this study concerns resilience engineering, defined as an organization's capacity to adapt to critical events or threats [9].

FLP and resilience can face several strategic challenges; unfortunately, the research that refers to the interaction of both subjects is scarce [10]. This study aims to analyze the resilience factors and variables used in the industry, followed by a synthesis of the most relevant indicators in the textile sector, pointing out its variables and elements. Finally, the relationship between resilience, textile sector indicators, and FLP is studied to obtain indicators applicable to facility layout design (FLD) with a resilient approach.

2 Methodology

The present work uses the Fink methodology, consisting of seven steps to select relevant information systematically [11]. In this manner, it is ensured that the research encompasses the resilience factors and indicators applied in the FLD of textile MSMEs.

The first step refers to selecting the research questions for which the PICO strategy was used [12], determining the Population, Intervention, Comparison, and Outcomes related to the research (Table 1). Thus, three questions were obtained: a) What variables or indicators are considered in a resilient model in the textile industry? b) What variables or indicators are used in the textile industry? c) What variables or indicators can be used in a resilient approach for an FLD in MSMEs in the textile sector?

As for Step 2, the definition of database sources, Scopus, and Web of Science were used for academic articles and LA Reference and RRAAE repositories for theses. For Step 3, selection of search terms, 12-character strings were selected to respond to the research questions: "Resilience" AND "Industry", ("Industry" AND "Textile" AND "Indicators") OR ("Indicador" OR "Indicadores") AND "Textil", "Industry" AND "Resilience" AND "Indicators", "Distribution problems" AND "Indicators" AND "Industry", "Resilience" AND "Industry" AND "Textile", "Resilience" AND "Indicators" AND "Facility layout problems", "Resilience" AND "Facility layout problems", "Indicators" AND "Facility layout problems", "Resilient facility location", "Standard" AND ("Facility Layout" OR "Plant Design" OR "Plant Layout") AND Textile", "Standard" AND ("Facility Layout" OR "Plant Design" OR "Plant Layout"), ("Facility Layout" OR "Plant Design" OR "Plant Layout") AND ("Textile" OR "Confection").

Table 1. Elements of the PICO strategy.

Element	Detail
Population	-Textile companies that need to follow up and monitor their plant distribution, considering demand, quality, costs, or worker safety
Intervention	-Application of appropriate indicators to follow up the plant, considering resilience to anticipate, face, and overcome business turbulences
Comparison	-The comparison is made between the different indicators found, their benefits, application methodology, and scope
Results	-Indicators for the FLD in the textile sector's MSMEs with a resilient approach

Applying practical and methodological selection criteria (Steps 4 and 5) facilitates the searches to provide relevant results. Thus, the study guarantees that articles and theses collected are pertinent and current. The following criteria were considered for their inclusion: 1) theoretical or applied articles and theses written in Spanish and English; 2) scientific journals articles; 3) articles and theses published in 2010–2021; 4) subject areas: Engineering, Management, Production, Mathematics, Decision Sciences, Economics, and Econometrics. In addition, the following exclusion criteria were used: 1) irrelevant articles, 2) duplicate articles, and 3) articles with comments from academic publications. The documents satisfying the exclusion criteria answered the following questions: Has the research design internal and external validity? Is reliability and validity present in the databases used? Are the analytical methods adequate to the characteristics and quality of the research data? The methodological quality of the documents decreased if they did not answer one or more questions.

The qualitative review of the documentation (Step 6) was performed using the Atlas.ti software, in which 53 codes were established to highlight and classify the information. The codes are described in detail in Appendix 1. Finally, the results were synthesized in Step 7. The number of documents obtained after the review and application of the search methodology is summarized in Fig. 1. As can be seen, 99 research studies were analyzed (Appendix 2), of which 59 correspond to articles and 40 to theses. Also, 26 studies have been identified that addressed issues related to textile industry indicators, 27 to FLP, and 46 referring to resilience in the industry.

3 Results and Discussion

The 99 selected research documents analysis is reported and discussed in two sections: a meta-analysis and descriptive analysis.

3.1 Meta-analysis

A 4W analysis (When, Who, What, and Where) was developed to better detail the results. The starting point is the "When" field, for which a temporal distribution was used (Fig. 2).

Notably, 2014 and 2020 reflect the most significant number of research studies carried out, with 15 and 13 studies, respectively. In addition, an increase in publications from 2014 to 2020 can be observed since the average number of published documents is 10.7, with a rise of 6 studies compared to 2010–2013 (4 papers).

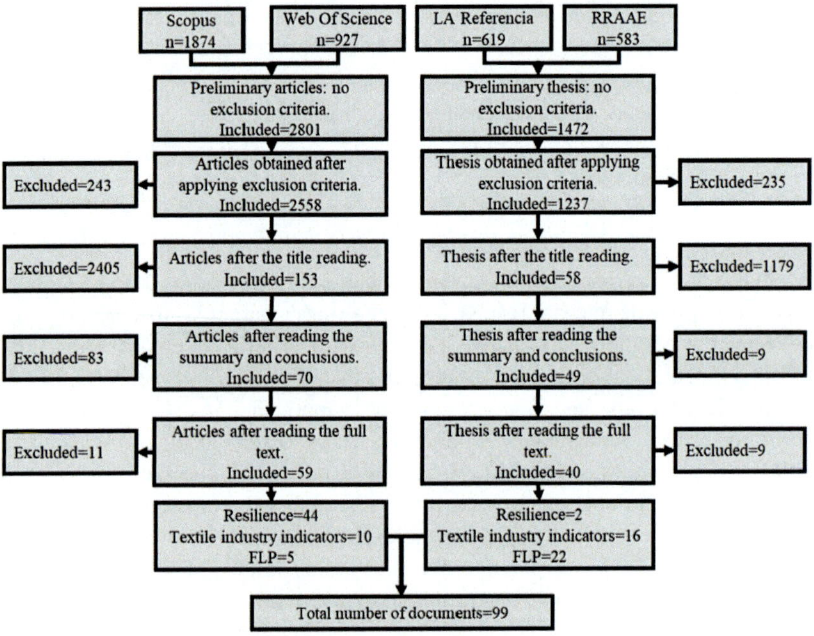

Fig. 1. Flowchart of the application of practical and methodological selection criteria.

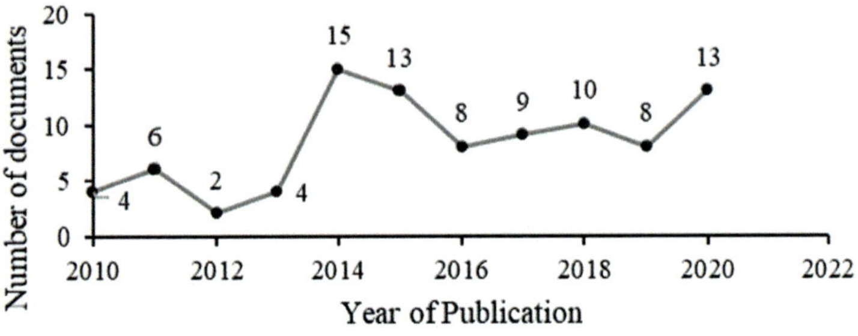

Fig. 2. Time distribution of sample papers.

At the highest peak of the sample, in 2014, only 33.3% of the 15 published studies talk about resilience, while the remaining 66.7% focused on textile sector indicators and FLD. Of the latter, most indicators are related to business management, lead times, failures, productivity, and environment [13]. Although there is a decrease of two investigations in the second peak (2020) compared to 2014, the studies on resilience in the industry amounted to 53.84%, i.e., only two more investigations on resilience in 2014. The studies on this topic remain centered on resilience measurement methodologies, resilience factors in the industry, and decision-making for resilient supply chains [14]. Regarding the studies on indicators of the textile sector and FLP in 2020, the approach is like 2014. The perspective of the indicators studied in the textile sector for FLD resides in the financial and productive scope. Therefore, these studies emphasize production costs, fulfillment of safety regulations, and environmental factors to a lesser extent [15]. Since the data collection period was carried out during 2021, this year is not considered in this analysis to avoid bias.

For the "Who" aspect, it is necessary to know who publishes the studies. Only 59 journal articles were considered. Figure 3 shows the journals with the highest number of publications, considering the three thematic areas: textile industry indicators, FLP, or Resilience. Appendix 3 shows in detail the mentioned information. Next, the "What" aspect aims to know what is developed in the studies, obtaining that 6.7% correspond to FLP studies, 16.94% to textile industry indicators, and 76.36% to resilience. However, if the 99 documents were considered, these figures change, i.e., 27.27% corresponds to FLP, 26.26% to textile sector indicators, and 46.46% to resilience.

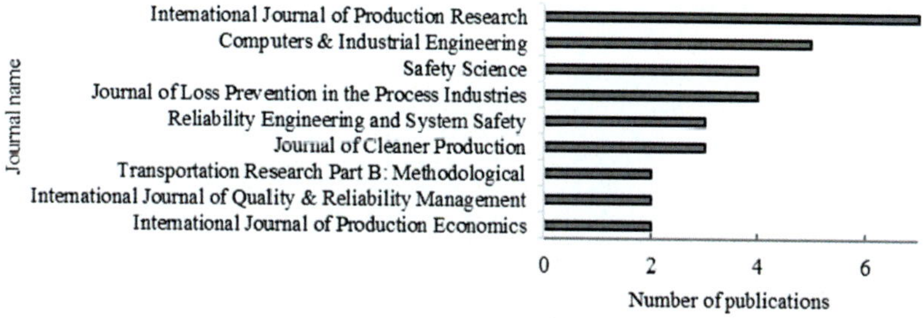

Fig. 3. Articles per journal.

The top journals with the most publications related to the study topics are the International Journal of Production Research, Computers & Industrial Engineering, the Journal of Loss Prevention in the Process Industries, and Safety Science. In these journals, the focuses of the publications, almost entirely, are on resilience, emphasizing the detection of disruptive events, the study of competitive strategies, and the construction of methodological frameworks for business resilience building.

The "Where" field is answered in the final step, i.e., geographic origin. As for the articles, the results indicate that studies have been conducted in 25 countries detailed in Fig. 4. Furthermore, four countries with the highest number of publications account for more than 50% of the sample, i.e., 31 out of the 59: USA, Iran, UK, and India. The studies in the USA speak almost entirely about resilience, presenting an increase from 2018. This research offers a common approach to measuring and implementing resilience to face risks. For example, Cameron and Chao [16] propose a decision-making model under uncertainty. It considers resilience factors to perceive and tolerate disruptive events, denoting main characteristics, the evaluation of resilience as a function of time, the robustness and restoration of the system, the association of resilience and the future benefit of the organization, and finally, an optimization method to solve decision-making models.

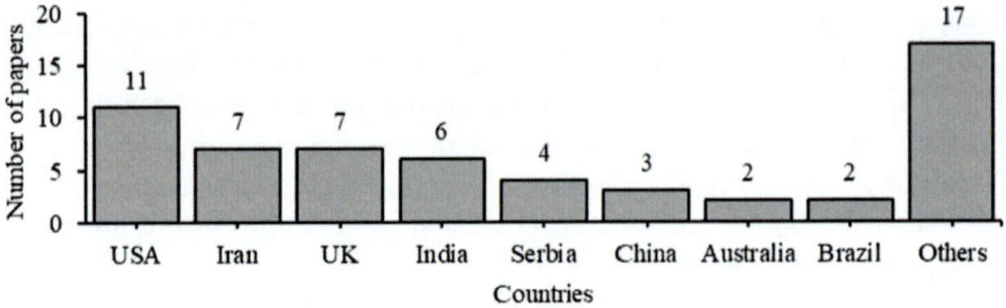

Fig. 4. Distribution of sample papers by country.

As for the UK and Iran, there has been an increase in publications since 2013. The main topic is resilience, with nine studies and three on FLP in the textile industry. Their trend in terms of resilience is repeated with the USA, i.e., methodologies for improvement and conceptual frameworks for measuring resilience [17]. The FLP studies suggest methods for optimization in FLD and algorithms for plant selection [18].

3.2 Descriptive Analysis

This section focuses on answering the three research questions posed. The first question corresponds to the *variables or indicators to be considered in a resilient industry model*. To this end, 46 studies (articles and theses) have been analyzed. These studies describe evaluation methods and strategies to identify resilience factors and indicators in the industry. For example, Aleksić et al. [19] propose a method to assess organizational resilience through a fuzzy mathematical model, proposing internal, external, and resilience factors for subsequent evaluation. This resilient measurement facilitates learning and improvement of various aspects since it is possible to report externally and demonstrate a certain level of performance, efficiently controlling and monitoring processes. Furthermore, according to the authors, a high resilience potential with minimized costs allows increasing the effectiveness of the operations.

The studies mention no single set of resilience elements and factors in the industry. Conversely, these should be used according to the organization; thus, the quantification of resilience is adequate since they contribute significantly to the competitiveness of MSMEs. Therefore, any weakness in the factors would cause inefficient and inadequate use of resources, leading to frustration and waste [20]. Nevertheless, through all this research on resilience, it is possible to synthesize the information, obtaining as a result, a set of indicators in which most of the authors agree (Table 2).

To answer the second research question, *What indicators are used in the textile industry?* It is necessary to know the indicators' different perspectives to understand the industrial sector trend clearly. In most investigations, indicators are applied or analyzed in the production and management areas since they are of interest to know the state of the primary operations of the companies. Their importance lies in the ease of obtaining a global vision of the company and, thus, knowing the most critical parameters and the causes of the existing problems. For example, Montava et al. [21] propose a series of indicators from different perspectives. The authors highlight financial, quality, and innovation perspectives, presenting indicators that consider the consequences of business management strategies and critical success factors. To a lesser extent, the indicators related to the environment focus their attention on water consumption, global warming, or resource use. Reyes et al. [22] developed a method for selecting life cycle assessment indicators in the textile industry in France. This method is adapted as an environmental learning engine since the approach mentioned is one of the most difficult to implement in the textile sector due to the nature of its processes. Therefore, the importance of this type of indicator lies in calculating the environmental impact of various textile products. Most of the indicators analyzed refer to the activities of the production area or top management administration, considering factors such as the quality of the environment. However, the supply chain and logistic indicators are left aside in most cases. Table 3 summarizes the indicators and variables.

Table 2. Industry resilience indicators and variables.

Indicator	Variables
Adaptive capacity	-Access to information -Communications and relationships -Information and knowledge -Innovation -Leadership -Decentralized decision making
Planning	-Planning strategies -Proactive posture -Recovery priorities -External resources
Key vulnerability management	-Severity of occupational accidents -Frequency of occupational accidents -Participation in drills -Process safety -Unplanned shutdowns per year -Accident learning
Just culture	-Understanding and perception of errors -Awareness -Fault tolerances
Information flow efficiency	-Management communication -Control capacity
Managerial efficiency/Top management commitment	-Rules and procedures -Violations of regulations -Work pressure and stress -Self-organization -Information culture -Teamwork
System safety efficiency	-Safety policies -Safety equipment -Safety and physical problems

Table 3. Textile industry indicators and variables.

Perspective	Indicator	Variable
Financial	-Financial position	-Liquidity -Solvency -Cash flow
	-Operating efficiency ratios	-Inventory turnover
	-Profitability ratios	-Gross margin percentage -Profitability on sales
Commercial	-Customer indicators	-Distribution costs -Billing -Non-payments
	-Commercial network	-Achievement of budgeted sales
Human resources	-Staff satisfaction	-Absenteeism -Occupational accidents -Occupational accident costs
Innovation	-Innovation indicators	-New article benefits -Innovation efficiency -R&D&I projects
Production systems	-Quality and service of the manufactured product	-Cost of non-quality -Returns -Failed delivery deadlines -Percentage of defects and waste -Ordering services
	-Environmental efficiency	-Cleaner production knowledge -Compliance with environmental standards -Resource management (air, energy, water, chemicals) -Solid waste management
	-Time	-Lead time for customer service -Lead time for incoming orders -Supplier response time -Average transit time -Operational times

Finally, the study answers, "*What are the variables or indicators that can be used in an FLD in MSMEs in the textile sector with a resilient approach*?" Initially, it can be said that no research has been found in which these two areas interact. Thus, the benefits of resilience in FLD have not yet been thoroughly studied. This means that information is not at the fingertips of organizations. However, the constant need to adapt to new, highly competitive markets generates circumstances in which companies must effectively know the current situation of their processes, operations, and performance in general. In this manner, MSMEs can understand the areas and variables generating problems, and thus their responses are quick and appropriate.

Most of the FLP studies in the textile sector include indicators that consider elements such as the materials flow, transport time, travel distances, or the use of machines. However, these criteria are not the only ones to consider since other indicators are essential to ensure optimal FLD and monitoring. For instance, Azadeh et al. [23] propose a multivariate fuzzy approach to find the appropriate strategy for the distribution of facilities with ambiguity. The authors use operational, qualitative, and dependent indicators such as distance, adjacency, and shape ratio. In this manner, the efficiency and accuracy of the models are greater, obtaining a correct arrangement of the work areas and machinery, thus achieving an economic reduction and at the same time a safer and fairer distribution for employees [18]. Seven factors synthesize the information present in an FLD to have a resilient approach and respond to the FLP principles (Table 4).

Table 4. Proposed indicators and variables for resilient plant layout design

Factor	Indicator	Variable
Material	-Material Handling Cost	-Traveled distance between activities
	-Material movement time	-Duration of travel between activities
	-Material volume moved	-Volume of material between activities
	-Time spent to move material	-Loading/Unloading time
Equipment	-Average time between failures	-Overall operation time in the period -Overall number of failures
	-Meantime to failure	-Stoppage times -Frequency of failures
	-Overall Equipment Effectiveness	-Availability -Efficiency
Waiting	-Time	-Average waiting time between queues -Lead time for order entry -Average transit time
	-WIP	-Average queue length
Service	-System safety efficiency	-Emergency equipment design -Safety routes -Safety and physical problems

(continued)

Table 4. (*continued*)

Factor	Indicator	Variable
Building/Facilities	-Adjacency	-Index of proximity between activities -Perimeter of contact between activities
	-Space sufficiency and utilization	-Space utilization efficiency -Productive area utilization
Change	-Facilities layout flexibility	-Expansion flexibility -Volume flexibility
	-Building expansion	-Available outdoor area -Current area
Human Resource	-Staff satisfaction	-Absenteeism -Occupational accidents
	-Training and development	-Cost of training provided -Operators with degrees or training
	-Fair culture	-Understanding and perception of errors

4 Conclusions

Through the literature review, 99 research studies were identified, of which 59 articles and 40 theses were analyzed systematically. The most common indicators and variables in terms of business resilience, FLP, and the textile industry were determined from the results obtained. The analysis of studies on resilience, indicators in the textile industry, and FLP denotes a rising trend. Indeed, in 2015–2021, there was a rate of published research of 9.71, higher than the period between 2010 and 2014, which was 6.2. In both periods, research on this topic focused on resilience measurement methodologies, resilience factors in the industry, and decision-making predominated.

Resilience research is superior to the other two topics analyzed as it represents 46.46% of the total sample. Although they focus almost entirely on evaluating and classifying resilience factors in the industry, no studies have examined the interaction between FLP and resilience. Together, these topics can address several strategic challenges generated in the wake of the Covid-19 pandemic. In addition, using indicators other than the traditionally used in FLD could allow better integration of flexibility.

As for the studies related to indicators in the textile industry, 84.62% were focused on costs, integrating them with other aspects such as commercial, material, or production. The analyzed indicators mainly apply to the production area and top management administration, highlighting the importance of organizations' global vision. However, these criteria are not the only ones considered since the research on FLP in the textile industry evidenced the use of indicators that evaluate the material flow, transit times, distance traveled, machine utilization, or the geometry of the building.

Finally, a series of indicators have been proposed for FLD, considering a resilient approach. These indicators respond to the FLP principles based on seven main factors: material, machinery, waiting, service, facilities, change, and human resources.

Despite certain limitations of this type of research, it is expected that the work will serve as a key tool for future studies in which the interaction between FLP and resilience is studied. For instance, one of the research limitations lies in obtaining the sample since it depends on the search terms and the database used. The subjectivity of the study is another limitation. The approach of the FLD indicators with a resilient attitude depends mainly on judgment, experience, and the number of researchers involved. Nevertheless, the systematic literature review provides a broad overview of the current trends of the aspects analyzed.

Acknowledgments. The authors would like to thank Corporación Ecuatoriana para el Desarrollo de la Investigación y Academia - CEDIA for the financial support given to the present research, development, and innovation work through its CEPRA program, especially for the CEPRA XV-2021-01-Resil-TEX. This research was partially funded by the University of Cuenca, the Technical University of Ambato, and the University of Azuay.

Appendix 1 – List of Codes Utilized in the Literature Review

The codes used for the systematic literature review using Atlas.ti software are detailed in the following link: https://imagineresearch.org/flp-rev-appendix-1/.

Appendix 2 – List of References Considered in the Literature Review

References of the 99 studies analyzed in the literature review can be found online at: https://imagineresearch.org/flp-rev-appendix 2/.

Appendix 3 – Distribution of Articles Per Journal

The journals and the number of articles published in them, as well as their respective subject, are detailed in: https://imagineresearch.org/flp-rev-appendix-3/.

References

1. Moosavi, J., Hosseini, S.: Simulation-based assessment of supply chain resilience with consideration of recovery strategies in the COVID-19 pandemic context. Comput. Ind. Eng. **160**, 107–593 (2021). https://doi.org/10.1016/j.cie.2021.107593
2. Taçoğlu, C., Ceylan, C., Kazançoğlu, Y.: Analysis of variables affecting competitiveness of SMEs in the textile industry. J. Bus. Econ. Manag. **20**, 648–673 (2019). https://doi.org/10.3846/jbem.2019.9853
3. Ugail, H., et al.: Social distancing enhanced automated optimal design of physical spaces in the wake of the COVID-19 pandemic. Sustain. Cities Soc. **68** (2021). https://doi.org/10.1016/j.scs.2021.102791

4. Ates, A., Bititci, U.: Change process: a key enabler for building resilient SMEs. Int. J. Prod. Res. **49**, 5601–5618 (2011). https://doi.org/10.1080/00207543.2011.563825
5. Zhao, L., Kim, K.: Responding to the COVID-19 pandemic: practices and strategies of the global clothing and textile value chain. Cloth. Text. Res. J. **39**, 157–172 (2021). https://doi.org/10.1177/0887302X21994207
6. Hosseini-Nasab, H., Fereidouni, S., Fatemi Ghomi, S.M.T., Fakhrzad, M.B.: Classification of facility layout problems: a review study. Int. J. Adv. Manuf. Technol. **94**(1–4), 957–977 (2017). https://doi.org/10.1007/s00170-017-0895-8
7. Flores, P., Siguenza, L., Lema, F., Tigre, F., Vanegas, P., Aviles, J.: A systematic literature review of facility layout problems and resilience factors in the industry. Int. J. Adv. Sci. Eng. Inf. Technol. **11** (2021)
8. Thoma, K., Scharte, B., Hiller, D., Leismann, T.: Resilience engineering as part of security research: definitions, concepts and science approaches. Eur. J. Secur. Res. **1**(1), 3–19 (2016). https://doi.org/10.1007/s41125-016-0002-4
9. Steen, R., Aven, T.: A risk perspective suitable for resilience engineering. Saf. Sci. **49**, 292–297 (2011). https://doi.org/10.1016/j.ssci.2010.09.003
10. Navazi, F., Tavakkoli-Moghaddam, R., Memari, P.: Layout optimization of injection process by considering integrated resilience engineering: a fuzzy-DEA approach. Int. J. Model. Simul. **41**, 52–66 (2021). https://doi.org/10.1080/02286203.2019.1670325
11. Fink, A.: Conducting Research Literature Reviews: From the Internet to Paper. SAGE Publications, Los Angeles (2019)
12. da Costa Santos, C.M., de Mattos Pimenta, C.A., Nobre, M.R.C.: The PICO strategy for the research question construction and evidence search. Revista Latino-Americana de Enfermagem **15**, 508–511 (2007). https://doi.org/10.1590/S0104-11692007000300023
13. Azadeh, A., Nazari, T., Charkhand, H.: Optimisation of facility layout design problem with safety and environmental factors by stochastic DEA and simulation approach. Int. J. Prod. Res. **53**, 3370–3389 (2015). https://doi.org/10.1080/00207543.2014.986294
14. Sundarakani, B., Pereira, V., Ishizaka, A.: Robust facility location decisions for resilient sustainable supply chain performance in the face of disruptions. Int. J. Logist. Manage. ahead-of-print (2021). https://doi.org/10.1108/IJLM-12-2019-0333
15. Costa, I., Martins, F.G., Alves, I.: Ecological Footprint as a sustainability indicator to analyze energy consumption in a Portuguese textile facility. Int. J. Energy Environ. Eng. **10**(4), 523–528 (2018). https://doi.org/10.1007/s40095-018-0268-6
16. MacKenzie, C.A., Hu, C.: Decision making under uncertainty for design of resilient engineered systems. Reliab. Eng. Syst. Saf. **192**, 106171 (2019). https://doi.org/10.1016/j.ress.2018.05.020
17. Mehrjerdi, Y.Z., Shafiee, M.: A resilient and sustainable closed-loop supply chain using multiple sourcing and information sharing strategies. J. Clean. Prod. **289**, 125141 (2021). https://doi.org/10.1016/j.jclepro.2020.125141
18. Vitayasak, S., Pongcharoen, P., Hicks, C.: A tool for solving stochastic dynamic facility layout problems with stochastic demand using either a genetic algorithm or modified backtracking search algorithm. Int. J. Prod. Econ. **190**, 146–157 (2017). https://doi.org/10.1016/j.ijpe.2016.03.019
19. Aleksić, A., Stefanović, M., Arsovski, S., Tadić, D.: An assessment of organizational resilience potential in SMEs of the process industry, a fuzzy approach. J. Loss Prev. Process Ind. **26**, 1238–1245 (2013). https://doi.org/10.1016/j.jlp.2013.06.004
20. Gunasekaran, A., Rai, B.K., Griffin, M.: Resilience and competitiveness of small and medium size enterprises: an empirical research. Int. J. Prod. Res. **49**, 5489–5509 (2011). https://doi.org/10.1080/00207543.2011.563831
21. Montava, I., García, R., Bonet, A., Díaz, P.: Textile industry indicators for management. Total Qual. Manag. Bus. Excell. **21**, 1–9 (2010). https://doi.org/10.1080/14783360903492470

22. Reyes, T., Gouvinhas, R.P., Laratte, B., Chevalier, B.: A method for choosing adapted life cycle assessment indicators as a driver of environmental learning: a French textile case study. Artif. Intell. Eng. Des. Anal. Manuf. **34**, 68–79 (2020). https://doi.org/10.1017/S0890060419000234
23. Azadeh, A., Moghaddam, M., Nazari, T., Sheikhalishahi, M.: Optimization of facility layout design with ambiguity by an efficient fuzzy multivariate approach. Int. J. Adv. Manuf. Technol. **84**(1–4), 565–579 (2015). https://doi.org/10.1007/s00170-015-7714-x

Financial Inclusion Through Digital Banking: The case of Peru

Jorge Julião[1](✉), Tonny Ayllon[1], and Marcelo Gaspar[2]

[1] Católica Porto Business School, Universidade Católica Portuguesa, Porto, Portugal
jjuliao@ucp.pt

[2] School of Technology and Management, Polytechnic Institute of Leiria, Leiria, Portugal

Abstract. Peru is currently considered one of South America's key emerging markets, with a significant recent history of economic stability. Nevertheless, the same country is also known for having one of the lowest financial inclusion rates in the region. To fight this latter aspect, the Peruvian government developed an interoperable mobile money platform, based on a digital electronic wallet for the unbanked, called the Billetera Móvil (BIM). Based on this mobile wallet, the purpose of this study aims to debate how financial inclusion may be facilitated by the banking sector's digital revolution. It discusses the fundamentals of FinTech, with a particular emphasis on financial inclusion and associated concerns. Moreover, current research aims to contribute to the understanding of how the recent digital transition can benefit disadvantaged target groups. Field observation and a mixed qualitative and quantitative approach allowed collecting empirical data to support the research findings and discussions. It was found out that, even though there is an effective potential for the digital transition to be impactful in the financial inclusion of the unbanked, a long path has yet to be crossed to achieve a successful result with this inclusive e-wallet.

Keywords: Digital transformation · Banking · Financial inclusion · FinTech · BIM Peru

1 Introduction

The ongoing digital revolution can be considered a period of economic, social, and technological transition. According to Gupta [1], the digital economy is a key driver to promote higher productivity, develop new markets, enhance innovation, promote economic growth, and boost the world economy. Without a doubt, it is more about developing products and new business models than it is about increasing productivity [2]. Current research [3–6] has demonstrated that digital technology fosters the economy and promotes the creation of new jobs.

When considering the banking sector, a full range of traditional products are now being disputed by FinTech [7]. Thus, as highlighted by Vasiljeva and Lukanova [8], FinTech's mission aims at providing the quickest, cheapest, and most adequate financial solutions possible, based on innovation and technology. Therefore, FinTech may promote

J. Machado et al. (Eds.): icieng 2022, LNME, pp. 294–304, 2023.
https://doi.org/10.1007/978-3-031-09360-9_24

new solutions for individual empowerment, allowing for increased transparency and increased accessibility to financial services.

FinTech now promises to reimagine the financial industry by enhancing (1) financial services efficiency, (2) cost reduction, (3) market creation, and (4) promoting more diverse, democratic, and stable financial growth [9]. On the other hand, what is happening with money as we know it? Sharma and Kulshreshtha [10] mention that the use of virtual cash or cashless transactions is a growing trend that has accelerated in recent years. It is also acknowledged that cashless payments are becoming increasingly popular in almost every sector. In this regard, the World Bank Group [11] refers that electronic money has the potential to reduce corruption, increase accountability, and provide a faster response time in the event of a natural disaster, pandemic crisis, or other emergencies. Additionally, electronic money allows for expanding consumer markets, improving bank access to the unbanked, boosting macroeconomic efficiency, and stimulating entrepreneurial activity [12]. For that reason, researchers [10] focused on the recent use of mobile wallets in Singapore, thus discussing challenges and benefits resulting from their adoption by the locals. That research allowed highlighting three critical characteristics that contribute to e-wallet acceptance and preference: ease of use, convenience, and security. In the same light, current research aims at discussing the challenges and benefits resulting from the recent adoption by the Peruvian government of digital electronic wallets for the unbanked, called the Billetera Móvil (BIM). However, as there is no significant published research focusing on the user's perception towards the use of such mobile wallets on the Peruvian market, current research aims at implementing exploratory research on the subject using a mixed qualitative and quantitative research approach. Moreover, this research aims to address the impact that the ongoing digital transition may have on the financial inclusion (FI) of disadvantaged target groups.

2 Financial Inclusion in the World and Latin America Landscapes

According to the World Bank Group, FI is a term that refers to the availability of affordable financial services to individuals and small businesses to address their specific needs in a responsible and sustainable way. Additionally, people may be considered financially included if they own or are able to contract formal financial products [13]. Also, FI entails expanding people's access to financial services, which allows for reducing poverty and fostering economic growth [14]. Finally, FI may be defined as a means to promote productive insertion, which encompasses efforts and initiatives aimed at increasing access to formal financial services for individuals [15].

According to the Global Findex database [11], 69% of adults currently own a bank account, up from 62 percent in 2014. Conversely, adults in high-income economies own 94% of bank accounts, whereas in developing economies' countries, that number is just 63% [11]. Actually, there are still significant levels of financial exclusion at a global scale, as around 1.7 billion adults worldwide can still be considered unbanked. Nevertheless, more than thirty emerging markets have seen a recent significant increase in FI rates, owing to concurrent growth the active use of electronic money, primarily via mobile wallets [11].

Nowadays, those who lack access to financial services rely on cash, which is insecure, risky, and even difficult to manage. Additionally, some studies conclude that some unbanked population may refuse to participate in the financial system voluntarily because they (1) dislike dealing directly with banks, (2) are unaware of the associated benefits, or (3) lack education on how to use digital platforms [14].

The analysis of data on FI in Latin America reveals that the region is characterized by a low and unequal level of household access to the formal financial system. Additionally, there are a limited number of instruments and mechanisms available to aid those participating in the formal financial system in improving their financial integration [11]. South America has one of the lowest rates of FI of any region. It is acknowledged that there is less than half (45.8%) of over-15 years of age population in the region to actually having financial access. This is lower than the global worldwide average (61%) and significantly lower than the 91% of adults in high-income countries [16]. According to the Global Findex 2014 in Latin America, the most frequently cited reasons for not having an account are (1) a lack of funds, (2) accounts are prohibitively expensive, and (3) respondents do not require an account [17].

2.1 Challenges and Opportunities

Combining high rates of financial exclusion, widespread mobile use, and technological innovation presents a significant opportunity to the FinTech sector. Globally, approximately 1.1 billion unbanked people own a mobile phone, accounting for roughly two-thirds of all unbanked adults. Following that, mobile financial services represent a means to reach unbanked individuals and a new revenue stream for mobile network operators [18]. Additionally, over 20% of unbanked adults worldwide (over 400 million people) are believed to receive their earnings or government grants through cash transfers [17]. By converting to electronic payrolls, for example, these workers could gain access to the formal financial system and realize its benefits.

As acknowledged, mobile technology combined with an appropriate and inclusive strategy has the potential to reach unbanked users by leveraging mobile phone penetration and Internet access [19]. Conversely, according to Beck *et al.* [20], increased FI cannot be achieved solely through digital technology. Digital technology, in and of itself, can act as a placebo, thus postponing the changes required to address the underlying issues [21].

Additionally, financial services must be tailored to the needs of consumers from disadvantaged groups. Special care must be taken with those who are utilizing financial services for the first time and may have limited literacy and numeracy skills [22]. FinTech operators also must be familiar with local customers' trends and preferences to develop services that are suitable for the low-income segment. All of this becomes more difficult due to the scarcity of data related to the product requirements for the poor [23]. Indeed, recent initiatives to increase financial inclusion, particularly in developing countries, have emphasized the use of technology, such as that of mobile wallets, electronic payments, and FinTech start-ups [24].

2.2 Best Practice M-Pesa

It is recognized that FI has been successful in Africa. Between 2012 and 2016, the continent added 7.2 million new users of digital financial services, a 250% increase over the baseline period [25]. All of this demonstrates that FI serves as a catalyst for rational development and equitable economic growth. Indeed, Kenya, in which M-Pesa was developed and implemented, is one of the best examples of success.

M-Pesa exemplifies how FinTechs can successfully address some of the problems faced by those living in developing countries [26]. In Kenya, where cash is king and both credit and debit cards are seldom used, citizens have opted out of card payments in favor of mobile payments [27]. Additionally, as a result of the widespread adoption of e-wallets, digital payments are nearly ubiquitous among account holders in Kenya [11].

In 2007, Vodafone's subsidiary "Safaricom" launched M-Pesa as a mobile payment digital platform [25]. M-Pesa enables users to deposit, send, and receive money, as well as pay for goods and services. M-Pesa has already brought all the benefits of a banking product to millions of impoverished Kenyans who previously lacked access to banking services through an unmatched mobile money infrastructure.

According to the World Bank Group [11], mobile payments are now available to 93 percent of Kenyans. Additionally, Kenyans can exchange money for virtual currency through 120,000 M-Pesa agents. M-Pesa operates directly over the telephone network. Sending money is similar to text messaging. Currently, people can also access such services through a mobile app.

2.3 Peru's Financial Inclusion

During seven years in a row Peru has ranked as the country with the best microfinance and financial inclusion environment, owing such leadership largely to the numerous initiatives implemented [11]. However, significant challenges remain for FI as despite significant efforts by the sectors, enormous gaps still persist in Peru [11].

For example, the amount of bank accounts tallied in Peru continues to be lower than the average for Latin America. Over 10 million users with more than 18-years of age do not have a bank account. Conversely, Peru is known to a high rate of mobile penetration throughout the country, including remote areas, of more than 70 percent.

Three major factors contribute to Peru's low FI rates: (a) Inadequate competition in the financing industry; (b) Inadequate penetration of payment infrastructure, namely point-of-sale (POS) terminals; (c) Inadequate financial education and consumption culture [11]. As a result, the Peruvian government promoted two key initiatives in recent years. The first initiative took place in 2013, with the introduction of the electronic DNI. The second is the "Modelo Peru" project, an interoperable payment platform launched in 2016 by the public and private sectors. Modelo Peru is backed by every financial institution, every telecommunications company, and the Peruvian government.

Modelo Peru is the result of a collaborative effort between numerous institutions that fostered the development of Peru's first digital payment platform. The Modelo Peru project stakeholders established an independent company called Pagos Digitales Peruanos to manage the country's electronic money exclusively (PDP). As a result, PDP introduced the BIM product (Billetera Movil). BIM was previously accessed via a USSD platform, which ensured that it was accessible from any type of phone or network. However, they are now accessing the service solely through the app and messenger. Currently, the BIM's primary challenges are expanding their network and establishing themselves as a conduit for government payments such as subsidies, pensions, and taxes. This is critical because, as Shahid and Razaq [28] point out, the mobile wallet that enables government payments has resulted in an increase in wallet account numbers in other countries.

3 Methodology

The qualitative method was the primary research tool in this study. This approach is characterized by a thorough data analysis, which has enabled a detailed examination of how banking technology may contribute to the promotion of FI. Furthermore, such a method was selected as it contributes to the detailed literature review characterization and the dedicated content analysis with the goal of discussing critical components and dimensions. Furthermore, in accordance with Saunders *et al.* [29], such a qualitative approach enables the testing of a theory in conjunction with its context, the comparison of concepts, and the explanation of outcomes. Additionally, qualitative research is suitable because it takes a novel research methodology, seeks to collect and discuss data in a context that is sensitive to the subject, and analyzes data both inductively and deductively, allowing for the establishment of patterns or themes [30].

Additionally, the business case study tool was used to aid in the comprehension of concepts studied in practice and within the Peruvian context. To accomplish this, in-depth observation and a review of the literature enabled analysing and discussing the challenges related to the use of e-wallets in Peru. This enabled the analysis of advantages and disadvantages, comparison of situations, and identification of study opportunities. According to Mfinanga *et al.* [31], the case study methodology can provide additional information about a specific event, as well as promote the analysis of a given case. For instance, it may include a specific narrative and characterization of a specific activity, making it a good candidate for BIM case discussion.

As there were no precedent and no comparable studies available, exploratory interviews were conducted to infer about the fundamentals of BIM and its acceptance. Based on the information gathered during such exploratory phase, a dedicated survey was designed to gain a better understanding of the solution from the users' perspective and to ascertain why they do not use or are uninterested in an electronic wallet despite its apparent benefits.

Subsequently to the explorative phase, a single in-depth interview was carried out with an executive who was responsible for implementing the BIM ecosystem. The interview was conducted in Spanish. The interview's purpose was to ascertain the status of the company PDP, as everything appears to suggest that they were not meeting expectations or are falling short of the results seen in cases such as M-Pesa. Five open-ended questions have been considered to conduct the interview, as previous research has demonstrated that such open-ended requests are preferable to closed-ended requests because they allow people to think freely and are not constrained by the research's frame of reference [32].

Following that, an online survey was made available to elicit not only the perspectives of BIM users but also those who are not. As a result, this research includes a questionnaire that complements the conducted research, with the primary objectives being (1) validated users' general knowledge and attributes about mobile wallets, (2) compare mobile wallets to cash and traditional financial services, (3) understand why people use or do not use mobile wallets.

Schierz *et al.* [33] and Vasiljeva and Lukanova [8] developed similar questions based on previous research. The scale was developed as described in [10] with some of the questions being used directly from that research, while others have been modified to reflect the current research's objectives and goals. The related data were collected using a seven-point Likert scale ranging from 1 to 7 [29]. The questionnaire was electronically distributed to 108 randomly selected respondents in Peru, and same number of valid responses were collected.

Finally, given the qualitative approach taken in this exploratory research, the NVIVO QDA software was used to support the analysis of the collected data.

4 Results

4.1 Interview Results

A single semi-structured, in-depth interview with a BIM executive who was responsible for the ecosystem's development covered five major points, as summarized below:

What Is BIM´s Primary Goal? BIM began as a financial sector strategy to accelerate FI processes in the country through collaboration and competition among financial intermediaries. BIM's primary objective is to make financial services available to more than ten million people (adults over the age of 18) who currently lack access to the financial system. Nowadays, many people lack trust in banks for a variety of reasons, according to our research: poor experience (perhaps the agencies mistreat them), or they believe that the related commissions and interests are hidden (even in the free savings accounts).

Why Is It Considered Key Inviting Financial Institutions to Participate in this Project? It may be considered that reaching faster to the segment of Peruvians that are out-cast from the financial system one would have to leverage the existing network of large banks. BIM enables the financial sector to expand its presence in areas where agencies do not operate. As a result, dedicated collaboration was required to have access to their network of agents and to coordinate communication efforts in a centralized way. 7.

What Have Financial Institutions to Gain as BIM Shareholders? BIM services are not provided for free. As such, it is expected to generate revenue through financial operations. Additionally, the 25 institutions that are currently connected to BIM, thus having the same opportunity for any Peruvian to access their BIM and safeguard their virtual money. Nevertheless, one such process is not always welcomed by large banks. Today, some banks lack a clear vision of their involvement in BIM; they cannot see the advantages of being a part of the operation.

5 Why Did BIM Change Its Method of Accessing Platforms from SMS to Only App and Messenger?

We were compelled to launch the app due to the prohibitively high cost of accessing the USSD. The telecom companies refused to offer us a discounted rate, even though being part of this same project. Additionally, we had to save money to make the project sustainable. If the Peruvian government truly wishes to reduce the use of cash (*e.g.*, to help reduce informality, theft, and increase Peruvians' productivity), it should use solutions such as a social tariff on the USSD to encourage the use of electronic money. Such measures haven't yet been implemented, and it is important to high-light that PDP is not a governmental entity; it was formed on its own initiative.

Is There Any Kind of Conflict of Interest Inside This Project? Indeed, an effective conflict of interest does exist. Particularly, when one of your shareholders, BCP (Peru's largest bank), declares that they expect to have ten million customers and compete with cash through their e-wallet "Yape," with similar goals to those of BIM.

5.1 Survey Results

Based on an online questionnaire, more than hundred valid answers were collected and analysed using the NVIVO QDA software. The main results of this survey are summarized next:

Demographics. Among those participating on the survey, 87% of the male respondents mentioned owning a bank account, whereas only 80% of the female respondents referred to having a bank account.

Mobile Wallet Related Information. It was observed that the respondents living in the country's capital mentioned having more information about the mobile wallet, in op-position to those living outside Lima. Particularly, 47% of those living outside the capital slightly disagree with the statement of having enough information compared to 26% in Lima. It can be settled that in Peru there is not enough information about the services and benefits of mobile wallets, with special incidence in the provinces when compared to those living in the capital city.

Timesaving from Using a Mobile Wallet. On what refers to the time savings perceived by potential users of the mobile wallets, the respondents highlighted this feature as being the most valued. When analyzing these results by gender, it was observed that 85% of the male respondents had a positive, or very positive outlook on such a feature, whereas only 72% of the female respondents perceived the same way that aspect.

Mobile Wallets Vs Cash and Banks. When analysing the appreciation of mobile wallets against traditional banking services and cash, two indicators stand out: 48% of the users strongly agree that the mobile wallet is more practical than carrying cash, and 42% agree that the services of a mobile wallet are cheaper than a bank. As a result, in all comparisons made, respondents generally agreed with and preferred the services of mobile wallet users over cash and banks. The components evaluated in these questions were based on the research by Sharma and Kulshreshtha [10], who conducted a study on mobile wallet adoption and concluded that three critical characteristics contribute to final users' needs and wants: usability, convenience, and security.

Why Do They Utilize or Intend to Utilize the Wallet Service? The last section of the online survey focused on the attributes that actual, or potential mobile wallets users value the most. Users appreciate the speed and ease of use of their mobile wallets, while potential users say that they would use them because they perceive them being convenient and fast to use. On the other hand, both groups coincide in placing on the last position the attribute of affordability.

6 Discussion and Conclusion

The current mixed quantitative & qualitative research combined both field observation with a survey based on a dedicated in-depth interview and a digital questionnaire. This study allowed identifying and discussing some of the main findings related to the potential of digital electronic wallets in Peru to promote FI. These will be highlighted next:

It is acknowledged that Peru is a commercially open country with the region's most stable and rapid economic growth. Furthermore, Peru has experienced more than 50% reduction in poverty rates over the last decade. Nevertheless, about a decade ago, the percentage of banking in Peru was among the lowest in the world. However, significant progress has been made in recent years. As Antón and Conde [34] mention, a growing number of people in Peru have currently access to financial services. As explained in the Global Findex questionary, this increase was fueled by a significant increase in rural accounts, which increased 14 percentage points from 29% to 43% between the years of 2014 to 2017 [11].

In Peru, although financial services have significantly increased in recent years (particularly the number of agents), when extrapolated to the amount per 1000 km, financial services remain insufficient.

Another distinguishing feature of Peruvian banking has been its significant costs of financial services, operations, and basic services. For instance, in some financial institutions, obtaining a physical statement could cost over $3, and cash withdrawals from bank agencies are subject to a $2.71 fee.

6.1 Main Findings

Some of the main findings of this study may be summarized as follows:

Perceptions on Traditional Banking. This research allowed identifying that there is an increasing trend toward avoiding traditional banking services, primarily because they are perceived as expensive, complex, and time-consuming.

FinTech vs. Traditional Banking. It was a consensual understanding that FinTech firms have the potential to outperform traditional banking since they develop low-bureaucracy internal processes, use improved technology, and have a greater capacity for innovation.

Advantages of FinTech. It was also observed that not only do individuals value the benefits of FI, but also society and governments. It was also noted that the perceived benefits ranged from increasing savings to assisting individuals in escaping poverty.

Consumer Profiles. This study highlights the need for special considerations towards the case of providing FI services, since the profile of consumers located outside the capital city, Lima, are not synonymous with those living in urban areas, who refer earning high incomes and having effective financial literacy, to name a few characteristics.

Impact of Banking Costs. Apart from the traditional banking low penetration rate, in Peru there is a significant cost differential between providers of basic financial services. As a result, there are numerous opportunities for FinTech to deliver faster services at lower costs and with greater accessibility for more people.

6.2 Discussion

According to the interview, PDP has not the full support of Peruvian governance or all of its shareholders (banks + telecommunications). Even though theory and successful examples such as M-Pesa demonstrate that government support is critical for this project. Additionally, Bayero's research [13] concluded that FI requires significant resources, efforts, and confidence in the concept of FI on the part of governance and all involved institutions.

The results of the survey allowed observing that most of the potential users do not have significant information about or are unaware about the e-wallet's advantages. Local observation and dedicated research enable the identification of financial education as a factor that needs to be improved in Peru. Dedicated signals confirm that Peruvian governance is attempting to improve it; nevertheless, much remains to be done. According to Cole *et al.* [22], new and compelling evidence demonstrates that financial literacy is a critical predictor of financial behavior in developing countries.

The business case analysis enabled an appreciation of the Peruvian government's efforts to promote financial inclusion and education (*e.g.*, back in 2014 financial literacy was about 28%). However, no additional proposals in the remaining areas as suggested by the World Bank Group [17] in that study have been made. As an example, the digitalisation of G2P payments, which was identified a success story in other countries, can be mentioned.

As per the Global Findex, 36% of Peruvians cite cost as a barrier to financial inclusion. This conclusion was confirmed by an analysis of the prices of the major financial products, which revealed that cost transactions are prohibitively expensive, even when conducted via Internet banking. That's why Ozili [14] observed in his study that traditional banks are being rejected as being expensive, complicated, and time-consuming.

6.3 Conclusions

The significant impact of the ongoing digital revolution was observed and discussed throughout this research study. As no relevant published research focusing on the user's perception towards the use of such mobile wallets on the Peruvian market was available, the current research aimed at implementing exploratory research on the subject using a mixed qualitative and quantitative research approach. It highlighted and discussed the perceptions of users and non-users of such digital banking tools in that emerging market.

This exploratory research also addressed the impact of the digital transition on the FI of disadvantaged target groups, with new financial services for those who have been denied access to the financial system's benefits for many years.

Additionally, the current research enabled an understanding of how the banking sector's technological revolution has resulted in the emergence of new companies known as FinTech, which are much more dynamic, adaptable, and capable of reaching segments that traditional banking neither desired nor was able to serve. Thus, a significant benefit of FinTech's scope is that it can serve the unbanked with more efficient and lower-cost technologies that enable them to compete with banks.

The highlight of this research is the significance and value of FI; prior research in a variety of contexts and countries confirms the positive relationship between increased FI and nation-building. As a result, governance should not hesitate to exert maximum effort to reduce financial exclusion rates; they should invest in building robust financial education programs, as well as promoting strategic partnerships to carry out FI projects.

Finally, based on the case of BIM Peru, one can conclude that BIM is not achieving the anticipated success. Several reasons could be that it has failed to cultivate an ecosystem that encourages and promotes mobile wallet use. Additionally, the strategy of gathering more than 20 private financial institutions may have not been the best choice, as it is reasonable to assume that such companies, as well as telecom operators, are more concerned with profit generation than with social responsibility. Nonetheless, it was demonstrated in this research study that Financial Inclusion can be promoted and achieved via digital banking.

References

1. Gupta, M.: Digital economy and the role of information technology. ANWESH Int. J. Manag. Inf. Technol. **3**, 53–58 (2018)
2. Ivanova, I., Sceulovs, D.: Identifying elements of the digital economy ecosystem. J. Bus. Manag. **16**, 5-14.5–15 (2018)
3. Kaplinsky, R., Kraemer-Mbula, E.: Innovation and uneven development: the challenge for low- and middle-income economies. Res. Policy **51**, 104394 (2022)
4. Dieppe, A., Francis, N., Kindberg-Hanlon, G.: Technological and non-technological drivers of productivity dynamics in developed and emerging market economies. J. Econ. Dyn. Control **131**, 104216 (2021)
5. Gaspar, M., Julião, J., Cruz, M.: Organizational strategies induced by the fourth industrial revolution: workforce awareness and realignment. In: Machado, J., Soares, F., Veiga, G. (eds.) HELIX 2018. LNEE, vol. 505, pp. 330–336. Springer, Cham (2019). https://doi.org/10.1007/978-3-319-91334-6_45
6. de Andrade Régio, M.M., Gaspar, M.R., do Carmo Farinha, L.M., de Passos Morgado, M.M.A.: Industry 4.0 and Telecollaboration to promote cooperation networks: a Pilot survey in the Portuguese Region of Castelo Branco. Int. J. Mech. Appl. Mech. **1** 243–248 (2017)
7. Lee, I., Jae, Y.: Fintech: ecosystem, business models, investment decisions, and challenges. Bus. Horiz. **61**, 35–46 (2018)
8. Vasiljeva, T., Lukanova, K.: Commercial banks and fintech companies in the digital transformation: challenges for the future. J. Bus. Manag. **11**, 25–34 (2016)
9. Wiradinata, T.: Mobile payment services adoption: the role of perceived technology risk. In: 2018 International Conference on Orange Technologies, ICOT 2018, pp. 1–5 (2018)

10. Sharma, G., Kulshreshtha, K.: Mobile wallet adoption in India: an analysis. J. Bank Manag. **18**(2019)
11. Demirgüç-Kunt, A., Klapper, L., Singer, D., Ansar, S., Hess, J.: The Global Findex Database 2017: measuring financial inclusion and opportunities to expand access to and use of financial services. World Bank Ec. Rev. **34**, S2–S8 (2020)
12. Drasch, B.J., Schweizer, A., Urbach, N.: Integrating the 'Troublemakers': a taxonomy for cooperation between banks and Fintechs. J. Econ. Bus. **100**, 26–42 (2018). https://doi.org/10.1016/j.jeconbus.2018.04.002
13. Bayero, M.A.: Effects of cashless economy policy on financial inclusion in Nigeria: an exploratory study. Proc. Soc. Behav. Sci. **172**, 49–56 (2015)
14. Ozili, P.K.: Impact of digital finance on financial inclusion and stability. Borsa Istanbul Rev. **18**, 329–340 (2018). https://doi.org/10.1016/j.bir.2017.12.003
15. Cepal, N.U.: La nueva revolución digital La nueva revolución digital. eLAC 2018 La revolucion digital (2015)
16. Caldentey, E.P., Titelman, D.: La inclusión financiera para la inserción productiva y el papel de la banca de desarrollo (2018)
17. WBG: Global Findex Database 2014: Measuring Financial Inclusion around the World (2014)
18. Kpodar, K., Andrianaivo, M.: ICT, financial inclusion, and growth evidence from African Countries. IMF Working Papers **11**, 1 (2011)
19. WBG: The Global Findex Database Latin America & Caribbean (2017)
20. Beck, T., Demirgüç-Kunt, A., Peria, M.S.M.: Banking services for everyone? Barriers to bank access and use around the world. World Bank Econ. Rev. **22**, 397–430 (2008)
21. Graham, M.: Digital Economies at Global Margins. MIT Press, Cambridge (2019)
22. Cole, S., Sampson, T., Zia, B.: Financial literacy, financial decisions, and the demand for financial services: evidence from India and Indonesia. Harvard Business School Working Paper 09-117, 1–37 (2009)
23. Chetty, K., et al.: Review of Fintech Strategies for Financial Inclusion in Sub-Saharan Africa (2019)
24. Koker, L.D.E., Berlin, D.I.W.: Financial inclusion and financial integrity: aligned incentives? World Dev. **44**, 267–280 (2013)
25. IFC: Digital Finance: The Future of Financial Inclusion in Africa (2018)
26. Haddad, C., Hornuf, L.: The emergence of the global fintech market: economic and technological determinants. Small Bus. Econ. **53**, 81–105 (2019)
27. Maldonado, L.: Los medios de pago, un paisaje en movimiento **72** (2015)
28. Shahid, Q., Razaq, L.: Demonetisation for changing payment behaviour and building platforms. J. Paym. Strat. Syst. **11**, 158–168 (2017)
29. Saunders, M., Lewis, P., Thornhill, A.: Research Methods for Business Students (2000). https://doi.org/10.1108/qmr.2000.3.4.215.2
30. Creswell: Five Quality Approaches to Inquiry (1997)
31. Mfinanga, F., Mrosso, R., Bishibura, S.: Comparing case study and ethnography as qualitative research approaches. Int. J. Latest Res. Human. Soc. Sci. **2**, 51–56 (2019)
32. Saris, W.E., Gallhofer, Irmtraud, N.: Wiley Series in Survey Methodology: Design, Evaluation, and Analysis of Questionnaires for Survey Research, John Wiley & Sons, Inc. (2014)
33. Schierz, P.G., Schilke, O., Wirtz, B.W.: Understanding consumer acceptance of mobile payment services: an empirical analysis. Electron. Commer. Res. Appl. **9**, 209–216 (2010)
34. Antón, P., Conde, T.: Modelo Peru Unique Model, Unique Challenges, Bright Future. Institute of International Finance (2017)

The Influence of COVID-19 on the Hospital Supply Chain – The Case of Portuguese NHS

José Sequeiros[1], Maria Teresa Pereira[1,2](✉), Marisa Oliveira[1,2], and Fernanda A. Ferreira[3]

[1] ISEP - School of Engineering, Polytechnic of Porto, 4200-072 Porto, Portugal
mtp@isep.ipp.pt

[2] INEGI - Instituto de Ciência e Inovação em Engenharia Mecânica e Engenharia Industrial, 4200-465 Porto, Portugal

[3] School of Hospitality and Tourism of Polytechnic Institute of Porto, Applied Management Research Unit (UNIAG), Rua D. Sancho I, 981, 4480-876 Vila do Conde, Portugal

Abstract. Available studies and reports, as well as real-time experiences, attest to profound, extensive, long-lasting effects on the supply chain caused by the pandemic. It is thus paramount to rethink the supply chain for medication and health care supplies to build a more resilient and adaptable management process. Data obtained from the NHS portal monthly reports from January 2017 to August 2020 – indicates that at the hospital level the impact was felt the most on the scheduling of medical appointments and prescriptions. Two forecasting methods were selected for this study: Simple Linear Regression and Holt-Winters with the trend and additive seasonality. There was a change in the behavior of hospital patients in the first year of COVID-19 pandemics. This change was shown in two main indicators, the number of hospital consultations and medication expenses. This changed behavior resulted in a decrease in demand for scheduled hospital services, 1.9M less than would be expected under normal conditions, and a foreseeable increase of €432M in the consumption of medicines in the last quarter of 2020, €103M more than in the same period. from the previous year.

Keywords: COVID-19 · Supply chain · Consumer behavior · Forecast methods

1 Introduction

Logistics should be seen as a value creation process [1–3]. The existence of competitive markets makes it necessary to optimize the chain of activities by all those involved. In this way, the way to increase competitiveness must pass through the optimization of the supply chain, efficiently integrating all the processes that compose it, with a focus on satisfying the required service levels [4]. The World Health Organization (WHO) declared COVID-19 a pandemic on March 11, 2020, just three months after the first infected people emerged in Wuhan, China. Two months later, the disease had already infected more than five million people, and 300,000 people had died [5]. At the economic level, and to date, the impacts can be translated into an average loss of GDP, with estimates

J. Machado et al. (Eds.): icieng 2022, LNME, pp. 305–315, 2023.
https://doi.org/10.1007/978-3-031-09360-9_25

of around 8.4% for the USA and the Eurozone. In the worst-case scenario designed by the OECD, global growth could fall by 50% [6].

Worldwide, and according to previous episodes, the occurrence of epidemics or pandemics poses enormous risks to the global supply chain as we know it. Three components stand out [7]:

- there is a long-term interruption of the chains, with an unpredictable duration,
- the disruption of the supply chain occurs simultaneously,
- there is also the interruption and alteration of the supply, demand, and logistics infrastructure.

This article aims to understand how the supply chain affected hospitals, how they responded to these changes, and how patients' habits changed. In analytical terms, how these changes materialized in two specific cases, the variation in the number of first and subsequent consultations and the consumption of medicines in the Portuguese National Health Service.

2 Literature Review

2.1 Hospital Supply Chain

The sustainability of the public healthcare system [8] has been a constant concern for years. Recent events have led to a political mindset shift, a review of the role of stakeholders, and an awareness of the necessity for better planning and forecasting tools.

Changes in economic circumstances, which placed added strain on the system - either due to natural evolution of offer and demand, or due to the pandemic context - enhanced the necessity for optimizing resources, achieving maxim efficiency, and a more adequate management. The importance of logistics in patient care support became paramount, particularly regarding the provision of supplies (inventory, purchase, distribution, and delivery to point of care). This became self-evident during the recent pandemic context which proved to be far more challenging than those of epidemic outbreaks SARS, MERS, Ebola, and Swine flu [11]. The supply chain of healthcare logistics faces strong scrutiny - products must be where they are needed when they are needed to avoid disruption and life loss [10] despite competing demands and lack of human resources to carry out distribution and patient care. With hospitals – and intensive care units in particular) far exceeding their capability, there was a shift in the form of the collaborative approach to the management of resources as governance, interdepartmental, inter-hospital collaboration prevailed even going as far as redesigning the procedures and interactions between healthcare units/services, government bodies, NGOs, and stakeholders [5, 12].

2.2 Global and Local Impacts

Although the supply chain disruption was triggered in China [11], other countries were soon just as affected – or more – by the shortage of medical equipment and supplies. The disruption of businesses, industry, manufacturers, distribution services challenged the response capacity of healthcare services and posed unforeseen challenges that led to the search for new strategies to contain the ripple effect [7, 13]. The resilience of systems was just as menaced as that of individuals and governments [14]. The 2020 pandemic hatched a black swan that impacted heavily on global value chains and triggered the questioning of practices and procedures. The decrease of demand for goods and services – as a result of confinement and mandatory isolation directives – resulted in several phenomena of adaptation/redesign of supply chains that have been the object of study and analysis:

- Hospitals resorted to inactive/idle resources to face the pandemic strain on their services.
- Flexibility was achieved by the possibility of eliminating processes.
- Human resources were given the training to adjust to the specificity of the pandemic and its demands [5].
- Facilities, such as laboratories, were converted to best serve national strategies, and technicians adapted to the requirements [5].
- Infectious disease units were allocated ICU beds from other units and were given priority over existing resources.
- The acquisition of goods and services was simplified as bureaucratic processes were eased – in response to emergencies, purchases could be made to any suppliers regardless of existing long-term binding contracts.

The Portuguese healthcare service provision context was quickly adapted to the growing number of hospital admissions (infections, casualties, hospitalization days) caused by the SARS-CoV-2 epidemic and the epidemiological infection by COVID-19. The nature of the disease – and the actions implemented by national authorities to restrain it – required a very specific response by healthcare services to avoid a nationwide public calamity. Despite the change of context, the pandemic data still displays a high territorial heterogeneity [15, 16].

The death toll for Portugal, as of March 2020, was higher than that recorded for the same reference period of 2019. The first half of July saw 99% of existing companies operating with a reported reduction in turnover and employees compared to an analogous period pre-pandemic. Price increase in essential commodities – particularly personal hygiene and prescription medication – was instantaneous and there was a lack of alcohol (gel and liquid) nationwide.

2.3 Challenges and Long-Term Changes in the Supply Chain

Available studies and reports, as well as real-time experiences, attest to profound, extensive, long-lasting effects on the supply chain caused by the pandemic. It is thus paramount to rethink the supply chain for medication and health care supplies to build a more resilient and adaptable management process. The supply chain management should be lean and agile, resorting to regular risk analysis, simulation, and optimization [6, 11, 17]. As social isolation resulted in a change in both consumer and patient behavior patterns, this caused a direct impact on the service sector as it affected the demand in all activities requiring a direct person to person interaction (Fig. 1).

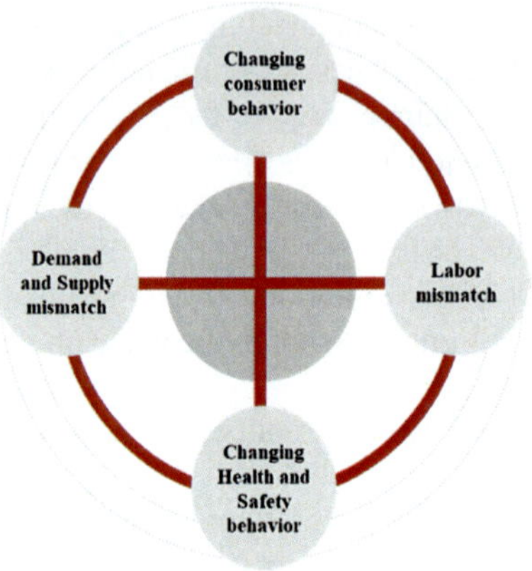

Fig. 1. Hospital's supply chain themes were identified as Interrelated during the COVID-19 (Adapted from [11]).

The consequences of this change will also have an impact on supply. This impact will be all the more significant depending on the level of digitization, automation, and human-machine interaction – these are prominent in both the industrial and agribusiness sectors and less relevant in the service sector, particularly in developing countries/economies.

The details of the aforementioned variable are outlined in Fig. 2. It describes how the change in consumer patterns will impact the future supply chain and how labor mismatch will affect service-providing procedures, specifically on health services.

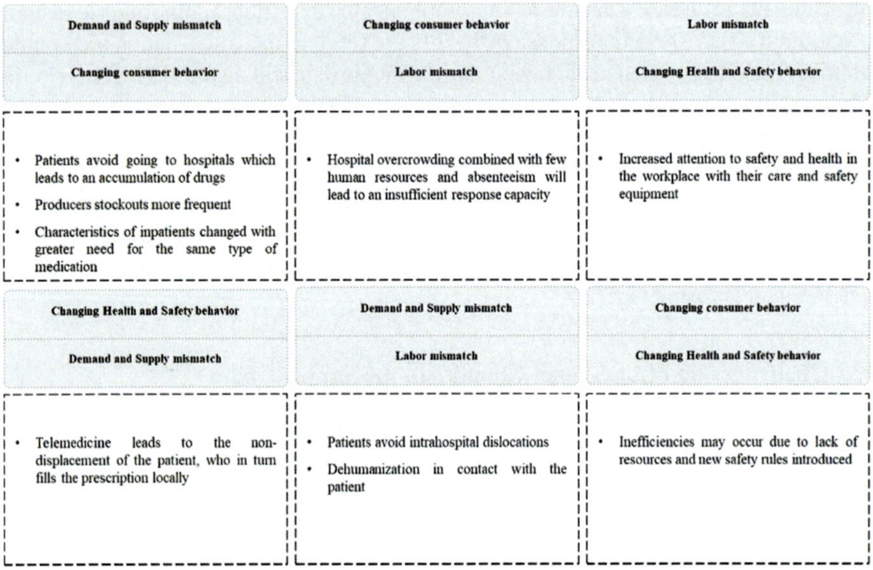

Fig. 2. Hospital's supply chain themes details identified as interrelated during the COVID-19 (Adapted from [11]).

3 Methodology

Data obtained from the NHS portal monthly reports by the Regional Health Administration Services – from January 2017 to August 2020 – indicates that at the hospital level the impact was felt the most on the scheduling of medical appointments and prescriptions.

We have selected two forecasting methods for this study: one simple method and one statistically more relevant more complex method. These methods were chosen for their ability to model trends and seasonal fluctuations present in economic health data, such as the number of medical appointments and prescriptions - both of which are addressed in this paper.

3.1 Simple Linear Regression

Regression analysis represents the minimum square distance of each observation to the trend line that represents the relationship of two variables. This is the simplest mathematical method to characterize quantitative data in a bivariate sample. It is used when the time series exhibits a long-term linear trend. Simple linear regression is only one independent variable X. When more than one independent variable X is considered for the model, then it is multiple linear regression. The simple linear regression model analyzes the relationship between two quantitative variables X and Y, whose trend is represented by an adjusted straight line.

Equation 1 expresses the model

$$\mathrm{Y_i = b.X_i + a + \varepsilon_i} \tag{1}$$

where:

Y is the dependent variable
X is the independent variable
b is the slope of the line
a is the intercept of the line with the Y-axis
ε_i is the residual random variable

The correlation coefficient is represented by R, dimensionless, and the Determination coefficient is represented by R^2, which represents the proportion of variation of the dependent variable Y explained by the independent variable X in the model, dimensional, that can support the forecast accuracy.

3.2 Holt-Winters with Trend and Additive Seasonality

The simple exponential smoothing model uses a weighted average of past time series values as the forecast for the most recent data. It is suitable for locally stationary series but is unsuitable for series with other characteristics, namely when trend and seasonality exist. Exponential Smoothing with Trend shows a consistent increase or decrease all the time in the series. Component of exponential smoothing (S_t) and trend (T_t) for period t are used to trend adjustment. The forecast accuracy and robustness of exponential smoothing methods such as Holt-Winters have led to widespread use in applications where many series necessitate an automated procedure, such as inventory control [18]. Equation 2 represents the Level component, Eq. 3 represents the Trend and Eq. 4 represents the Seasonality component, and Eq. 5 the forecast for the next period.

$$\textbf{Level} \quad S_t = \alpha(A_t - I_{t-L}) + (1 - \alpha)(S_{t-1} + T_{t-1}) \tag{2}$$

$$\textbf{Trend} \quad T_t = \beta(S_t - S_{t-1}) + (1 - \beta)T_{t-1} \tag{3}$$

$$\textbf{Seasonality} \quad I_t = \gamma(A_t - S_t) + (1 - \gamma)I_{t-L} \tag{4}$$

$$\textbf{Forecast for period } t+1 \text{ and } t+m \quad F_{t+1} = S_t + T_t + I_{t-L+1} \text{ then } F_{t+m} = S_t + mT_t + I_{t-L+m} \tag{5}$$

S_t is the value of the smoothed forecast for period t
T_t is the estimate of the trend value for period t
I_t is the seasonality value for period t
A_t is the real value observed in period t
α is the exponential smoothing constant, $0 \leq \alpha \leq 1$
β is the trend smoothing constant, $0 \leq \beta \leq 1$
γ is the seasonality smoothing constant, $0 \leq \gamma \leq 1$
F_{t+1} is the forecast value in period $t + 1$
F_{t+m} is the forecast value in the period $t + m$
L spacing of seasonality - number of periods in the seasonal cycle (must be constant in the time and the variation must be greater than the random variations or "noise"

3.3 Errors

Forecasting errors are used to calculate forecast accuracy. The error associated with the forecast is the difference between the value observed in the series time and the forecast. The mean square error (MSE), Eq. 6, it is the most popular measure of synthesis of the forecast errors used in the evaluation of the performance of each method. The MSE is the best method for this study for two main reasons, its wide usage and the series are evaluated under different conditions, consequently having the same measurement level as the underlying series [19].

$$MSE = \frac{1}{n}\sum_{i=1}^{n}(Y_i - \hat{Y}_i)^2 \tag{6}$$

4 Results

4.1 Medicine Expenses in NHS Hospitals

Figure 3 shows monthly data regarding the monitoring of expenses with medication prescribed/dispensed by NHS public management hospital institutions.

The data reports on current consumption for prescription drugs are covered by the National Hospital Medicines Code. These include medicines for human use with Marketing Authorization and Exceptional Use Authorization. The analysis of the graphic in

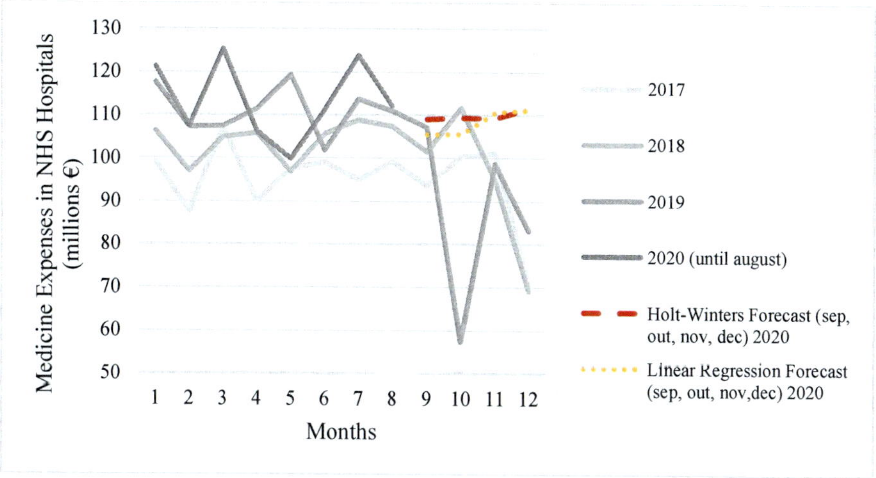

Fig. 3. NHS hospitals medicine expenses – linear regression and Holt-Winters forecast for (Sep, Out, Nov, dec) 2020 [20].

Fig. 3 enables the understanding of the seasonality of the consumption of prescription drugs in the NHS hospitals - peaks occur in March and July. In 2020 there was a sharp drop in prescriptions during the months of confinement - the values were close to those of 2017 (approximately 20M €). The values recorded for September, October, November, and December were estimated using the two methods described previously: Linear Regression and Holt-Winters with the trend and additive seasonality, Table 1.

Table 1. Results obtained for medicine expenses in NHS Hospitals considering different two methods.

Method	Sep 2020	Oct 2020	Nov 2020	Dec 2020	MSE
Simple Linear Regression	108,990,673	109,271,026	109,192,406	111,371,109	10,233,834
Exponential Smoothing with Trend and Additive Seasonality	105,489,547	105,441,421	110,210,993	110,816,570	10,970,268

The results on Table 1 were obtained with:

Linear Regression a: 83,991,021.87 b: 380,279.00
Holt-Winters α: 0.04 β: 0.14 γ: 0.48

A self-feeding method inputted the forecast for a given month to the following month. In certain months, such as October and December, linear regression displayed results with lower MSE. Still, the analysis of the evolution curves denotes both trend and seasonality of these time series, making the Holt-Winters method the most suitable for analysis. As a result, the additional cost of prescription drugs for the last quarter of 2020 should amount to € 432M - € 103M higher than 2019. A sharp drop in spending at the end of the year is expected, because of the need for budgetary compliance with expenses already defined at the beginning of the year.

4.2 Consultations in Portuguese NHS Hospitals

Since 2013, the figures for the total number of consultations have not been so low (data not shown in the graph, but available in [23]). The historic low recorded in April 2020, data that explains the mandatory confinement, and the general cancellation of non-urgent medical appointments. In another study, carried out by two responsible bodies in Portugal, and published in September 2020, interesting results were presented, which corroborate what was described above regarding the change in habits of users of public hospital services. The main cause for this change in attitude is the fear of contagion. More than half of the Portuguese had difficulty accessing health care, there was the impossibility of scheduling 692 thousand medical appointments in the period March-September 2020; and, despite needing health care, around 50% of the Portuguese would only use health services in situations of high severity [22].

The data used in this work were obtained from the Portuguese NHS monitoring platform, data reported by the hospitals themselves, monthly [23]. Changing habits influence the number of consultations carried out, whether in the first contact with a particular specialty or in a follow-up consultation or therapeutic prescription.

The graph in Fig. 4 represents the monthly distribution of the number of consultations carried out on the Portuguese NHS, since 2017. By observing it, it is possible to visualize the seasonality in this distribution, with an increase in the coldest months (October, November, December, and January) and a decrease in the warmer months (June, July, and August). The average number of consultations is around 950 thousand monthly consultations. The study was concerned with the prediction of the number of consultations that would be carried out until the end of the year, namely, for October, November, and December 2020. Two methods were used, Linear Regression and Holt-Winters, additive, given the trend and seasonality of the values. Table 2 aggregates the values obtained from the two methods, as well as the MSE of each one. The method with the best results is the Holt-Winters, additive, given the lower MSE in both cases. Compared to 2019, there were 1,915,499 fewer consultations on the Portuguese NHS.

The results in Table 2 were obtained using the following parameters.

Table 2. Prediction of the number of consultations performed in Portuguese NHS, October, November, and December 2020, using simple linear regression and exponential smoothing with trend and additive seasonality.

Method	Type of consultation	Oct 2020	Nov 2020	Dec 2020	MSE
Simple Linear Regression	First	270,809.8	269,747.0	263,135.9	35,610.7
	Subsequent	701,939.3	692,659.5	682,242.9	79,003.8
Exponential Smoothing with Trend and Additive Seasonality	First	248,233	227,448	210,296	32,656
	Subsequent	518,837	568,706	580,731	66,156

For first consultations:

Linear Regression a: 302651.95 b: −548.83
Holt-Winters α: 0.33 β: 0.00 γ: 0.50

For subsequent consultations:

Linear Regression a: 735693.80 b: −742.37
Holt-Winters α: 0.21 β: 0.00 γ: 0.38

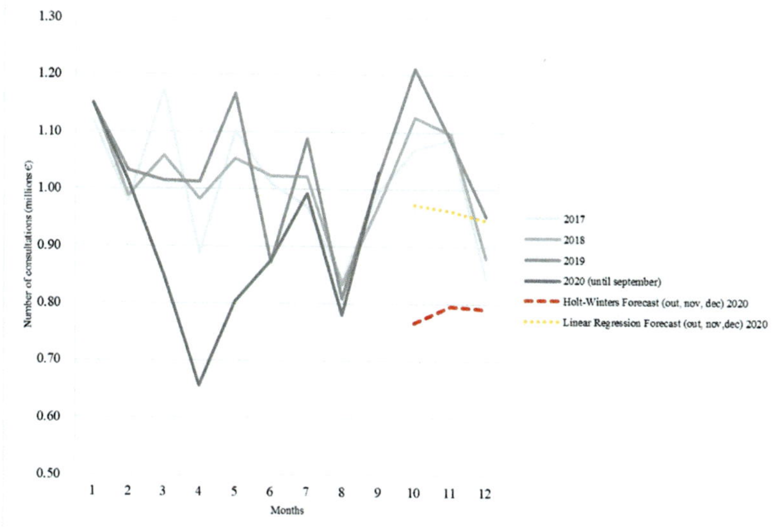

Fig. 4. A total number of consultations (first and subsequent) by year in all NHS hospitals, Linear Regression and Holt-Winters forecast for (Out, Nov, Dec) 2020 [23].

The results are in line with the work described at the beginning of the chapter, highlighting the difficulties of access and fear of moving to institutions.

5 Conclusions

The sustainability of the Portuguese NHS has been a relevant topic when the need to seek maximum efficiency [8]. This sustainability is compromised by two opposing factors, the increased demand for healthcare, which contrasts with the economic reality that reduces investment in the Portuguese NHS [1].

However, in the first year of the COVID-19 pandemic, there was a change in the behavior of hospital patients. This change was shown in two indicators typically reported by hospital entities, the number of hospital visits and medication expenses. This resulted in a decrease in demand for scheduled hospital services, 1.9M less than would be expected under normal conditions, and a foreseeable increase of €432M in the consumption of medicines in the last quarter of 2020, €103M more than in the same period. from the previous year.

Acknowledgments. This work is funded by the European Regional Development Fund through the Regional Operational Program North 2020, within the scope of Project GreenHealth-Digital strategies in biological assets to improve well-being and promote green health, Norte-01-0145-FEDER-000042. The authors are grateful to the FCT Portugal for financial support by national funds FCT/MCTES to UNIAG, under Project no. UIDB/04752/2020.

References

1. Brunero, S., Lamont, S.: The process, logistics, and challenges of implementing clinical supervision in a generalist tertiary referral hospital. Scand. J. Caring Sci. **26**(1), 186–193 (2012)
2. Aguilar-Escobar, V.G., Bourque, S., Godino-Gallego, N.: Hospital Kanban system implementation: evaluating satisfaction of nursing personnel. Investig. Eur. Dir. y Econ. la Empres. **21**(3), 101–110 (2015)
3. Sequeiros, J.: Analysis of the Supply Chain Management in a Portuguese Public Hospital: a Case Study. Dissertaion thesis. Instituto Superior de Engenharia do Porto (2020)
4. Rais, A., Alvelos, F., Figueiredo, J., Nobre, A.: Optimization of logistics services in hospitals. Int. Trans. Oper. Res. **25**(1), 111–132 (2018)
5. Id, A., Órfão, R.: Resposta da Anestesiologia Portuguesa à Pandemia por Anaesthesiology Response to the COVID-19 Pandemic in Portugal 29 (2020)
6. Lozano-Diez, J.A., Marmolejo-Saucedo, J.A., Rodriguez-Aguilar, R.: Designing a resilient supply chain: an approach to reduce drug shortages in epidemic outbreaks. EAI Endorsed Trans. Pervas. Heal. Technol. **6**(21), 1–12 (2020)
7. Ivanov, D.: Predicting the impacts of epidemic outbreaks on global supply chains: a simulation-based analysis on the coronavirus outbreak (COVID-19/SARS-CoV-2) case. Transp. Res. Part E Logist. Transp. Rev. **136**, 101922 (2020)
8. Nunes, A.M.: Gestão Hospitalar: Quatorze Anos de Empresarização em Portugal. Rev. Gestão em Sist. Saúde **6**(2), 154–162 (2017)

9. Moreira, I.M.D.: Comparação de resultados: Prestação de Cuidados de Saúde Hospitalares Públicos e Privados. Dissertation thesis. Universidade do Porto (2011)
10. Stank, T.P., Pellathy, D.A., In, J., Mollenkopf, D.A., Bell, J.E.: New frontiers in logistics research: theorizing at the middle range. J. Bus. Logist. **38**(1), 6–17 (2017)
11. Mollenkopf, D.A., Ozanne, L.K., Stolze, H.J.: A transformative supply chain response to COVID-19. J. Serv. Manag. **32**(2), 190–202 (2020)
12. Søreide, K., et al.: Immediate and long-term impact of the COVID-19 pandemic on delivery of surgical services. Br. J. Surg. **107**(10), 1250–1261 (2020)
13. Queiroz, M.M., Ivanov, D., Dolgui, A., Fosso Wamba, S.: Impacts of epidemic outbreaks on supply chains: mapping a research agenda amid the COVID-19 pandemic through a structured literature review. Ann. Oper. Res. 1–38 (2020)
14. Murphy, J.F., Jones, J., Conner, J.: The COVID-19 pandemic: Is it a 'Black Swan'? Some risk management challenges in common with chemical process safety. Process Saf. Prog. **39**(2), 8–10 (2020)
15. ERS: Impacto da pandemia COVID-19 no Sistema de Saúde – período de março a junho de 2020, p. 46 (2020)
16. Instituto Nacional de Estatística: Indicadores de contexto para a pandemia COVID-19 em Portugal (2020)
17. Govindan, K., Mina, H., Alavi, B.: A decision support system for demand management in healthcare supply chains considering the epidemic outbreaks: a case study of coronavirus disease 2019 (COVID-19). Transp. Res. Part E Logist. Transp. Rev. **138**, 101967 (2020)
18. Bastos, J.A.: Forecasting the capacity of mobile networks. Telecommun. Syst. **72**(2), 231–242 (2019). https://doi.org/10.1007/s11235-019-00556-w
19. van der Laan, E., van Dalen, J., Rohrmoser, M., Simpson, R.: Demand forecasting and order planning for humanitarian logistics: an empirical assessment. J. Oper. Manag. **45**, 114–122 (2016)
20. SNS - Portal da Transparência: Despesa com Medicamentos nos Hospitais do SNS (2020)
21. Saúde - Gabinete da Ministra, Despacho n.º 3219/2020. 2020. pp. 102–105 (2020)
22. Associação Portuguesa dos Administradores Hospitalares (APAH) and Ordem dos Médicos. Acesso a cuidados de saúde em tempos de pandemia (2020)
23. SNS - Portal da Transparência: Consultas Médicas Hospitalares (2020)

Author Index

J. Machado et al. (Eds.): icieng 2022, LNME, pp. 317–318, 2023.
https://doi.org/10.1007/978-3-031-09360-9